Environmental Remediation Technologies, Regulations and Safety

nova science publishers
New York

Environmental Remediation Technologies, Regulations and Safety

Investigations into the Decarbonisation of Railways
Inara Watson, PhD
2021. ISBN: 978-1-68507-250-6 (eBook)

Who Will Save Amazonia? World Heritage or Full Destruction
Marcílio de Freitas
Marilene Corrêa da Silva Freitas, PhD
(Editors)
2021. ISBN: 978-1-53619-893-5 (Hardcover)
2021. ISBN: 978-1-53619-910-9 (eBook)

Modeling Forest Landscape Dynamics
Bharath Setturu
K.S. Rajan
T. V. Ramachandra, FIE, FIEE (UK), FNIE
2021. ISBN: 978-1-53619-594-1 (Hardcover)
2021. ISBN: 978-1-53619-651-1 (eBook)

The Future of Phytoremediation
James C. Flores (Editor)
2021. ISBN: 978-1-53619-625-2 (Hardcover)
2021. ISBN: 978-1-53619-645-0 (eBook)

Environmental Performance: An Analysis of its Determinants
Haroon A. Khan
2020. ISBN: 978-1-53618-525-6 (Softcover)
2020. ISBN: 978-1-53618-594-2 (eBook)

More information about this series can be found at
https://novapublishers.com/product-category/series/environmental-remediation-technologies-regulations-and-safety/

Tamara Lazarević-Pašti, PhD
Editor

Organophosphates

Detection, Exposure and Occurrence

Volume 1:
Impact on Health and the Natural Environment

nova
science publishers
New York

DOI: https://doi.org/10.52305/IMSO3553

NOTICE TO THE READER

Library of Congress Cataloging-in-Publication Data

ISBN: 978-1-68507-652-8

Published by Nova Science Publishers, Inc. † New York

Contents

Preface

Organophosphates are widely used nowadays. They have applications as pesticides, drugs, plasticizers, flame retardants or chemical warfare agents. Their acute toxicity is ascribed to the inhibition of acetylcholinesterase, a key enzyme in the transmission of nerve impulses in animals. Their toxic effects are manifested by the accumulation of acetylcholine and can lead to paralysis or death. Organophosphates also inhibit the other enzymes playing important roles in various biochemical processes. Organophosphate poisoning therapy is based on the application of specific antidotes as well as non-specific and symptomatic procedures.

Organophosphates: Detection, Exposure and Occurrence. Volume 1: Impact on Health and Natural Environment begins with a review of organophosphate pesticides' overall impact on human health, presenting the current state of knowledge and future perspectives regarding their use, regulations, routes of exposure, and the connected impacts on human health. Next, the authors provide insight into the cytotoxic and genotoxic effects of organophosphate pesticides in humans, their metabolism, mechanisms of action, and the potential treatment strategies to minimize their possible impact on the environment and human health. Further, recent findings associating a low-level, chronic exposure to organophosphates with neurological and cognitive impairments that can persist long after the normalization of AChE levels are reviewed, suggesting that mechanisms of action for acute and chronic organophosphate poisoning may differ. One such mechanism is the contribution of chronic organophosphate poisoning to the development of synaptic vulnerability that can further lead to neurodegenerative and neuropsychiatric pathologies such as Alzheimer's disease, Parkinson's disease, and cognitive impairments in children. Next, the relationship between organophosphate exposure and depression development is assessed. Epidemiological studies have reported higher incidences of depression in particular groups of individuals with chronic health conditions and individuals in specific occupations, such as farming, fishing and forestry. Also, several studies have reported that people exposed to organophosphates in occupational and environmental settings are more prone, compared with the general population, to develop cancer. Thus, the authors aim to link epidemiologic studies to available evidence from basic science research to better understand the role that organophosphate exposures play in the molecular pathogenesis of different types of cancer. Further, organophosphates' fate in the environment, biomonitoring, analysis and legislation are reviewed. In addition, the key accomplishments and future challenges to the role of ionic liquids in organophosphate analysis and degradation are discussed, with particular emphasis on the more benign ionic liquids while foreseeing the development of more sustainable processes. Moreover, an overview of the literature data concerning the electrochemical oxidation of several widely used organophosphate pesticides is presented. The influence of the electrode materials on selected organophosphates, the choice of an appropriate electrode, supporting electrolyte, and knowing the electrolysis parameters such as current density, temperature, and oxidation rate were also elucidated. Additionally, the use of carbon materials in organophosphate remediation from the environment was reviewed. The application of metal-organic frameworks for the detection of selected organophosphate compounds was discussed as well. The special emphasis was on metal-organic-frameworks-based sensors, their sensitivity, selectivity, advantages and disadvantages in the OPs detection process. Finally, the use of organophosphates as acaricides and flame retardants is discussed in more detail.

Chapter 1 - Although an extensive legal framework regulates the development of organophosphate pesticides (OP), serious concerns have been raised considering human health risks resulting from consumption and acute or/and chronic exposure. Most pesticides and their residues are toxic to both the environment and humans, so international and national regulatory bodies have been established to control and enforce pesticide regulations and the maximum concentrations of residues in food and feed that are considered safe for humans. However, excessive pesticide application increases the risk of combined exposures to multiple residues that may lead to serious cumulative toxicity and adverse effects on human health. The health hazards can range from short-term to chronic conditions. For most of the OPs, toxicological data are collected from numerous studies covering a wide range of toxicological endpoints; however, the risk of health hazards due to pesticide exposure depends not only on how toxic the compounds are but also on the dose level at which harmful effects might be observed, and the frequency, timing and levels of exposure. This chapter presents the current state of knowledge and future perspectives regarding OPs use, regulations, routes of exposure, and the concomitant impacts on human health.

Chapter 2 - Pesticides are used in almost all food production areas and represent pollutants of worldwide importance. The utilization of organophosphate (OP) pesticides is considered a major global health problem. The OP pesticides toxicity is closely related to their metabolism, oxidative stress induction, mitochondrial dysfunction, and inhibition of the enzyme acetylcholinesterase (AChE). OP pesticides affect cellular and molecular targets, which may lead to systemic disturbances and, finally, the development of various human diseases. The extent and severity of OPs-induced adverse effects depend not only on the specific pesticide properties, exposure levels, and route of absorption but also on polymorphisms in the main genes involved in OP metabolism. This chapter aims to provide insight into the cytotoxic and genotoxic effects of OP pesticides in humans, their metabolism, mechanisms of action, and the potential treatment strategies to minimize their possible effects on the environment and human health.

Chapter 3 - Organophosphates (OPs) are the main components of herbicides, insecticides, and pesticides used worldwide. In addition, OPs are used in the production of chemical weapons such as nerve gasses, sarin, and soman. The main mechanism of OP action in the case of lethal poisoning is the inhibition of acetylcholine esterase (AChE) activity resulting in excess of acetylcholine (ACh), and the induction of "cholinergic crisis," a clinical condition that often ends in coma and death. Thus, the levels of AChE in the blood are used as the primary biomarker for evaluating OP-induced poisoning. Normal levels of AChE in the blood have verified that low-level chronic OP exposure is relatively safe. However, recent findings associated a low-level, chronic OP exposure to neurological and cognitive impairments that can persist long after the normalization of AChE levels suggesting that mechanisms of action for acute and chronic OP poisoning may differ. One such mechanism is the contribution of chronic OP poisoning to the development of synaptic vulnerability that can further lead to neurodegenerative and neuropsychiatric pathologies such as Alzheimer's disease, Parkinson's disease, ALS, ADHD, and cognitive impairments in children. The potential mechanisms of action during low-level chronic OP exposure in various preclinical models and paradigms were discussed. In addition, the need for the development of new biomarkers in order to reevaluate systemic, neural, and neurodevelopmental toxicity, and thus the applicative doses, of various OPs that exist today is underscored. Finally, potential prophylactic measures that can be applied as the countermeasures for the increased synaptic vulnerability are suggested.

Chapter 4 - Organophosphates are the most widely used group of pesticides globally and are considered one of the most hazardous to vertebrates. The neurotoxic effects of high-level acute poisoning are well established and involve changes in peripheral, autonomic and central nervous system function. These events result in a constellation of physical, cognitive and psychiatric symptoms. The impact of long-term low-level exposure to organophosphates on human health is less known. Organophosphates disrupt many other neurotransmitters, which are involved in mood regulation, such as serotonin and acetylcholine. Depression is a common mental health condition affecting millions of people globally, and its etiology is connected with disturbances in neurotransmitter functioning and genetic factors. Epidemiological studies have reported higher incidences of depression in particular groups of individuals with chronic health conditions and individuals in specific occupations, such as farming, fishing and forestry. The reason behind the high risk of mood disorder in farming populations is indistinguishable. Research showed that individuals suffering from depression have elevated levels of the neurotransmitter acetylcholine. Since organophosphates are potent inhibitors of acetylcholinesterase, the enzyme responsible for acetylcholine's hydrolysis, there are strong indications that acetylcholine level elevation could be the crucial cause of the symptoms of clinical depression in exposed individuals.

Chapter 5 - Organophosphorous compounds have been widely used as pesticides in agricultural, commercial and residential settings. A number of studies have reported that people exposed to these compounds in occupational and environmental settings are more prone, compared with the general population, to develop cancer. However, a direct link in terms of causality between organophosphate exposure and carcinogenesis in humans has not been established, mainly because the exact biologic mechanisms underlying these associations are still unclear. This chapter aims to link epidemiologic studies to available evidence from basic science research to provide a better understanding of the role organophosphate exposures play in the molecular pathogenesis of different types of cancer.

Chapter 6 - The presence of organophosphorus (OP) compounds in soil, atmosphere, groundwater, surface water, and river sediments originates from using those compounds in pest management on vegetation and animals, as well as in disposal of cattle and sheep dips on farmland. The environmental persistence of OP pesticides is thought to be stricken by transporting methods combined with chemical and organic degradation methods. The introduction of OP residues in remote environments demonstrates that those pesticides may be transported long distances via air and water. Today, it is known that the fate of the OP in the environment is affected by transport (sorption, volatilization, leaching, runoff), chemical (solubility, oxidation/reduction, hydrolysis, photolysis), and biological processes. Biological tracking is an important factor in any complete evaluation of exposure to OP. Historically, biological impact tracking (BIM) through the extent of the reduction of blood cholinesterase activity became the method of choice. It is involved the measurement of plasma cholinesterase (ChE) and erythrocyte acetylcholinesterase (AChE), reflecting the influence of OPs' intake on inhibition of mentioned blood enzymes as surrogates of an AChE in neural tissue and neuromuscular junctions. An alternative or complementary approach to the biological control of OPs is based on the analysis of metabolites in the urine. These methods detect specific metabolites for the particular OP or non-specific metabolites common for various OPs – the dialkyl phosphates (DAPs). Once ingested, OPs no longer accumulate substantially in the human body, and over 75% is rapidly metabolized to DAPs. After, they are excreted through the urine 6 to 24 hours after exposure. Depending on the type of sample, different methods of analysis are used. Most analytical methods for the analysis of OP pesticides are useful for determination in the plant, water, soil, or food. However, they are not directly applicable to biological samples taken in an emergency or during an autopsy. A solid-phase microextraction (SPME) method has been used as a variable solvent-free alternative to conventional liquid-liquid extraction. The Quick, Easy, Cheap, Effective, Rugged, and Safe method coined the "QuEChERS" today is also the best known and most widespread method for the analysis of pesticide residues. Legislation related to pesticides can be divided into two parts: (1) part relating to registration (permission to place into circulation and use); (2) the part related to the regulation of the maximum allowed amount of pesticide residues in food.

No plant protection product can be registered in the EU until it is scientifically determined that it:

- has no adverse effects on consumers, farmers, or passive observers;
- does not cause adverse effects on the environment;
- has sufficient effectiveness against targeted pests.

Chapter 7 - Organophosphorus pesticides (OPs) are highly effective pesticides whose usage accounts for approximately 30% of the pesticides used worldwide. However, overusing OPs in addition to their difficult degradation is threatening environmental and human health even at very low concentrations that are below detection limits of the standard analytical techniques. For that reason, great efforts have been placed in developing rapid and effective preparation methods for OPs prior to analysis, as well as in improving the performance of different analytical techniques for OPs detection from various samples. Much of the recent progress in terms of efficiency and sustainability has come from incorporating ionic liquids (ILs) as solvents or in the creation of IL-supported materials for OPs extraction purposes. ILs are liquid molten salts typically composed of bulky and unsymmetrical organic cations and various anions. ILs are non-flammable and non-volatile, have excellent (electro)chemical stability and solvation ability for a wide range of compounds and materials. Owing to a large number of ion combinations, ILs have been referred to as tunable designer solvents, the feature that overcomes the limited selectivity and efficiency of common organic solvents and allows the development of more effective extraction platforms. This review will summarize the main results achieved by the use of IL-based processes in the extraction of OPs as a part of sample preparation (liquid-liquid microextraction and solid-phase microextraction) or in designing different nanocomposites to improve electrochemical performance for the detection of different OPs. Moreover, this review will outline and discuss achievements in nucleophilic degradation of OPs where IL can act as both the nucleophile and solvent. The key accomplishments and future challenges to the role of ILs in OPs analysis and degradation will be discussed, with particular emphasis on the more benign ILs while foreseeing the development of more sustainable processes.

Chapter 8 - Electrochemical oxidation is in general considered to be among the best methods for removing organic pesticides from wastewater. The oxidation of organophosphates induces the formation of a great number of various more or less toxic short-time living intermediates and finally leads to their complete degradation. This chapter presents an overview of the literature data concerning the electrochemical oxidation of several widely used organophosphate pesticides (malathion, diazinon, chlorpyrifos, dimethoate, methyl parathion). The influence of the electrode materials on selected organophosphates, the choice of an appropriate electrode, supporting electrolyte, and knowing the electrolysis parameters such as current density, temperature, and oxidation rate were elucidated. The theoretical calculation enabled us to identify the phenomena engaged near the electrode and to completely determine the structures of the products of electrochemical oxidation formed during the degradation. The comparison of the results relating to the various electrodes indicates that these materials have the power to reduce the quantity of the organic matter in the electrolyzed solution.

Chapter 9 – Organophosphates are considered some of the most toxic substances synthesized by man due to their high neurotoxicity. Their wide use in modern agriculture and other areas results in an urgent need for their efficient removal from the environment. Among numerous removal strategies, adsorption is considered a promising approach for organophosphate pesticide removal because it is simple, economical, and environmentally friendly. Carbon-based materials are a very good choice for this purpose because of their properties, including high specific surface, various pore structures and rich surface chemistry. Furthermore, carbon materials not only have a high adsorption capacity towards organophosphates, but they are generally cheap, and their properties could be easily tuned. When analyzing the adsorption of organophosphates on carbon materials, the key is to understand the mechanisms of these interactions. However, available literature shows a great diversity in used approaches and sometimes diverges from basic theoretical principles. In this chapter, we critically address the analysis of organophosphates adsorption on carbon materials. First, we present the properties of organophosphate pesticides, followed by the classification of carbon materials and an overview of techniques used in their characterization. This is essential to understand how organophosphates can interact with carbon surfaces and what are the key factors determining adsorption performance. Next, we summarize thermodynamic and kinetic descriptions of adsorption processes and outline critical points in the analysis of adsorption thermodynamics. This is followed by an overview of organophosphate-carbon interactions and experimentally obtained adsorption capabilities of different carbons reported in the literature. We conclude with the suggestions to standardize adsorption measurements, making them easier to compare and suitable for extracting the key carbon properties for developing advanced superior adsorbents for organophosphate pesticides.

Chapter 10 - Organophosphate compounds (OPs) are highly acute toxic compounds used in agriculture production as pesticides. They bind to acetylcholinesterase leading to nerve paralysis and organ failure in humans. Hence, it has become mandatory on a global level to detect and remove organophosphate compounds from the environment. There are many conventional methods (nanoscale materials or sophisticated instruments) available to detect these compounds. However, these methods are time-consuming, expensive, and require an analytical laboratory with skilled staff. These constraints were the driving force for developing sensing platforms capable of detecting pesticides with excellent sensitivity. Metal-organic frameworks (MOFs) belong to porous coordination polymers fabricated from metal clusters/ions and multifunctional organic ligands. The tunable size and shape of pores, stable framework and the opportunity for desirable post-synthetic modifications classify them as promising candidates for potential use in detecting OPs at low nanomolar levels. This chapter will discuss the application of MOFs for the detection of selected organophosphate compounds. The special emphasis will be on MOF-based sensors, their sensitivity, selectivity, advantages and disadvantages in the OPs detection process.

Chapter 11 - Organophosphate insecticides are a group of chemicals widely used in agriculture, horticulture, and veterinary medicine. They act in the same way by blocking acetylcholinesterases and rapidly undergo hydrolytic degradation in sunlight, which makes them a plausible alternative to other groups of pesticides, especially organochlorine and carbamate ones. Today, organophosphates are extensively used throughout the world due to their efficiency. However, they can also affect beneficial soil invertebrates that have an important role to play in the food web, decomposition of organic substances, and maintenance of soil structures. Among such soil invertebrates are spiders. Comprising one of the most abundant and diverse groups of natural enemies, spiders inhabit various types of agroecosystems and take part to a significant extent in pest control. Unfortunately, extensive use of pesticides can disturb the role of spiders, so the purpose of this chapter would be to provide an overview of data published about the effect exerted on these soil invertebrates by organophosphates, which—like their overall ecotoxicology—is a subject that has received rather limited attention to date.

Chapter 12 - The application of organophosphorus flame retardants (OPFRs) is continuously increasing over the years. There is a growing concern for OPFRs' safety due to their high production. Unavoidable, large quantities of OPFRs enter the environment. Sewage and solid waste (particularly e-waste) treatment plants are the major sources of OPFRs released into the environment. OPFRs are extensively detected in air, dust, water, soil, sediment, and sludge. This chapter discusses and compares the most advanced analytical procedures of OPFRs, including extraction, purification, gas, and liquid chromatography determination techniques. The environmental occurrence of OPFRs was analyzed. Finally, the accumulation and metabolic pathway of OPFRs on living organisms are overviewed.

Chapter 1

Organophosphate Pesticides and Human Health: Current Knowledge and Future Prospects

Sandra Petrović*, **PhD**

Department of Physical Chemistry, "Vinča" Institute of Nuclear Sciences - National Institute of the Republic of Serbia, University of Belgrade, Belgrade, Serbia

Abstract

Although an extensive legal framework regulates the development of organophosphate pesticides (OP), serious concerns have been raised considering human health risks resulting from consumption and acute or/and chronic exposure. Most pesticides and their residues are toxic to both the environment and humans, so international and national regulatory bodies have been established to control and enforce pesticide regulations and the maximum concentrations of residues in food and feed that are considered safe for humans. However, excessive pesticide application increases the risk of combined exposures to multiple residues that may lead to serious cumulative toxicity and adverse effects on human health. The health hazards can range from short-term to chronic conditions. For most of the OPs, toxicological data are collected from numerous studies covering a wide range of toxicological endpoints; however, the risk of health hazards due to pesticide exposure depends not only on how toxic the compounds are but also on the dose level at which harmful effects might be observed, and the frequency, timing and levels of exposure. This chapter presents the current state of knowledge and future perspectives regarding OPs use, regulations, routes of exposure, and the concomitant impacts on human health.

---

* Corresponding Author's Email: sandra@vin.bg.ac.rs.

In: Organophosphates
Editor: Tamara Lazarević-Pašti
ISBN: 978-1-68507-652-8

Keywords: organophosphate pesticides, human health, future prospects

Introduction

With the continuous growth of the world population, global food demand is rapidly increasing, intensifying the use of pesticides. Pesticides are used most in agricultural production to reduce crop yield losses and increase agriculture productivity and product quality (Antonini and Argilés-Bosch 2017; Damalas and Eleftherohorin 2011). The increased quantity and frequency of pesticide applications consequently increased their impact on the environment and human health. Exposure to pesticides might be occurred directly from occupational, agricultural and household use and indirectly through the diet. Besides agricultural activities, pesticides are also highly applied in public health actions to control vector-borne diseases (e.g., malaria, yellow fever, dengue, etc.) and unwanted plants in public places, parks and gardens (Kim, Kabir, and Jahan, 2017; Zikankuba et al. 2019). However, uncontrolled pesticide applications and mistreatments may cause environmental pollution and adverse human health effects (Savić et al. 2019). Pesticide residues and corresponding metabolites can contaminate soils and water, enter the food chain, and finally cause adverse effects on human health. Therefore, the development and application of pesticides have to be controlled by strict regulation processes raised by international organizations.

Organophosphate pesticides (OPs) are broad-spectrum, very effective insecticides that are highly applied worldwide in agriculture, homes, gardens and veterinary practices for preventing, destroying, or controlling pests (Zikankuba et al. 2019). Over 100 organophosphorus compounds are currently in commercial use as insecticides to control pests in agricultural food commodities. Some of the most commonly used compounds such as chlorpyrifos, malathion, diazinon, dichlorvos, phosmet, azinphos-methyl, parathion (no longer registered for any use), methyl parathion fenitrothion, tetrachlorvinphos and azamethiphos are included in this class (Adeyinka, Muco, and Pierre 2020). As neurotoxicants, the OPs are toxic not only to insects but to most animal species and humans. Numerous studies have demonstrated that OPs also affect the immune response and display teratogenic, cytotoxic and genotoxic effects *in vitro* and *in vivo* (Čolović et al. 2010; Li 2007; Lu et al. 2012; Ojha et al. 2013; Savić et al. 2019). In addition, biomarkers of exposure to OPs have been extensively detected in human populations. Exposure to OPs that includes inhalation, ingestion and direct contact might be associated with an increased risk of various chronic illnesses, including cancer, respiratory, metabolic and neurologic disorders (Kim, Kabir, and Jahan, 2017; Terry 2012). Although current regulations covering pesticide application stipulate that pesticides should have no harmful effects on human health, exposure to the OPs by various routes may cause cumulative effects and lead to severe additive toxicity (Roberts and Reigart 2013). Therefore, this chapter is organized to provide a brief overview of pesticide regulations, OPs application, toxicity, routes of exposure, and impacts on human health.

Pesticide Regulations

The use of pesticides is a widespread agricultural practice that can be harmful to humans and the environment. Therefore, the potential benefits of pesticide application must be balanced against potential harms (Donley 2019). Accordingly, international and national regulatory bodies have been established to control and enforce pesticide regulations and the maximum concentrations of residues in food and feed that are considered safe for humans (maximum residue levels (MRLs)). The Joint Meeting on Pesticide Residues (JMPR) is an expert body established mutually by the Food and Agriculture Organization of the United Nations (FAO) and the World Health Organization (WHO) in order to harmonize the risk assessment on the pesticide residues following a Good Agricultural Practice (GAP) (Zikankuba et al. 2019). The JMPR encompasses the WHO Core Assessment Group and the FAO Panel of Experts on Pesticide Residues in Food and the Environment. The WHO Core Assessment Group is responsible for estimating Acceptable Daily Intakes (ADI), acute reference doses (ARfDs) and other pesticide toxicological data. The FAO Panel is responsible for assessing pesticide residues data and estimating supervised trials median residue values (STMRs), highest residues (HRs) and MRLs in food and feed. In the next step, the MRLs recommended to the Codex Committee on Pesticide Residues (CCPR) for consideration should be finally adopted by the Codex Alimentarius Commission (CAC) that acts as a risk manager (Yamada 2017). However, on the global level, these limits remain variable, especially among developed and developing countries. The world's largest agricultural producers and pesticide users are the European Union (EU), United States of America (USA) and China, each having different pesticide regulatory systems established to protect humans and the environment. In the European Union, a large body of EU legislation regulates the use of pesticides and their residues in food and feed. The

European Commission (EC) and the Member States evaluate and authorize plant protection products following Regulations 1107/2009 and 396/2005 (Donley 2019). The European Food Safety Authority (EFSA) makes proposals regarding the setting of MRLs and gives scientific advice to the EC on potential risks associated with the presence of pesticide residues in food and feed and their effects on human or animal health (Zikankuba et al. 2019). The EC regulation 396/2005 directly concerns public health, providing a system of setting MRLs for all food treated with pesticides and their residues (Brancato et al. 2018). In the USA, the Environmental Protection Agency (EPA) is responsible for the pesticide registration, regulations and establishment of MRLs in food and feed following the Federal Food, Drug, and Cosmetic Act (FFDCA) and the Federal Insecticide, Fungicide, and Rodenticide Act (FIFRA) (Donley 2019). In China, the Institute for the Control of Agrochemicals under the Ministry of Agriculture is the main pesticide regulatory body responsible for pesticide registration and management (Donley2019).

The risk assessment of the effects of pesticides on the environment and human health is complex and includes several factors: the types of pesticides used, the dosage applied, the periods and level of exposure, and the environmental characteristics of the locality where pesticides are applied (Damalas and Eleftherohorin 2011). Although pesticides are developed to meet regulatory compliance and standards requirements and to act with minimal risk to human health and the environment, they certainly might pose a significant risk to humans through acute and/or chronic exposure and consumption (Damalas and Eleftherohorin 2011). Over the past years, the main concern was related to the potential risk of combined exposure to multiple pesticide residues in the diet and dose addition of these compounds, which may lead to cumulative toxicity. As a result, EFSA announced several methodologies to calculate the cumulative risk on groups of pesticides that, causing the same toxic effects in tissues, organs and physiological systems, can produce cumulative toxicity, even if they do not exhibit similar modes of action (Crivellente et al. 2019). Therefore, in 2019, EFSA established cumulative assessment groups (CAGs) of pesticides for the following effects on the nervous system: brain and/or erythrocyte acetylcholinesterase (AChE) inhibition, functional alterations of the motor, sensory and autonomic divisions and histological neuropathological changes in neural tissue (Crivellente et al. 2019).

The CAG included 47 pesticide active substances, of which 11 were N-methyl carbamate insecticides and 36 were OPs associated with the brain and/or erythrocyte acetylcholinesterase inhibition. The investigated substances were characterized by no observed adverse effect levels for short- and long-term cumulative exposure/risk assessment using all information regarding the brain or erythrocyte AChE inhibition (Crivellente et al. 2019). In 2020, EFSA announced the first two reports on retrospective (for the years 2014 – 2016) cumulative risk assessments (CRAs) of dietary exposure to pesticide residues, which concerned acute effects on the nervous system and chronic effects on the thyroid gland for ten population groups from different countries and different age classes (Craig et al. 2020a,b). In 2021, EFSA published a report on retrospective (for the years 2016 – 2018) CRAs of dietary exposure to pesticide residues conducted for chronic inhibition of acetylcholinesterase (Anastassiadou et al. 2021). It was concluded that cumulative exposure to pesticides contributing to the chronic inhibition of acetylcholinesterase did not exceed the threshold established by risk managers. However, the non-dietary routes of exposure to pesticides were not considered in the assessment (Anastassiadou et al. 2021). Recently, the Report from the Commission to the European Parliament and the Council concluded that a methodology for CRA covering simultaneous exposure to pesticide residues was more complex than initially expected and that faster progress is needed. In this regard, the EU Commission and EFSA developed an Action Plan to accelerate work on developing other methods for cumulative risk assessment and their subsequent implementation (European Commission 2021). In 2021, the EU Member States adopted this Action Plan under document SANTE/10178/2021 (EFSA-SANTE/10178/2021 2021 (EC)). As reported, for the years 2022-2030, the plan will focus on assessing human health risks from dietary and non-dietary exposure to pesticides. In this regard, the applicability of CAGs established for dietary CRA to non-dietary CRA will be evaluated. The priority will be the establishment of new CAGs; so far, the CAGs were derived for the nervous system and the thyroid gland, and they will need to be established for 15 organ systems. The EFSA developed a prioritization method that will enable: (i) identification of pesticides that may present the highest risk in terms of dietary exposure and (ii) identification of organ systems. New CRAs will be carried out for pesticides that have: 1) chronic effects on the nervous system (2021), 2) acute effects on craniofacial malformations (2022) and 3) effects on other organ systems in accordance with the establishment of new CAGs (2022-2030). In addition, CAGs will be periodically updated in order to include new substances, and accordingly, the CRAs will comprise new active substances in terms of their approval and new product authorization (European Commission 2021).

Toxicity, Routes of Exposure and Mechanism of OPs Action

Toxicity of Pesticides

The toxicity of pesticides is estimated by laboratory testing on animals such as rats, mice and rabbits. The acute toxicity of pesticides is measured by LD_{50} and LC_{50} values. According to the World Health Organization & International Programme on Chemical Safety (2010), LD_{50} (lethal dose, 50 percent) is the statistical estimate of the dose of a pesticide (the number of milligrams (mg) of toxicant per kilogram (kg) of body weight) required to kill 50% of the population of test animals. Thus, a pesticide with a smaller LD_{50} is more toxic than a pesticide with a higher value because it takes less pesticide to kill 50% of the test animals. The LC_{50} is the average concentration (expressed as milligram (mg) per liter (l) of pesticide dust, vapor, gas, mist or fume that is capable of killing 50% of the test animals over a given period of exposure (Deadman 2017). Classification of pesticides based on the acute oral and dermal toxicity to the rat (World Health Organization and International Programme on Chemical Safety 2010) is presented in Table 1.

Table 1. Classification of pesticides based on the acute oral and dermal toxicity to the rat

LD_{50} for the rat (mg/kg body weight)		**Class**	
Oral	**Dermal**		
less than 5	less than 50	Ia	Extremely hazardous
5–50	50–200	Ib	Highly hazardous
50–2,000	200–2,000	II	Moderately hazardous
Over 2,000	Over 2,000	III	Slightly hazardous
5,000 or higher		U	Unlikely to present acute hazard

Some commonly used OPs are classified as extremely hazardous and highly hazardous; for example, parathion and parathion methyl are classified as WHO class Ia, while phosphamidon and methamidophos are classified as WHO class Ib. According to FAO, almost half of all WHO class Ia and Ib pesticides are OPs. As a result, some OPs have been banned, or their use has been restricted in many countries, especially in developing countries where farmers often are untrained or do not have the equipment to handle pesticides safely (Zikankuba et al. 2019). For example, parathion has been severely restricted globally since the 1980s; diazinon has been restricted in the EU, USA, and Canada but is still in use in Brazil; tetrachlorvinphos is banned in the EU but continues to be used in the USA; malathion is in continued global use in agriculture and public health actions for mosquito control (Pluth, Zanini, and Battisti 2019).

Routes of Exposure to OPs

As for all other types of pesticides, the exposure to OPs can occur directly from occupational (agricultural workers, greenhouse workers, pesticide manufacturing) and non-occupational use (residential exposure, exposure in public places), and indirectly through the diet (pesticide residues in food and drinking water) (Kim, Kabir, and Jahan, 2017). In addition, exposure might be accidental or intentional. Routes of direct exposure include dermal (contact), respiratory (inhalation), oral (ingestion), and ocular (eye) pathways (Kim, Kabir, and Jahan, 2017). Dermal and respiratory exposures are the most common routes of exposure to OPs, which may occur during mixing, loading, disposing, and cleaning of pesticides (Salvatore et al. 2008). Solid forms of pesticides (powders, dust, and granules) are not absorbed through the skin as easily as liquid formulations. The hazard from dermal absorption depends on the toxicity of the pesticide, the duration of the exposure, and the body part affected (Damalas and Koutroubas 2016). Pesticide inhalation, which can occur due to the presence of volatile components, may cause serious damage to the nose, throat, and lung tissues if inhaled in sufficient amounts. Vapors and very small particles pose the most serious risks (Damalas and Koutroubas 2016). Oral exposure to OPs may cause severe injuries or even death. It usually arises by accident but can also be intentional (self-poisoning). The majority of cases of oral exposure to OPs are related to carelessness, such as swallowing of pesticides by mistake when they are improperly stored in food containers, poor hygiene after pesticide handling when hands are not properly washed before eating or smoking, etc. (Damalas and Eleftherohorin 2011). Absorption of pesticides, particularly vaporous, dust, or aerosols, by the eyes may cause local effects on the smooth muscles of the eyes, severe injuries, blindness, or even death. Therefore, eye protection is needed when handling and spraying pesticides (Kim, Kabir, and Jahan, 2017).

Dietary (indirect) exposure to OPs is related to the pesticide residues that can remain in/on food and might pose a serious risk to human health through consumption. As mentioned above, the risk assessment of dietary exposure to residues in food and feed is regulated by a very extensive legal framework, ensuring consumer safety (Crivellente et al. 2019). The appropriate analyses have demonstrated that dietary intakes of OPs as determined in total diet samples are generally below the ADI values for each specific pesticide analyzed. The exceptions are the propetamphos and phosalone in infants, for the years 1995-1996 when exposures up to twice the ADI were reported (Crivellente et al. 2019).

Mechanism of OPs Action

The OPs are esters of phosphoric acid that share a common mechanism of toxicity comprising irreversible acetylcholinesterase inhibition in the nervous system with subsequent accumulation of the acetylcholine (ACh) and the continuous stimulation of ACh receptors (Adeyinka, Muco, and Pierre 2020; Lazarević-Pašti et al. 2017). The AChE plays a critical role in controlling nerve impulse transmission from nerve fibers to smooth and skeletal muscle cells, secretory cells and autonomic ganglia and within the central nervous system (CNS) (Bondžić et al. 2020; Roberts and Reigart 2013). The AChE breaks down the ACh into choline and acetate by hydrolytic process, thus terminating its effect on the muscarinic and nicotinic receptors. Nicotinic and muscarinic receptors are widely distributed throughout the body. Central nicotinic receptors (N2) function within the CNS and can also be found in the sympathetic, parasympathetic ganglia of the peripheral nervous system (PNS) and the adrenal medulla. Peripheral nicotinic receptors (N1) function at the neuromuscular junction (Adeyinka, Muco, and Pierre 2020). Muscarinic receptors (M1-M5) are present in the CNS and function in both the PNS and CNS, mediating innervation to visceral organs (Adeyinka, Muco, and Pierre 2020). OPs irreversibly bind to AChE and prevent the breakdown of ACh, which leads to overstimulation of the muscarinic receptors located at cholinergic junctions in skeletal neuromuscular junctions, the nicotinic receptors in autonomic ganglia, and receptors in the CNS (Jaga and Dharmani 2003; Rusyniakand Nañagas 2004). Stimulation of N1 nicotinic receptors causes fasciculation and muscular weakness, whereas stimulation of N2 receptors can cause hypertension and tachycardia (Adeyinka, Muco, and Pierre 2020). Stimulation of muscarinic M1-M5 receptors in the CNS causes seizure, anxiety and agitation, while stimulation of M2 and M3 receptors in the gastrointestinal system leads to abdominal cramps, drooling, and salivation. In the respiratory system, stimulation of M2 and M3 receptors causes bronchospasm, bronchorrhea and rhinorrhea, while in smooth muscles of internal organs, it leads to abdominal cramps and urinary urgency. In the eyes, stimulation of M2 and M3 receptors causes miosis, whereas stimulation of M2 receptors in the heart can cause hypotension bradycardia (Abrams et al. 2006).

The OPs inhibit both the red blood cell (RBC) AChE activity and the butyrylcholinesterase (BuChE) activity. The BuChEis inhibited more quickly and restored to the normal levels within 60 days, indicating acute OPs poisoning. On the other side, the RBC AChE is depressed less rapidly and is returned to the normal levels in several weeks to several months, being an indicator of chronic OPs exposure (Sullivan and Blose 1992).

Impact of OPs on Human Health

Exposure to OPs may pose a significant risk to human health. Based on the length of exposure, the adverse effects upon exposure to OPs may be (i) acute (arise within minutes to 24 hours), (ii) subacute (arise between 24 hours and two weeks) and (iii) chronic (exposure beyond weeks to years) (Adeyinka, Muco, and Pierre 2020).

Acute Effects

Acute toxicity assigns to the effects of a single exposure or frequent exposure to OPs over a short time. Poisoning is the major effect of acute exposure to OPs. The toxicokinetics and toxicodynamics of OPs poisoning depend on the route and extent of exposure and also the chemical structure and dose of the compound (Kwong 2002). Acute exposure to OPs affects the parasympathetic, sympathetic, and central nervous system via previously described OPs mechanism of action (Jaga and Dharmani 2003). In general, irrespective of the route of absorption, systemic effects are similar. According to the degree of the severity of poisoning, the symptoms may be mild, moderate or severe. The standard treatment of OPs poisoning includes atropine, which blocks the ACh receptors, thereby preventing the action of ACh at appropriate sites, and pralidoxime, which reverses the AChE inhibition by the reactivation of the enzyme (He 2000; Jaga and Dharmani 2003). Signs and symptoms of OPs poisoning can be divided into nicotinic, muscarinic, and CNS effects (Roberts and Reigart 2013).

From a clinical perspective, the manifestation of these effects may be described as follows:

- Cardiovascular system- arrhythmias, ventricular tachycardia, bradycardia, hypertension, hypotension, prolonged QTc, cardiac arrest (Karki et al. 2004)
- Respiratory System- bronchoconstriction, aspiration pneumonia, respiratory failure from respiratory muscle weakness, severe bronchospasm, pulmonary edema, asphyxia as an immediate cause of death in OPP poisoning (Hulse et al. 2014)
- Central Nervous System- psychosis, seizure, hallucination, anxiety, restlessness, insomnia, headache, apathy, confusion, change in mental status (Jamal 1997)
- Gastrointestinal and Metabolic Systems- increased gastrointestinal tone and peristalsis, nausea, vomiting, abdominal cramps, diarrhea, pancreatitis, hyperglycemia, low bicarbonate, temporary liver damage (Hamaguchi et al. 2006)
- Renal System- acute renal failure (Agostiniand Bianchin 2003)

The severity of the toxic effects of OPs mainly depends on their metabolism and detoxification. The OPs such as parathion, chlorpyrifos, and diazinon undergo oxidative metabolism to active oxon forms, which then undergo hydrolysis to inactive metabolites. Significant individual variability in the activity of the key enzymes involved (for example, paraoxonase responsible for the hydrolysis of many oxons) was observed (Mutch, Blain, and Williams 1997). Accordingly, considering the adverse effects, substantial interindividual variability in response to OPs was observed (Mutch, Blain, and Williams 1999).

Rarely, some OPs, such as mipafox, trichlorphon, leptophos, merphos, and chlorpyrifos, were reported to induce a delayed and persistent neuropathy unrelated to AChE activity (Hayes and Laws 1991). This syndrome, termed OPP-induced delayed neuropathy (OPIDN), is manifested by weakness or paralysis and paresthesia of the extremities and may persist for weeks to years. The mechanism underlying OPIDN is the inhibition of the neuropathy target esterase (NTE) enzyme (Jamal 1997). Besides OPIDN, an intermediate syndrome, which occurs after initial improvement, approximately 24–96 hours after exposure, has been described. This syndrome is characterized by muscular weakness and acute respiratory paresis, and like OPIDN, it lacks muscarinic symptoms (Roberts and Reigart 2013). The OPs reported to produce this syndrome are methyl parathion, fenthion and dimethoate (Roberts and Aaron 2007). In addition, several reports indicated that after massive acute exposures to OPs, the long-term effects might occur. Symptoms that were usually reported included neuropsychiatric sequelae, depression, memory problems, persistent headaches and motor weakness, which in rare cases have persisted for months to years (Gershon and Shaw 1961; Metcalf and Holmes 1969).

Besides inhibition of AChE or NTE, for the longer-term effects that follow acute poisoning by OPs, several other mechanisms are suggested to be responsible. These mechanisms include the phosphorylation of proteases, esterases or proteins involved in cell signaling, excessive calcium influx in cells at nerve endings, interaction with cytoskeletal proteins, prolonged receptor stimulation at nerve endings leading to muscle fasciculation and necrosis, hypoxic brain damage, and psychological stress from the acute poisoning producing post-traumatic stress disorder (Abou-Donia 1993; Baker and Sedgwick 1996; Gupta and Abou-Donia 1995).

Chronic Effects

As mentioned above, the persistent effects as a response to OPs exposure might be those arising after acute poisoning and those occurring as the consequence of long-term or repeated low-level or subacute pesticide exposures over time (Roberts and Reigart 2013). As for acute toxicity, the effects of chronic toxicity are dose-related as well. The repeated exposure to OPs at levels insufficient to cause acute effects may not cause immediate injury but can significantly increase the risk of chronic adverse effects (Adeyinka, Muco, and Pierre 2020). In general, the link between pesticide exposure and disease is not always easy to establish. Nevertheless, the data from animal models and evidence from epidemiological studies support observations that chronic exposure to OPs may produce chronic health conditions. The chronic health conditions associated with the long-term OPs exposure include neurological effects, particularly neurodevelopmental abnormalities, cancer, metabolic disorders, endocrine disruptions, reproductive disorders, and respiratory effects (Roberts and Reigart 2013). As previously described, the acute toxicity of OPs is mainly related to inhibition of cholinesterase activity; however, this mechanism cannot alone account for the broad range of adverse effects arising following exposure to OPs, especially those related to the chronic low-dose exposure to pesticides (Terry

2012). In addition to cholinesterase, some of the non-cholinesterase targets of OPs are papain, carboxylesterase, adenylyl cyclase, acylpeptide hydrolase, fatty acid amide hydrolase, cannabinoid CB1 receptors, albumin transferrin, kinesin, ATP Synthase, and tubulin (Terry 2012).

Many registered OPs such as chlorpyrifos and diazinon are specifically toxic to the nervous system. Neurologic symptoms associated with chronic exposure to OPs include peripheral neuropathy, manifested by paresthesia, pain, anesthesia, and paresis and ataxia (Roberts and Reigart 2013). The CNS effects usually include depression, anxiety, memory impairment, and some other symptoms, such as insomnia, headaches, and neurobehavioral disorders (Lockridge and Schopfer 2020). Chronic low-dose OPs exposure is also associated with the increased risk of Parkinson's disease (Wang et al. 2014) and Alzheimer's disease (Yan et al. 2016). The mechanisms leading to neurotoxicity and dementia upon long-term exposure to OPs still need to be elucidated. A recent study has suggested that tubulin, as one of the OPs targets, can be modified by chlorpyrifos oxon leading to the formation of high-molecular-weight protein multimers. The cross-linked tubulin and other proteins in the brain might disrupt the microtubule function leading to neurotoxicity (Lockridge and Schopfer2020). The major concern with the neurotoxicity of OPs is their effects on the immature nervous system. In infants and children, the effects of exposure to the OPs can be permanent, while effects in adults may be temporary and reversible (Lim and Bolstad 2019). It could be explained by the greater sensitivity and susceptibility of the rapidly developing nervous system (Lim and Bolstad 2019). Several studies on prenatal exposure to OPs and cognitive capabilities in children have reported abnormal reflexes in newborns, mental and motor delays in preschoolers, lower IQ, poorer working memory and perceptual reasoning among elementary school-age children. In addition, prenatal exposures also increased the risk of attention-deficit/hyperactivity disorder (ADHD) and autism spectrum disorder (ASD) (Hertz-Picciotto et al. 2018).

Numerous epidemiologic studies have been reported a link between pesticide exposure and cancer. Tumors of the prostate, kidney, pancreas, and breast are among the most steadily reported outcomes (Roberts and Reigart 2013). Considering the OPs exposure, as comprehensively discussed by Lerro et al. (2015) and Pluth, Zanini, and Battisti (2019), diazinon, chlorpyrifos, and terbufos use has been linked to lung cancer; diazinon, terbufos, fonofos, and malathion use has been linked to leukemia; terbufos use has been linked to non-Hodgkin lymphoma and lymphohematopoietic cancer; azinphosmethyl, diazinon, terbufos, fonofos, and malathion use has been linked to prostate cancer. The majority of studies mentioned above have been focused on occupational exposure among farmers in predominantly male populations. However, considering the female population, the results of the Agricultural Health Study cohort (n = 30,003) have shown an increased risk of hormonally-related cancers, including thyroid, ovary, and breast cancers in female spouses of pesticide applicators (Lerro et al. 2015). Namely, malathion use was associated with an increased risk of thyroid cancer, while diazinon use was associated with ovarian cancer (Lerro et al. 2015). The association between OPs exposure and breast cancer has been observed for malathion and parathion, which were reported to alter estrogen activity and upregulate genes associated with carcinogenesis, in some cases in combination with estrogen (Calaf and Roy 2008). In addition, some studies linked chlorpyrifos use with an increased risk of breast cancer (Kim, Kabir, and Jahan, 2017).

The pediatric cancer types reported to be associated with pesticide exposure are leukemia and brain tumors (Roberts and Reigart 2013). In the majority of cases, the OPs exposure was related to household insecticide use during pregnancy. The strongest association was found for pediatric acute lymphoblastic leukemia (ALL). Recently, a case-control study of children newly diagnosed with ALL and their mothers (n = 41 child-mother pairs) has shown significant differences between children with ALL and controls for the organophosphate metabolites diethylthiophosphate and diethyldithiophosphate (Soldin et al. 2009). However, further studies are warranted to confirm the observed association.

Numerous studies have shown that pesticides may act as endocrine-disrupting chemicals that interfere with the hormone systems even at low doses, affecting the normal development and function of multiple organs (Leemans et al. 2019). Much experimental evidence suggests that pesticides may act as thyroid disruptors disturbing the central regulation, iodine uptake, and production and distribution of thyroid hormones (Boas, Feldt-rasmussen, and Main 2012). The association between exposure to OPs and thyroid function can be evaluated by measuring the urinary levels of dialkyl phosphates (DAP) produced during the metabolism of OPs. Since most commonly used OPs could be metabolized into six DAPs, the concentrations of DAPs in urine are often used as the biomarkers of OPs exposure. An epidemiological study conducted on children in Indonesia (n = 66) has shown that the mean TSH level in DAP-positive children was significantly higher than that observed in children with no detectable levels of DAP (Suhartono et al. 2018). In the adult population, the results of the study conducted in China (n = 325) have shown that DAP levels in the urine of pregnant women were positively correlated with FT4 concentration, indicating the possible impairment of thyroid function during pregnancy as the consequence of OPs exposure (Leemans et al. 2019). Furthermore, the Agricultural Health Study results of male pesticide applicators demonstrated that the use of malathion was associated with an increased prevalence of hypothyroidism (Lerro et al. 2015). Besides, chlorpyrifos was also found to disturb the adrenal and thyroid gland homeostasis and inhibit sex hormones (Yang et al. 2019).

Several studies have demonstrated that long-term exposure to OPs may lead to metabolic dysregulations such as hypertension, hyperglycemia, overweight or dyslipidemia (Leonel Javeres et al. 2021). The results of the studies performed in Cameroon (n = 220) and Pakistan (n = 220) on participants that were exposed to malathion, parathion and chlorpyrifos directly (farmers and sprayers) or indirectly (people living in the pesticide-spraying areas) have shown dysregulated liver and kidney function regardless of gender and age groups (Leonel Javeres et al. 2021). Furthermore, a growing body of evidence specifies the significant association between the OPs exposure and risk of obesity and type 2 diabetes mellitus (Czajka et al. 2019). The results of the Agricultural Health Study encompassing the licensed pesticide applicators (n = 31,787) in the USA have shown that long-term exposure to coumaphos, phorate, terbufos and trichlorfon was associated with an increased risk of diabetes (Montgomery et al. 2008).

Considering the effects of OPs exposure on the reproductive system, animal studies have shown that OPs can disturb serum reproductive hormone concentrations and semen quality parameters (Dirican and Kalender 2012). However, the adverse effects of OPs on human reproductive health have not been fully characterized. Focusing on male infertility problems, in the study conducted in Spain, the associations between urinary concentrations of six DAP metabolites and reproductive parameters were investigated in men (n = 162) seeking infertility evaluation (Melgarejo et al. 2015). It was found that sperm concentration and total sperm count were inversely associated with dimethyl phosphate, dimethyl thiophosphate, and dimethyl dithiophosphate metabolite levels. Thus, the study results have shown that exposure to OPs might be associated with decreased sperm count and altered reproductive hormone levels in men (Melgarejo et al. 2015). Nevertheless, further studies are needed to support these findings.

The epidemiologic studies on the association between pesticide exposure and respiratory health in children and adults have reported mixed results. Several studies have demonstrated that OPs exposure might lead to the development or exacerbation of asthma. It appears that OPs may contribute to the exacerbation of asthma by irritation, inflammation, immunosuppression, or endocrine disruption (Kim, Kabir, and Jahan, 2017). In a cohort of women and their children (n = 359) conducted in the USA, it was found that early exposure to OPs led to respiratory symptoms associated with childhood asthma (Raanan et al. 2015). In adult populations, several examinations in the Agricultural Health Study have shown that exposure to chlorpyrifos, malathion and parathion were positively associated with wheeze in farmers (Roberts and Reigart 2013). Similar findings were obtained in the group of the commercial applicators exposed to chlorpyrifos, dichlorvos and phorate. In a study involving farm women (n = 25,814) in the USA, any use of pesticides was found to be related to atopic asthma (Hoppin et al. 2008).

The data obtained shows that evidence of toxicity arising from chronic, low-level exposure to OPs is accumulating, so both the public and private sectors should put effort into the development and implementation of pesticide use strategies to minimize the adverse health effects.

Future Prospects

In recent years, several strategies have been proposed to develop and manufacture new, effective, safe to humans, and environmentally friendly pesticides, and to, mandated by legislation globally, ensure the cumulative assessment of human health risks from exposure to currently used pesticides. As discussed above, according to the Action Plan for the years 2022-2030, established by the EU Commission and EFSA, new methods for cumulative risk assessment from dietary and non-dietary exposure to pesticides will be developed and implemented. The plan is based on current knowledge and will be flexible in responding to new scientific developments and experiences (European Commission 2021). Due to the pronounced scientific progress in chemistry, biology, and molecular biology, the way of searching for novel pesticides, the re-assessment of the safety of the already used pesticides, the development of new biosensors for pesticide detection, and methods for pesticide degradation into the less toxic products, have been improved continuously. New techniques with better reliability than those already existing will enable a better prediction of the potential hazards of pesticides and, hence, contribute to reducing the adverse effects on human health and the environment.

General strategies for creating new pesticides encompass (i) development of pesticides that are effective at extremely low doses, (ii) development of pesticides that are rapidly degradable and less residual in the environment, and (iii) development of selective toxic chemicals effective in the control of pest but not toxic against humans and beneficial organisms (Umetsu and Shirai2020). Together with the protection of human health and the environment, the development of novel pesticides is very important for avoiding pesticide resistance and enabling sustainable agricultural production. In the last decade, at least 105 chemical pesticides (fungicides, insecticides, nematicides, acaricides, and herbicides) have been launched or under development. However, although most of them appear safe to humans and the environment, only a few products have been developed for practical use (Umetsu and Shirai 2020). Chemical pesticides are frequently used for crop protection and are expected to be commonly used in the future. Nevertheless, pesticide development has been increasingly shifting from chemical pesticides to biological pesticides, RNAi pesticides, abiotic stress control agents, genetically modified crops, and seeds (Umetsu and Shirai 2020). Biological pesticides are reported to affect the environment to a lesser extent than chemical pesticides; it is predicted that they will equal chemical pesticides on the market by the 2050s (Damalas and Koutroubas 2018).

As already described, the use of OPs in agriculture is estimated to account for approximately 40% of all the pesticides used in this field (Thunga et al. 2010). Because of the OPs toxicity, many detection methods based on the biological, chemical and physical parameters have been engaged in order to identify even trace amounts of the pesticides. The traditional methods for pesticide detection are based on instrumental techniques such as gas chromatography (GC), high-performance liquid chromatography (HPLC), or chromatographic methods coupled with mass spectrometry (MS) detectors, which provide profuse qualitative and quantitative information of the residues with high accuracy. However, the main limitations of these methods are the time-consuming sample preparation and requirement for highly trained technicians and expensive equipment (Zamora-Sequeira et al. 2019). Common rapid methods, such as immunoassays, spectroscopic analyses, and electrochemical techniques that can be used complementarily to instrumental techniques, significantly improved the detection of pesticides. The development of sensors (electrochemical sensors, optical sensors including chemiluminescence, fluorescence and colorimetric sensors, and piezoelectric sensors) and biosensors represents a novel strategy for monitoring pesticide contamination. In addition, the introduction of nanomaterials into their construction enhances the efficiency of sensors and biosensors as analytical tools for the detection of pesticides (Zamora-Sequeira et al. 2019). The use of biosensors in pesticide detection enables efficient and precise analysis at a low cost. The AChE-based biosensor has been commonly used to detect OPs (diazinon, dimethoate, dichlorvos, chlorpyrifos, malathion, methyl parathion, glyphosphate, etc.). The main limitation of this method is the lack of stability and persistent need for a substrate for the quantification of the pesticide level. Therefore, numerous methods based on the AChE inhibition, including colorimetric and electrochemical assays, have been developed to improve the system's stability and enable a more efficient analysis with lower detection capabilities. Moreover, the immobilization of AChE with different nanocomposites (e.g., single-walled carbon nanotubes, multi-wall carbon nanotubes, gold nanorods, etc.) has been seen as a potent tool to increase the response of the biosensor in pesticides detection (Kaushal, Khatri, and Arya 2021).

Since food samples usually contain more than one pesticide, there is a growing need to develop sensitive multi-residue detection methods. As comprehensively discussed by Jia et al. (2020), the two strategies for rapid multi-residue detection methods have been proposed; the first strategy based on different recognition elements includes the use of antibodies, aptamers, and molecular imprinted polymers (MIPs), while the second strategy based on the inherent characteristics of pesticides includes the utilization of the enzymatic inhibition-based sensors, near-infrared (NIR) spectroscopy and a surface-enhanced Raman scattering (SERS) spectroscopy. The proposed methods and their further improvements will enable efficient and accurate multi-residue screening and detection.

Several abiotic and biotic degradation strategies have been developed and continuously improved to minimize human exposure to pesticides and their potential adverse effects on the environment (Kaushal, Khatri, and Arya 2021). The main goal is to eliminate pesticides completely or to a great extent from the environment. Abiotic strategy refers to the direct chemical or mechanical breakdown of OPs into nontoxic forms. Some of the methods include the use of oxidants such as ferrate (VI), manganese dioxide, manganese dioxide charged with bisulfite and nanostructured titania-iron mixed oxides, catalytic hydrolysis by molybdates and silver, and direct degradation of pesticides by gamma irradiation (Kaushal, Khatri, and Arya 2021).

The biotic strategy includes microbial and enzymatic degradation of pesticides. For example, *Bacillus aryabhattai* was found to be effective in the degradation of chlorpyrifos and parathion, while several bacterial species (*Pseudomonas putida, Sphingomonassp* and *Rhodoccocusrhodochrous*) were efficient in the degradation of malathion (Geed et al. 2016). Considering the enzymatic degradation of OPs, the enzymes extracted from bacterial colonies, such as the prolidase extracted from the marine bacterium *Pseudoalteromonas* sp. can be, with further protein engineering, used for the OPs degradation. In addition, the organophosphorus hydrolase from *Pseudomonas nitroreducens* AR-3 strain has been proved to be efficient in the elimination of chlorpyrifos (Aswathi, Pandey, and Sukumaran 2019). The combined use of abiotic and biotic degradation strategies for OPs might enable a higher rate of their elimination from the environment.

Conclusion

The OPs are efficient against pests and are widely used in agriculture to prevent crop damage and increase agriculture productivity, so their usage in the past decades has been intensified. Although numerous studies on the toxicity of OPs have been performed, there are still research gaps in the prediction of their long-term effects on human health and the environment. Since the harms to humans and non-target species can be widespread and severe, there is a continuous need to highlight the public health significance of OPs exposure. Nowadays, pesticide development has been increasingly shifting from chemical pesticides to biological pesticides and other less toxic formulations; however, the OPs are frequently used and are expected to be commonly used in the future. Due to the pronounced scientific progress in chemistry, biology, and molecular biology, several strategies, including advances in creating, detection and degradation of OPs, aimed at reducing their adverse health and environmental effects have been developed and continuously improved.

Acknowledgments

This research was funded by the Ministry of Education, Science and Technological Development of the Republic of Serbia, grant number 451-03-9/2021-14/200017.

References

Abou-Donia, Mohamed B. 1993. "The Cytoskeleton as a Target For Organophosphate Ester Induced Delayed Neuropathy (OPIDN)." *Chemico-Biological Interactions* 87:383-93. doi: 10.1016/0009-2797(93)90066-8.

Abrams, Paul, Karl-Erik Andersson, Jerry J. Buccafusco, Christopher Chapple, William C. de Groat, Alison D. Fryer, Gary Kay, Alan Laties, Neil M. Nathanson, Pankaj J. Pasricha, and Alan J. Wein. 2006. "Muscarinic Receptors: Their Distribution and Function in Body Systems, and the Implications for Treating Overactive Bladder." *British Journal of Pharmacology* 148 (5):565-78. doi: 10.1038/sj.bjp.0706780.

Adeyinka, Adebayo, ErindMuco, and Louisdon Pierre. 2020. *Organophosphates*. StatPearls [Internet]. Treasure Island (FL): StatPearls Publishing.

Agostini, Moreno, and Andrea Bianchin. 2003. "Acute Renal Failure from Organophospate Poisoning: A Case of Success with Haemofiltration." *Human & Experimental Toxicology* 22 (3):165-67. doi: 10.1191/0960 327103ht343cr.

Anastassiadou, M., J. Choi, T. Coja, B. Dujardin, A. Hart, A. F. Hernandez-Jerrez, S. Jarrah, A. Lostia, K. Machera, I. Mangas, A. Mienne, M. Schepens, A. Widenfalk, L. Mohimont, and European Food Safety Authority. 2021. "Cumulative Dietary Risk Assessment of Chronic Acetylcholinesterase Inhibition by Residues of Pesticides." *EFSA Journal* 19 (2):6392. doi: 10.2903/j.efsa.2021.6392.

Antonini, Carla, and Josep M. Argilés-Bosch. 2017. "Productivity and Environmental Costs from Intensification of Farming. A Panel Data Analysis Across EU Regions." *Journal of Cleaner Production* 140:796–03. doi: 10.1016/j.jclepro.2016.04.009.

Aswathi, Aswathi, Ashok Pandey, and Rajeev K. Sukumaran. 2019. "Rapid Degradation of the Organophosphate Pesticide – Chlorpyrifos by a Novel Strain of *Pseudomonas nitroreducens* AR-3." *Bioresource Technology* 292:122025. doi: 10.1016/j.biortech. 2019.122025.

Baker, D. J., and E. Michael Sedgwick. 1996. "Single Fibre Electromyographic Changes in Man after Organophosphate Exposure." *Human & Experimental Toxicology* 15 (5):369-75. doi: 10.1177/0960327196 01500501.

Boas, Malene, Ulla Feldt-rasmussen, and Katharina M. Main. 2012. "Molecular and Cellular Endocrinology Thyroid Effects of Endocrine Disrupting Chemicals." *Molecular and Cellular Endocrinology* 355:240–48. https://doi.org/10.1016/j.mce. 2011.09.005.

Bondžić, Aleksandra M., Tamara D. Lazarević-Pašti, Andreja R. Leskovac, Sandra Ž. Petrović, Mirjana B. Čolović, Tatjana N. Parac-Vogt, and Goran V. Janjić. 2020. "A New Acetylcholinesterase Allosteric Site Responsible for Binding Voluminous Negatively Charged Molecules – The Role in the Mechanism of AChE Inhibition." *European Journal of Pharmaceutical Sciences* 151. doi: 10.1016/j.ejps.2020.105376.

Brancato, A., D. Brocca, C. De Lentdecker, Z. Erdos, L. Ferreira, L. Greco, S. Jarrah, D. Kardassi, R. Leuschner, C. Lythgo, P. Medina, I. Miron, T. Molnar, A. Nougadere, R. Pedersen, H. Reich, A. Sacchi, M. Santos, A. Stanek, J. Sturma, J. Tarazona, A. Theobald, B. Vagenende, A. Verani, L. Villamar-Bouza, and European Food Safety Authority. 2018. "Setting of Import Tolerances for Flubendiamide in Apricots,

Peaches, Nectarines, Plums and Soya Beans." *EFSA Journal* 16 (1):5128. doi: 10.2903/j. efsa.2018.5128.

Calaf, G. M., and D. Roy. 2008. "Cancer Genes Induced by Malathion and Parathion in the Presence of Estrogen in Breast Cells." *International Journal of Molecular Medicine* 21 (2):261-68.doi: 10.3892/ijmm.21.2.261.

Craig, P. S., B. Dujardin, A. Hart, A. F. Hernandez-Jerez, S. H. Bennekou, C. Kneuer, B. Ossendorp, R. Pedersen, G. Wolterink, L. Mohimont, and European Food Safety Authority. 2020a. "Cumulative Dietary Risk Characterisation of Pesticides That Have Acute Effects on the Nervous System." *EFSA Journal* 18 (4):6087. doi: 10.2903/j.efsa.2020.6087.

Craig, P. S., B. Dujardin, A. Hart, A. F. Hernandez-Jerez, S. H. Bennekou, C. Kneuer, B. Ossendorp, R. Pedersen, G. Wolterink, L. Mohimont, and European Food Safety Authority. 2020b. "Cumulative Dietary Risk Characterisation of Pesticides That Have Chronic Effects on the Thyroid." *EFSA Journal* 18 (4):6088. doi: 10.2903/ j.efsa.2020.6088.

Crivellente, F., A. Hart, A. F. Hernandez-Jerez, S. H. Bennekou, R. Pedersen, A. Terron, G. Wolterink, L. Mohimont, and European Food Safety Authority. 2019. "Establishment of Cumulative Assessment Groups of Pesticides for Their Effects on the Nervous System." *EFSA Journal* 17 (9):5800. doi: 10.2903/j.efsa.2019.5800.

Czajka, Magdalena, Magdalena Matysiak-Kucharek, Barbara Jodlowska-Jedrych, Krzysztof Sawicki, Berta Fal, Bartłomiej Drop, Marcin Kruszewski, and Lucyna Kapka-Skrzypczak. 2019. "Organophosphorus Pesticides Can Influence the Development of Obesity and Type 2 Diabetes with Concomitant Metabolic Changes." *Environmental Research* 178:108685. doi: 10.1016/j.envres.2019.108685.

Čolović, Mirjana, Danijela Krstić, Sandra Petrović, Andreja Leskovac, Gordana Joksić, Jasmina Savić, Mladen Franko, Polonca Trebše, and Vesna Vasić. 2010. "Toxic Effects of Diazinon and Its Photodegradation Products." *Toxicology Letters* 193 (1):9-18. doi: 10.1016/j.toxlet.2009. 11.022.

Damalas, Christos A., and Ilias G. Eleftherohorin. 2011. "Pesticide Exposure, Safety Issues, and Risk Assessment." *International Journal of Environmental Research and Public Health* 8:1402-19. doi: 10.3390/ijerph 8051402.

Damalas, Christos A., and Spyridon D. Koutroubas. 2016. "Farmers' Exposure to Pesticides: Toxicity Types and Ways of Prevention." *Toxics* 4 (1):1. doi: 10.3390/ toxics4010001.

Damalas, Christos A., and Spyridon D. Koutroubas. 2018. "Current Status and Recent Developments in Biopesticide Use. *Agriculture* 8 (13):1-6. doi: 10.3390/agriculture 8010013.

Deadman, Michael L. 2017. "Sources of Pesticide Residues in Food: Toxicity, Exposure, and Risk Associated with Use at the Farm Level." In *Pesticide Residue in Foods*, edited by Mohidus S. Khan and Mohammad S. Rahman, 7-35. Cham: Springer International Publishing. https://doi.org/10.1007/978-3-319-52683-6_2.

Dirican, Enver Kerem, and Yusuf Kalender. 2012. "Dichlorvos-Induced Testicular Toxicity in Male Rats and the Protective Role of Vitamins C and E." *Experimental and toxicologic pathology* 64:821-30. doi: 10.1016/j. etp.2011.03.002.

Donley, Nathan. 2019. "The USA Lags Behind other Agricultural Nations in Banning Harmful Pesticides." *Environmental Health* 18 (1):44. doi: 10.1186/s12940-019-0488-0.

European Commission. 2021. *Cumulative Risk Assessment*. https://ec.europa. eu/food/plant/pesticides/max_residue_levels/cumulative_risk_en.

EFSA-SANTE/10178/2021 2021 (European Commission). https://ec.europa.eu/food/system/files/2021-03/pesticides_mrl_cum-risk-ass_sante-10178-2021.pdf.

Geed, S. R., M. K. Kureel, A. K. Shukla, R. S. Singh, and B. N. Rai. 2016. "Biodegradation of Malathion and Evaluation of Kinetic Parameters Using Three Bacterial Species." *Resource-Efficient Technologies* 2:S3-S11. doi: 10.1016/j.reffit.2016.09.005.

Gershon, S., and F. H. Shaw. 1961. "Psychiatric Sequelae of Chronic Exposure to Organophosphorus Insecticides." *Lancet* 277 (7191):1371-74. doi: 10.1016/s0140-6736(61)92004-9.

Gupta, Ram P., and Mohamed B. Abou-Donia. 1995. "Neurofilament Phosphorylation and Calmodulin Binding by Ca/Calmodulin Dependent Protein Kinase in the Brain Subcellular Fracion of DFP Treated Hen." *Neurochemical Research* 20 (9):1095-105. doi: 10.1007/BF00995565.

Hayes, Wayland J., and Edward R. Laws, eds. 1991. *Handbook of pesticide toxicology*. San Diego: Academic Press.

Hamaguchi, Mitsuhide, Akira Namera, Noriko Tsuda, Toshifumi Uejima, Katsuyuki Maruyama, Tohoru Kanai, and Ikuhiro Sakata. 2006. "Severe Acute Pancreatitis Caused by Organophosphate Poisoning." *Chudoku Kenkyu* 19 (4):395-99.

He, F. 2000. "Neurotoxic Effects of Insecticides—Current and Future Research Review." *Neurotoxicology* 21:839–45.

Hertz-Picciotto, Irva, Jennifer B. Sass, Stephanie Engel, Deborah H. Bennett, Asa Bradman, Brenda Eskenazi, Bruce Lanphear, and Robin Whyatt. 2018. "Organophosphate Exposures during Pregnancy and Child Neurodevelopment: Recommendations for Essential Policy Reforms." *Plos Medicine* 15 (10). doi: 10.1371/journal.pmed.1002671.

Hoppin, Jane A., David M. Umbach, Stephanie J. London, Paul K. Henneberger, Greg J. Kullman, Michael C. R. Alavanja, and Dale P.Sandler. 2008. "Pesticides and Atopic and Nonatopic Asthma among Farm Women in the Agricultural Health Study." *American Journal of Respiratory and Critical Care Medicine* 177 (1):11-18. doi: 10.1164/rccm.200706-821OC.

Hulse, Elspeth J., James O. J. Davies, A. John Simpson, Alfred M. Sciuto, and Michael Eddleston. 2014. "Respiratory Complications of Organophosphorus Nerve Agent and Insecticide Poisoning Implications for Respiratory and Critical Care." *American Journal of Respiratory and Critical Care Medicine* 190 (12):1342-54. doi: 10.1164/rccm.201406-1150CI.

Jaga, Kushik, and Chandrabhan Dharmani. "Sources of Exposure to and Public Health Implications of Organophosphate Pesticides." *Revista Panamericana de Salud Pública* 14 (3):171-85. doi: 10.1590/s1020-49892003000800004.

Jamal, Goran A. 1997. "Neurological Syndromes of Organophosphorus Compounds." *Adverse Drug Reactions and Toxicological Reviews* 16 (3):133-70.

Jia, Min, E. Zhongbo, Fei Zhai, and Xin Bing. 2020. "Rapid Multi-Residue Detection Methods for Pesticides and Veterinary Drugs." *Molecules* 25 (16):3590. doi: 10.3390/ molecules25163590.

Karki, P., J. A. Ansari, S. Bhandary, and S. Koirala. 2004. "Cardiac and Electrocardiographical Manifestations of Acute Organophosphate Poisoning." *Singapore Medical Journal* 45 (8):385-89.

Kaushal, Jyoti, Madhu Khatri, and Shailendra K. Arya. 2021. "A Treatise on Organophosphate Pesticide Pollution: Current Strategies and Advancements in Their Environmental Degradation and Elimination." *Ecotoxicology and Environmental Safety* 207:111483. doi: 10.1016/j.ecoenv.2020.111483.

Kim, Ki-Hyun, Ehsanul Kabir, and Shamin A. Jahan. 2017. "Exposure to Pesticides and the Associated Human Health Effects." *Science of the Total Environment* 575:525-35. doi: 10.1016/j.scitotenv.2016.09.009.

Kwong, Tai C. 2002. "Organophosphate Pesticides: Biochemistry and Clinical Toxicology." *Therapeutic Drug Monitoring* 24 (1):144-49. doi: 10.1097/ 00007691-200202000-00022.

Lazarević-Pašti, Tamara, Andreja Leskovac, Tatjana Momić, Sandra Petrović, and Vesna Vasić. V. 2017. "Modulators of Acetylcholinesterase Activity: from Alzheimer's Disease to Anti-cancer Drugs." *Current Medicinal Chemistry*24 (30):3283-309. doi: 10.2174/0929867324666170705123509.

Leemans, Michelle, Stephan Couderq, Barbara Demeneix, and Jean-Baptiste Fini. 2019. "Pesticides with Potential Thyroid Hormone-Disrupting Effects: A Review of Recent Data." *Frontiers in Endocrinology* 10:743. doi: 10.3389/fendo.2019.00743.

Leonel Javeres, Mbah Ntepe, Rabia Habi, Ngondi Judith Laure, Syed Tahir Abbas Shah, Martin Valis, Kamil Kuca, and Syed Muhammad Nurulain. 2021. "Chronic Exposure to Organophosphates Pesticides and Risk of Metabolic Disorder in Cohort from Pakistan and Cameroon." *International Journal of Environmental Research and Public Health* 18 (5):2310. doi: 10.3390/ijerph18052310.

Lerro, Catherine C., Stella Koutros, Gabriella Andreotti, Melissa C. Friesen, Michael C. Alavanja, Aaron Blair, Jane A. Hoppin, Dale P. Sandler, Jay H. Lubin, Xiaomei Ma, Yawei Zhang, and Laura E. Beane Freeman. 2015. "Organophosphate Insecticide Use and Cancer Incidence among Spouses of Pesticide Applicators in the Agricultural Health Study." *Occupational and Environmental Medicine* 72 (10):736-44. doi: 10.1136/Oemed-2014-102798.

Li, Qing. 2007. "New Mechanism of Organophosphorus Pesticide-induced Immunotoxicity." *Journal of Nippon Medical School* 74 (2):92-105. doi: 10.1272/ jnms.74.92.

Lim, Lori, and Heather M. Bolstad. 2019. "Organophosphate Insecticides: Neurodevelopmental Effects."In *Encyclopedia of Environmental Health,* edited byJerome Nriagu, 785-791. Elsevier. doi: 10.1016/B978-0-12-409548-9.10930-3.

Lockridge, Oksana, and Lawrence M. Schopfer. 2020. "The Cross-Linking Action of Organophosphorus Poisons; Implications for Chronic Neurotoxicity." In *Handbook of Toxicology of Chemical Warfare Agents,* edited byRamesh C. Gupta, 1027-1033. Academic Press. https://doi.org/ 10.1016/B978-0-12-819090-6.00059-3.

Lu, Xian T., Yun Ma, Cui Wang, Xiao F. Zhang, Da Q. Jin, and Chang J. Huang. 2012. "Cytotoxicity and DNA Damage of Five Organophosphorus Pesticides Mediated By

Oxidative Stress in PC12 Cells and Protection by Vitamin E." *Journal of Environmental Science and Health Part B-Pesticides Food Contaminants and Agricultural Wastes* 47 (5):445-54. doi: 10.1080/ 03601234.2012.663312.

Melgarejo, Maria, Jaime Mendiola, Holger M. Koch, Miriam Monino-Garcia, Jose A. Noguera-Velasco, and Alberto M. Torres-Cantero. 2015. "Associations between Urinary Organophosphate Pesticide Metabolite Levels and Reproductive Parameters in Men from an Infertility Clinic." *Environmental Research* 137:292-98. doi: 10.1016/j.envres.2015.01.004.

Metcalf, David R., and Joseph H. Holmes. 1969. "Toxicology and Physiology. EEG, Psychological, and Neurological Alterations in Humans with Organophosphorus Exposure." *Annals of the New York Academy of Sciences* 160 (1):357-65. doi: 10.1111/j.1749-6632.1969.tb15857.x.

Montgomery, M. P., F. Kamel, T. M. Saldana, C. R. Alavanja, and D. P. Sandler. 2008. "Incident Diabetes and Pesticide Exposure among Licensed Pesticide Applicators: Agricultural Health Study, 1993-2003." *American Journal of Epidemiology* 167 (10):1235-46. doi: 10.1093/aje/kwn028.

Mutch, Elaine, Peter G. Blain, and Faith M. Williams. 1997. "Cytochrome P450 isozymes involved in parathion metabolism by human liver." *Human and Experimental Toxicology* 16:409.

Mutch, Elaine, Peter G. Blain, and Faith M. Williams. 1999. "The role of metabolism in determining susceptibility to parathion toxicity in man." *Toxicological Letters* 107:177-87. doi: 10.1016/s0378-4274(99)00044-2.

Ojha, Anupama, Santosh K. Yaduvanshi, Satish C. Pant, Vinay Lomash, and Nalini Srivastava. 2013. "Evaluation of DNA Damage and Cytotoxicity Induced by Three Commonly Used Organophosphate Pesticides Individually and In Mixture, in Rat Tissues." *Environmental Toxicology* 28 (10):543-52. doi: 10.1002/tox.20748.

Pluth, T. Bremm, Lucas A. G. Zanini, and Iara D. E. Battisti. 2019. "Pesticide Exposure and Cancer: An Integrative Literature Review." *Saúde em Debate*, 43 (122):906–24. doi: 10.1590/0103-1104201912220.

Raanan, Rachel, Kim G. Harley, John R. Balmes, Asa Bradman, Michael Lipsett, and Brenda Eskenazi. 2015. "Early-life Exposure to Organophosphate Pesticides and Pediatric Respiratory Symptoms in the CHAMACOS Cohort." *Environmental Health Perspectives* 123 (2):179-85. doi: 10.1289/ehp.1408235.

Roberts, Darren M., and Cynthia K. Aaron. 2007. "Management of Acute Organophosphorus Pesticide Poisoning." *British Medical Journal* 334:629-34. doi: 10.1136/bmj.39134.566979.BE.

Roberts James R., and J. Routt Reigart. 2013. *Recognition and Management of Pesticide Poisoning: Overview*. 6th ed. Washington, DC: US Environmental Protection Agency Office of Pesticide Programs. https://www.epa.gov/ pesticide-worker-safety/ recognition-and-management-pesticide-poisonings.

Rusyniak, Daniel E., and Kristine A. Nañagas. 2004. "Organophosphate Poisoning." *Seminars in Neurology* 24 (2):197-04. doi: 10.1055/s-2004-830907.

Salvatore, Alicia L., Asa Bradman, Rosemary Castorina, José Camacho, Jesús Loopez, Dana B. Barr, John Snyder, Nicholas P. Jewell, and Brenda Eskenazi. 2008. "Occupational Behaviors and Farmworkers' Pesticide Exposure: Findings from a

Study in Monterey County, California." *American Journal of Industrial Medicine* 51 (10):782-94. doi: 10.1002/ajim. 20622.

Savić, Z. Jasmina, Sandra Ž. Petrović, Andreja R. Leskovac, Tamara D. Lazarević Pašti, Branislav J. Nastasijevic, Brankica B. Tanović, Slavica M. Gašić, and Vesna M. Vasić. 2019. "UV-C Light Irradiation Enhances Toxic Effects of Chlorpyrifos and Its Formulations." *Food Chemistry* 271:469-78. doi: 10.1016/j.foodchem.2018.07.207.

Soldin, Offie P., Hala Nsouly-Maktabi, Jeanine M. Genkinger, Christopher A. Loffredo, Juan A. Ortega-Garcia, Drew Colantino, Dana B. Barr, Naomi L. Luban, Aziza T. Shad, and David Nelson. 2009. "Pediatric Acute Lymphoblastic Leukemia and Exposure to Pesticides." *Therapeutic Drug Monitoring* 31 (4):495-501. doi: doi 10.1097/FTD.0b013e3181aae982.

Suhartono, Suhartono, Apoina Kartini, Hertanto W. Subagio, Budiyono Budiyono, Agustini Utari, Suratman Suratman, and Mateus Sakundarno. 2018. "Pesticide Exposure and Thyroid Function in Elementary School Children Living in an Agricultural Area, Brebes District, Indonesia." *International Journal of Occupational and Environmental Medicine* 9 (3):137-44. doi: 10.15171/ijoem.2018.1207.

Sullivan, John B., and J. Blose. 1992. "Organophosphate and Carbamate Insecticides." In *Hazardous Materials Toxicology: Clinical Principles of Environmental Health,* edited by John B. Sullivan and Gary R. Krieger, 1015–1026. Baltimore, Maryland, United States of America: Williams and Wilkins.

Terry, Alvin V. Jr. 2012. "Functional Consequences of Repeated Organophosphate Exposure: Potential Non-Cholinergic Mechanisms." *Pharmacology and Therapeutics* 134 (3):355–65. doi: 10.1016/j.pharmthera. 2012.03.001.

Thunga, Girish, Kishore Ganna Sam, Kanav Khera, Sureshwar Pandey, and Sudha Vidya Sagar. 2010. "Evaluation of Incidence, Clinical Characteristics and Management in Organophosphorus Poisoning Patients in a Tertiary Care Hospital." *Journal of Toxicology and Environmental Health Sciences* 2 (5):73-76. doi: 10.5897/JTEHS. 9000029.

Umetsu, Noriharu, and Yuichi Shirai. 2020. "Development of Novel Pesticides in the 21st Century." *Journal of Pesticide Science* 45 (2):54-74. doi: 10.1584/jpestics.D20-201.

Wang, Anthony, Myles Cockburn, Thomas T. Ly, Jeff M. Bronstein, and Beate Ritz. 2014. "The Association between Ambient Exposure to Organophosphates and Parkinson's Disease Risk." *Occupational and Environmental Medicine* 71 (4):275-81. doi: doi 10.1136/oemed-2013-101394.

World Health Organization & International Programme on Chemical Safety. 2010. *The WHO recommended classification of pesticides by hazard and guidelines to classification 2009.* World Health Organization. https://apps. who.int/iris/handle/ 10665/44271.

Yamada, Yukiko. 2017. "Importance of Codex Maximum Residue Limits for Pesticides for the Health of Consumers and International Trade." In *Food Safety Assessment of Pesticide Residues,* edited by Árpád Ambrus and Denis Hamilton, 269–282. Europe: World Scientific. doi: 10.1142/978178634 1693_0007.

Yan, Dandan, Yunjian Zhang, Liegang Liu, and Hong Yan. 2016. "Pesticide Exposure and Risk of Alzheimer's disease: A Systematic Review and Meta-Analysis." *Scientific Reports* 6:32222. doi: 10.1038/srep32222.

Yang, Fang-Wei, Yi-Xuan Li, Fa-Zheng Ren, Jie Luo, and Guo-Fang Pang. 2019. "Assessment of the Endocrine-Disrupting Effects of Organophosphorus Pesticide Triazophos and Its Metabolites on Endocrine Hormones Biosynthesis, Transport and Receptor Binding In Silico." *Food and Chemical Toxicology* 133:110759. doi: 10.1016/j.fct.2019.110759.

Zamora-Sequeira, Roy, Ricardo Starbird-Perez, Oscar Rojas-Carillo, and Seiling Vargas-Villalobos. 2019. "What are the Main Sensor Methods for Quantifying Pesticides in Agricultural Activities? A Review." *Molecules* 24 (14):2659. doi: 10.3390/molecules 24142659.

Zikankuba, V. Lwoga, Gaspary Mwanyika, Julius E. Ntwenya, and Armachius James. 2019. "Pesticide Regulations and Their Malpractice Implications on Food and Environment Safety." *Cogent Food & Agriculture* 5 (1):1601544. doi: 10.1080/23311932.2019.1601544.

Chapter 2

Organophosphate Pesticides: Cytotoxicity, Genotoxicity and Current Treatment Strategies

Andreja Leskovac*, PhD
Department of Physical Chemistry, "Vinča" Institute of Nuclear Sciences - National Institute of the Republic of Serbia, University of Belgrade, Belgrade, Serbia

Abstract

Pesticides are used in almost all food production areas and represent pollutants of worldwide importance. The utilization of organophosphate (OP) pesticides is considered a major global health problem. The OP pesticides toxicity is closely related to their metabolism, oxidative stress induction, mitochondrial dysfunction, and inhibition of the enzyme acetylcholinesterase (AChE). OP pesticides affect cellular and molecular targets, which may lead to systemic disturbances and, finally, the development of various human diseases. The extent and severity of OPs-induced adverse effects depend not only on the specific pesticide properties, exposure levels, and route of absorption but also on polymorphisms in the main genes involved in OP metabolism. This chapter aims to provide insight into the cytotoxic and genotoxic effects of OP pesticides in humans, their metabolism, mechanisms of action, and the potential treatment strategies to minimize their possible effects on the environment and human health.

Keywords: organophosphate pesticides, cytotoxicity, genotoxicity, treatment strategies

* Corresponding Author's Email: andreja@vin.bg.ac.rs.

In: Organophosphates
Editor: Tamara Lazarević-Pašti
ISBN: 978-1-68507-652-8

Introduction

Pesticides are used in almost all food production areas and affect virtually the entire population, making them a pollutant of global importance. According to WHO data, 1 to 5 million cases of serious intoxication and more than 300,000 lethal cases have been estimated worldwide per year due to pesticide poisoning (WHO 2016).

Organophosphates, a class of organophosphorus compounds, encompass the largest and the most frequently used group of pesticides with more than one hundred known individual compounds. According to their application, OPs are classified as insecticides, herbicides, and nerve agents. Although many OPs are characterized by rapid degradation in the environment and living organisms, most of them are biocides, which generally imply high toxicity for humans (Ferrer 2003).

OP pesticides can affect humans both directly and indirectly *via* various routes of exposure. Inhalation is the most common way of human exposure, but dermal contact and accidental ingestion of contaminated foods and water are also frequent (Argentin, Divizia, and Cicchetti 2015). The factors underlying the toxic effects of OPs mainly encompass the products of their metabolism, which induce damage at the cellular and molecular levels, including oxidative damage to macromolecules as a consequence of oxidative stress induction, mitochondrial dysfunction, inhibition of the enzyme acetylcholinesterase (AChE), and several other enzymes involved in key cellular processes. Therefore, exposure to OP pesticides has been linked to cytotoxicity and genotoxicity, and consequently neurotoxicity, endocrine disruption, reproductive disturbances, and immunomodulating effects, leading to chronic conditions such as neurobehavioral disorders, endocrine disorders, reproductive dysfunctions, and immunological disorders (Kim, Kabir, and Jahan 2017; Lushchak et al. 2018). In addition, pesticide exposure has been linked to an increase in the incidence of malignancies such as non-Hodgkin's lymphoma, multiple myeloma, soft tissue sarcoma, and lung sarcoma. Other cancers linked to OPs exposure include pancreatic, stomach, liver, bladder, and gall bladder cancer (Bolognesi et al. 2011).

The majority of OP pesticides induce well-known acute effects on human health. However, assessing the abovementioned chronic effects as a consequence of continuous long-term pesticide exposure still has much ambiguity and controversy. The chronic toxicity caused by long-term exposure to low doses of pesticides can manifest considerably later, complicating the assessment of their effects. Since long-term exposure to pesticides has severe effects on genome integrity and silently mediates cancer development, genetic vulnerability is likely to be a key factor in the observed outcomes.

The long-term effects of OP pesticides in human populations have been mainly assessed by conducting biomonitoring studies of occupationally exposed workers, which consider the impact of all endogenous and exogenous factors, metabolism, and inter-individual variability in detecting adverse pesticide effects. The commonly used agricultural OPs display different toxicity levels, and their specific risks were related to target organ cells. The cytotoxicity and genotoxicity of OPs might vary significantly depending on exposure levels, route of absorption, specific pesticide properties, and individual factors, such as age, gender, nutritional condition, general health, and polymorphisms in genes encoded xenobiotic-metabolizing enzymes (Angerer, Ewers, and Wilhelm 2007; Suarez et al. 2021; Kalyabina et al. 2021). For human health risk assessment, it is important to assess numerous events involved in pesticide toxicity and the extent of pesticide impacts on biological processes, which is crucial for predicting adverse effects. The adverse outcome pathway links critical events at different levels of biological processes that might induce detrimental consequences at the level of the whole organism and even subsequent generations of organisms in a population (Kalyabina et al. 2021).

Furthermore, when assessing pesticide mixtures' toxicity and health risk, it should be taken into account that they can affect each other following the concepts of additivity and interaction, decreasing (antagonism) or increasing (synergism) each other's effect (Silins and Högberg 2011). In addition, another important aspect of OPs, their potentiation, when a non-toxic substance enhances the impact of another substance, is important to consider when assessing their genotoxicity and cytotoxicity. Therefore, the effects of pesticide mixtures should be examined based on a cumulative risk evaluation rather than the summation of the effects of individual chemicals separately (Hernández et al. 2013).

This chapter aims to provide insight into the cytotoxic and genotoxic effects of organophosphates in humans, their metabolism, direct and indirect mechanisms of their actions, and highlight the importance of mitigating OPs effects through the development of appropriate treatment strategies.

Effects of OPs on Cellular and Molecular Targets - Basic Considerations

The toxic effects of OPs are based on their ability to trigger the processes that result in damage at the cellular and molecular levels. Cytotoxicity can be defined as the ability of a substance to cause damage to cells and consequently cell death (Vinken and Blaauboer 2017), while genotoxicity refers to the ability of a substance to cause damage to DNA in cells and/or the cellular apparatus that maintains the integrity of the genome (Ren et al. 2017). Cells with damaged DNA may or may not die. In fact, the persistence of cells with damaged DNA (mutations) that are not repaired underlies cancer development (Ren et al. 2017). Accordingly, OPs have the potential to cause both cytotoxicity and genotoxicity.

OPs interact directly with cell membranes, cause structural abnormalities, suppress the cellular antioxidant defense system, and trigger free radicals, which can cause damage to cell macromolecules (Mishra and Srivastava 2015). Furthermore, OP pesticides may cause direct genotoxic effects by forming structural alterations in DNA, resulting in higher levels of DNA damage and the subsequent expression of mutant gene products in various cell types. Since exposure to pesticides represents a genotoxic risk, different biomarkers are used to evaluate primary and permanent DNA damage in the biomonitoring of human populations. For the occupational risk assessment, the comet assay, the micronucleus (MN) assay, chromosomal aberration (CA) analysis, and sister chromatid exchange (SCE) analysis are used most frequently (Naravaneni and Jamil 2007). The elevated levels of these biomarkers, in general, imply that the investigated compound has substantial genotoxic effects.

Considering the long-term effects of OP pesticides in human populations, biomonitoring studies of occupationally exposed workers are extremely important for detecting any changes in their health conditions (Angerer, Ewers, and Wilhelm 2007). Agricultural workers are mainly exposed to pesticide mixtures with varying modes of action (Kapka-Skrzypczak et al. 2011). Therefore, it is not easy to estimate the effects of long-term exposure to OP pesticides on numerous aspects of human health. The main disadvantages of using data from cytogenetic biomonitoring studies for risk assessment are the difficulties in establishing a consistent causal exposure-diseases relationship, problems in obtaining reliable information on exposure levels in retrospective studies, concerns regarding overlap between contact with other chemicals and the problem of non-comparable populations among those exposed to varying levels of chemicals (Kapka-Skrzypczak et al. 2011). In addition, there is a large inter-individual exposure variability in the case of occupational contact with pesticides and complex combinations of a large number of chemicals that are commonly used (Jovicic et al. 2013). Nevertheless, despite all the mentioned problems, biomonitoring studies are priceless and irreplaceable because they provide insight into the real action of pesticides.

Numbered biomonitoring studies have been conducted in various countries to assess the cytotoxic and genotoxic consequences of pesticide exposure. Studies covering the last few years have shown mainly the effects of pesticide mixtures, including OPs. For example, in the study conducted in Mexico that included 80 participants, of which 54 were workers exposed to a complex mixture of pesticides, encompassing 47% OPs (diazinon, dimethoate, gusathion, lorsban, malathion, orthene, parathion, tamaron, and paraquat), all DNA damage parameters were significantly elevated in the exposed group compared to the control group. Furthermore, all parameters of primary DNA damage measured by comet assay were considerably increased in the exposed group compared to the control group. Besides, a significantly higher level of genomic instability in the exposed group was determined in the frequency of nuclear buds and the frequency of binucleated cells. However, they did not determine decreased AChE and butyrylcholinesterase (BuChE) in blood nor serum of exposed workers (Valencia-Quintana et al. 2021).

In a study performed in Greece, 2020. Potential genotoxic and cytotoxic effects caused by the occupational exposure of 24 farmers were evaluated. In 62.5% of blood samples, five pesticides (trifluralin, chlorpyriphos methyl, metolachlor, fenthion, and dimethoate) and three metabolites (fenthion sulfone, fenthion sulfoxide, and 4,4'DDE) were detected. According to their results, statistically significant differences in MN frequencies in peripheral blood lymphocytes were found among exposed and control groups, indicating a possible clastogenic and aneugenic effect of pesticides (Moshou et al. 2020).

Also, in a similar study conducted in India, 2019. The effects of exposure to pesticides on 77 women workers (exposed group) and 66 women who served as the control group (non-exposed group) were investigated. Among pesticide mixtures, at least five were organophosphates (dimethoate, phosalone, quinalphos, profenophos, and ethion). The results revealed that the exposed group had lower AChE and BuChE activities than those in the control group. In addition, the comet assay used to measure DNA damage has shown a significantly elevated level of DNA damage in exposed workers compared to non-exposed (Dhananjayan et al. 2019).

Worth mentioning is the first study in Brazil, 2019, that followed 12 male soybean growers exposed to different pesticides, including two organophosphate insecticides (chlorpyrifos and methamidophos). At the same time, the control group included 12 non-exposed male individuals. The comet assay evaluated DNA damage in whole peripheral blood, while BuChE activity in plasma was accessed to measure exposure to cholinesterase inhibitors. The study results showed that the period of high exposure to a mixture of pesticides significantly increased levels of genotoxic damage and inhibited BuChE activity. However, the increased level of DNA damage was persistent 28 days after exposure determined by repeated blood sampling, suggesting that not all DNA damage was repaired or that some new DNA lesions were induced due to low-level exposure to pesticides (Bernieri et al. 2019).

All studies mentioned above provide insight into the direct cytotoxic effects of organophosphates in humans *in vivo*. Moreover, the performed studies were complex and comprehensive because they considered the impact of all endogenous and exogenous factors, metabolism, and inter-individual variability.

General Aspects of OPs Toxicity

In general, at the cellular level, the toxicity of OPs is closely related to their metabolism. It encompasses alteration of oxidant/antioxidant balance, oxidative stress, mitochondrial dysfunction, and inhibition of the enzyme AChE. Thus, damages occurred at the molecular and cellular levels tend to accumulate, inducing subclinical manifestations for an extended period. Furthermore, impaired enzyme activities, DNA expression, and damaged membranes, when combined, will result in an irreversible disturbance of metabolism and organ system functioning (Kalyabina et al. 2021).

Metabolism of OPs

Metabolism of OPs is one of the key factors in determining their toxicity (Lushchak et al. 2018). In living organisms, OPs are metabolized in the liver through two phases. Phase I involves biotransformation (oxidations, reductions, or hydrolysis), while during phase II, conjugations of the polar groups of metabolites occur, and subsequently, they transform into more soluble and excretable forms. Enzymes involved in phase I metabolism include cytochrome P450-dependent monooxygenases (CYPs), flavin-containing monooxygenases (FMOs), paraoxonases, esterases, and hydrolases, among others, while enzymes usually involved in phase II include glutathione transferases, glucuronosyltransferase, sulfotransferase, and acetyltransferases (Tang et al. 2006).

Although these metabolic processes are expecting to be detoxication reactions, in some cases, enzymes transform substances into highly reactive forms, more reactive than the parental compound (Abass et al. 2010). This phenomenon, known as metabolic activation, occurs mostly during the oxidation of OP compounds in which desulphurization generates metabolites known as oxons (Sams, Mason, and Rawbone 2000).

In particular, it was determined that malathion metabolites and contaminants aggravate its toxicity. Two main products of malathion are malaoxon, the bioactivated form of malathion, and isomalathion, an isomer generated during the thermal or photochemical isomerization of malathion. Both have been considered responsible for insecticidal (anti-AChE) activity of malathion (Ojha and Gupta 2015). Furthermore, it was found that compared to pure malathion, malaoxon and isomalathion induced genetic damage in human blood lymphocytes in a dose-dependent manner (Blasiak et al. 1999). Also, the DNA damage generated by malaoxon was more extensive than the damage caused by isomalathion. Therefore, observed genotoxicity at higher concentrations of malathion could be due to its metabolic biotransformation to malaoxon or a consequence of the presence of malaoxon and/or isomalathion as well other unspecified impurities in commercial formulations of malathion (Blasiak et al. 1999). Another study proved that isomalathion was much more cytotoxic than malathion (Josse et al. 2014). At the same time, both compounds had comparable genotoxic effects in the human liver HepaRG (hepatocytes) cell line. In addition, isomalathion reduced cell viability and significantly induced caspase-3 activity in a dose-dependent manner (Josse et al. 2014).

Furthermore, chlorpyrifos and its metabolite 3,5,6-trichloro-2-pyridinol (TCP) were tested for cytotoxicity *in vitro* using two mammalian cell lines, HEK293 (human embryonic kidney) and N2a (mouse neuroblastoma). Results have shown that TCP is more hazardous than chlorpyrifos, inducing the concentration-dependent decrease of cell viability (Echeverri-Jaramillo et al. 2021).

Toxic effects of diazinon and its irradiated solutions (O-analog diazinon (diazoxon) solution and 2-isopropyl-6-methyl-4-pyrimidinol (IMP)) were investigated using cultivated human blood cells and skin fibroblasts. It was found that photodegradation treatment strengthened the prooxidative properties of diazinon and its potency to induce cytogenetic damage (Colovic et al. 2010). Besides, diazinon and its products increased the incidence of micronuclei (*via* a clastogenic mode of action) in a dose-dependent manner, considerably suppressed cell proliferation, enhanced malondialdehyde level, and reduced AChE and Na+/K+-ATPase activity at all concentrations tested (Colovic et al. 2010).

The impact of acephate pesticide, commonly used in agriculture and horticulture, and its selective toxicity toward insects is ascribed to faster conversion to its metabolite methamidophos in insects than in mammals. Both acephate and methamidophos have shown cytotoxic and genotoxic effects *in vivo* and *in vitro* (Ozkan et al. 2009).

Oxidative Stress and ROS-Mediated OPs Toxicity

The OPs-induced alteration of oxidant/antioxidant balance may affect crucial cellular processes, leading to enzyme inactivation, lipid peroxidation, and DNA damage that can induce cell death or promote carcinogenesis (Lushchak et al. 2018).

Namely, pesticides are shown to alter cellular redox balance by enhancing reactive oxygen species (ROS) and reactive nitrogen species (RNS), inducing lipid peroxidation and depletion of antioxidant defense (Lushchak et al. 2018). An increase in superoxide anion ($O_2^{\bullet-}$) production and decrease of the activity of cellular antioxidant enzymes, superoxide dismutase (SOD), catalase (CAT), reduced glutathione (GSH), and glutathione peroxidase (GPx) cause disturbances in oxidant-antioxidant balance and influence cell viability. Besides, pesticides alter the activities of several other enzymes, such as alanine transaminase (ALT), aspartate transaminase (AST), alkaline phosphatase (ALP), and lactate dehydrogenase (LDH) (Lushchak et al. 2018).

ROS generated through OPs-induced oxidative stress may affect cell membrane integrity, induce mitochondrial dysfunction, and cause DNA damage, resulting in genotoxic, mutagenic, and carcinogenic consequences (dos Santos et al. 2016). The imbalance between harmful reactive oxygen species and physiological antioxidant defense causes oxidative stress, which results in oxidative damage (Nita and Grzybowski 2016).

Oxidative stress has been proposed as a key mechanism involved in the pathogenesis of several diseases after acute, chronic, and developmental exposure to OPs in both animals and humans. Numerous *in vivo* and *in vitro* studies supported the oxidative stress theory (Zheng et al. 2020; Guo et al. 2013; Soltaninejad and Abdollahi 2009; Toledo-Ibarra et al. 2016; Yan, Wang, and Zhu 2013). A comprehensive analysis conducted by Soltaninejad and Abdollahi (2009) that reviewed 127 articles, of which 112 were experimental studies and 15 including humans, provides important evidence of the implication of oxidative stress in toxicity (Soltaninejad and Abdollahi 2009). It should be emphasized that the evaluation of oxidative stress biomarkers in blood was useful for monitoring exposed people (Soltaninejad and Abdollahi 2009).

Several studies have confirmed the association of OPs exposure and ROS-mediated OPs toxicity. For instance, glyphosate and its main metabolite (aminomethylphosphonic acid, AMPA) were reported to induce DNA damage (single- and double-stranded DNA formation) and increase $^{\bullet}OH$ level through ROS-mediated effects in human peripheral blood mononuclear cells (PBMCs) (Wozniak et al. 2018). Additionally, dimethoate, widely used to control crop insects, was found to inhibit some metabolic enzymes, alter neuro-physiological parameters in acute and subchronic applications and generate oxidative stress in different tissues (Ayed-Boussema et al. 2012).

Exposure to fenitrothion, an organophosphorus insecticide, promoted oxidative stress and induced numerous deleterious changes in various tissues. It was observed that fenitrothion inhibited AChE activity, increased GABA, serotonin, and dopamine levels in the brain. Besides, it induced DNA damage indicated by elevation of 8-hydroxy 2-deoxyguanosine and mRNA expression of pro-apoptotic genes, including Bax and p53, and reduced Bcl-2 expression (Alam et al. 2019).

Studies of chlorpyrifos have shown oxidative stress-induced cytotoxicity and genotoxicity in human lymphocytes as revealed by its high potency to induce lipid peroxidation, significantly enhancing micronuclei incidence and decreasing cell proliferation in a concentration-dependent manner (Savic et al. 2019).

Malathion is a non-systemic pesticide most extensively used to protect plants from pests and disease. Although it inhibits AChE, oxidative stress has been proposed as a crucial mechanism in both acute and chronic toxicity in humans (Badr 2020). Accordingly, malathion is reported to affect oxidants/antioxidants balance, induce DNA damage, and trigger apoptosis. Furthermore, based on comprehensive *in vivo* and *in vitro* studies, malathion has been identified as a mutagen, genotoxic and carcinogenic chemical compound (Blasiak et al. 1999). Moreover, *in vitro* studies on human peripheral blood lymphocytes exposed to malathion showed a dose-dependent decrease of mitotic index and increased chromosomal changes and sister chromatid exchanges (Balaji and Sasikala 1993).

According to Argentin et al. (2015), in human gingival fibroblasts (HGF) exposed to different concentrations of methyl parathion (mPT) for 24 h oxidative stress induction, cytotoxicity, and genotoxicity, and increment in NO levels through stimulation of inducible NO synthase expression were observed (Argentin, Divizia, and Cicchetti 2015). In addition, cell proliferation potential was significantly reduced at the highest investigated concentrations. Therefore, it was concluded that cell injuries produced by mPT were concentration-dependent (Argentin, Divizia, and Cicchetti 2015). According to WHO, methyl parathion is considered extremely toxic OP (Category I A according to WHO) that causes *in vivo* and *in vitro* cytotoxicity through damage to DNA and formation of ROS (Edwards, Yedjou, and Tchounwou 2013; Ojha et al. 2013). In HGF cultures, exposure to 50 or 100 g/ml of mPT caused apoptosis, whereas 10 g/ml mPT did not affect cell viability. These results indicated that pesticide-induced cell damage depends on application dose (Argentin, Divizia, and Cicchetti 2015).

The mechanisms of genotoxicity of the individual compounds and combination of OPs pesticides terbufos and fenthion were investigated on HepG2 (hepatocellular carcinoma) cell line and zebrafish embryos (Wahyuni et al. 2021). The neutral comet test and phosphorylation of H2AX (γH2AX) revealed that terbufos and fenthion induced DNA double-strand breaks (DSBs) in treated cells. Furthermore, in HepG2 cells, terbufos, fenthion, and their combination decreased expression of DNA HR repair gene Xrcc2 and expression of DNA non-homologous end joining (NHEJ) repair gene Xrcc6 (Wahyuni et al. 2021). In addition, fenthion and terbufos caused DNA damage and even apoptosis in hepatocellular carcinoma cells (Wu et al. 2011). Terbufos also induced ROS, apoptosis, and DNA damage in testicular cell lines (Hung et al. 2016). On the other side, fenthion increased lipid peroxidation, enhanced oxidative stress and micronuclei formation in *C. gariepinus* blood erythrocytes (Nwani et al. 2017), and generated chromosomal aberrations in human lymphocytes (Rani and Rao 1991).

Some studies indicate that oxidative stress is a major player in diazinon-induced cytotoxicity (Abdel-Daim, Abushouk, Bahbah, et al. 2020; Abdel-Daim et al. 2018). Diazinon, a highly toxic and commonly used pesticide for environmental pest control, was reported to generate free radicals, which interact with cellular macromolecules, causing lipid peroxidation, dysfunctional enzymes, and genotoxicity (Colovic et al. 2010). Also, diazinon inhibited key antioxidant enzymes and thus reduced the total antioxidant capacity of the cell (Toledo-Ibarra et al. 2016). Furthermore, it was found that diazinon increased serum concentrations of IL-6 and TNF-α (Abdel-Daim, Abushouk, Bahbah, et al. 2020) and the expression of cyclooxygenase-II and NF-κB, which is upstream of TNF-α (Proskocil et al. 2019). Therefore, increased TNF-α expression may directly affect diazinon intoxication due to oxidative stress (Abdel-Daim, Abushouk, Bahbah, et al. 2020). The abovementioned is in agreement with the fact that oxidative stress and TNF-α expression are linked; oxidative stress increases TNF-α expression, and TNF-α promotes oxidative stress (Fischer and Maier 2015).

OPs-Induced Mitochondrial Dysfunction

Mitochondria play important roles in metabolism, energy regulation, protein folding, and cell proliferation (Boskabady et al. 2018). Moreover, mitochondrial dysfunction can result in oxidative stress, and oxidative attack can lead to impaired mitochondrial function. Numerous data indicate a strong association between OP exposure and oxidative stress, mitochondrial dysfunction, and neurotoxicity (Guo et al. 2013). OPs affect oxidative phosphorylation by altering complexes I, II, III, IV, and V in the mitochondria, reducing mitochondrial adenosine triphosphate (ATP) synthesis and promoting oxidative and genotoxic damage. Accordingly, dysfunction and disruption of mitochondria trigger cell death *via* cytochrome C release and activation of caspases (Guo et al. 2013).

The first enzyme (complex I) in the mitochondrial electron transport chain, NADH dehydrogenase, catalyzes the transfer of electrons from NADH to ubiquinone in a reaction associated with proton translocation across the membrane while assisting in the formation of the electrochemical potential required for ATP production. It is found that some OPs can significantly decrease the activity of NADH dehydrogenase. Thus, the activity of the main enzyme of the mitochondrial electron transport chain, NADH-dehydrogenase, can be considerably reduced by some pesticides (Sharma, Lu, and Bai 2009).

For example, chlorpyrifos was found to mediate oxidative stress and neurotoxicity by disturbing NADH-dehydrogenase activity (Chen et al. 2017). The study conducted by Chen et al. (2017) has shown that NADH dehydrogenase activity was decreased in chlorpyrifos (500 μM) treated SH-SY5Y (neuroblastoma) cells as compared to controls (Chen et al. 2017). Similar results were obtained in the rat after dichlorvos administration.

Furthermore, mitochondrial NADH dehydrogenase activity was decreased by 62%-70.2% in dichlorvos treated animals compared to controls (Binukumar et al. 2010). In addition, chronic low-level dichlorvos exposure in adult rats has been found to cause apoptotic neurodegeneration by increasing mitochondrial Ca++ levels, compromising mitochondrial complexes I, III, and IV activities, decreasing the level of mitochondrial antioxidants glutathione and superoxide dismutase, and promoting oxidative stress (Kaur et al. 2007). Furthermore, besides reducing NADH dehydrogenase activity, chlorpyrifos induced mitochondrial fragmentation and ATP decline (Chen et al. 2017). In contrast, another OP pesticide trichlorfon (200 μM), did not affect NADH dehydrogenase activity nor other mitochondrial functions, suggesting that mitochondrial dysfunction did not underly the trichlorfon-induced cell toxicity (Chen et al. 2017).

Some studies implicated the involvement of mitochondria in the toxicity induced by low levels of OPs. It was observed that exposure of cultured cortical neurons to low concentrations of chlorpyrifos and its oxidative metabolite chlorpyrifos oxon resulted in alterations of mitochondrial dynamics (Terry 2012).

All changes in mitochondrial function, including mitochondrial DNA and mitochondrial dynamics, disturb mitochondrial homeostasis and function and may be linked to Parkinson's and Alzheimer's disease pathogenesis. In addition, oxidative stress related to the development of neurological dysfunction and neuronal death has a crucial pathogenic role in both disorders (Yan, Wang, and Zhu 2013).

OPs-Induced Inhibition of AChE

Synaptic AChE inhibition appears pivotal for toxicity and associated neurotoxicity after exposure to OPs (Naravaneni and Jamil 2007; Lazarevic-Pasti et al. 2017; Bondzic et al. 2020; Colovic et al. 2010). Ordinarily, AChE binds to the neurotransmitter acetylcholine (ACh) and has a critical role in the induction of nerve impulses at the nerve junction, effectively ending the stimulation of the next neuron. The rate and degree of AChE inhibition depend on the structure and nature of OPs. Its inhibition disrupts signal transmission, resulting in intoxication manifested as restlessness, hyperexcitability, convulsions, paralysis, and death (Lionetto et al. 2013). The OPs are considered to be functionally irreversible inhibitors of AChE. OPs inhibit AChE by phosphorylation of the serine hydroxyl group in the substrate-binding domain of the enzyme, which results in the accumulation of acetylcholine, overstimulation of acetylcholine receptors in the autonomic nervous system, central nervous system, and neuromuscular junction (NMJ) (Eddleston and Chowdhury 2016). Thus, some OP insecticides cause delayed neuropathy and long-term neurological consequences in chronically exposed workers (Kamel and Hoppin 2004).

In addition, some studies indicated that acetylcholinesterase enzyme inhibition is associated with the increase in reactive oxygen species, not only in workers who are exposed to OP pesticides but also in those exposed to other pesticides (Cortes-Iza and Rodriguez 2018).

Genetic Polymorphisms-Biomarkers of Susceptibility to OP Toxicity

Although the toxic effects of OPs might vary significantly depending on many factors, the biological aspects of their effects appear to be primarily influenced by genetic polymorphisms (Suarez et al. 2021). Variations in alleles of genes that encode enzymes involved in the cell cycle regulation, redox state, bioactivation, and detoxification of OPs, as well as the enzymes that these compounds target, could have a significant influence on their toxicity. Since proteins coded by different genotypes alter substrate biotransformation, monitoring genetic polymorphisms are of particular importance for risk assessment in the general population. The changes in genes that encode cytochrome P450, glutathione transferases, acetyltransferases, and paraoxonases are most significant in terms of individual susceptibility to OPs exposure and OPs-induced toxicity (Teodoro et al. 2019).

Cytochrome P450 (CYP)

The human genome comprises 57 functional, protein-coding cytochrome P450 (CYP) genes grouped according to their sequence homology into 18 families and 44 subfamilies (Nebert, Wikvall, and Miller 2013). The CYP proteins are involved in the chemical defense, synthesis, and metabolism of major biological signaling molecules. For example, although CYP catalyzes detoxication reactions, many pesticides can be bioactivated, converted by CYP into highly toxic oxon forms (Jan et al. 2016).

It was estimated that the majority of oxidative metabolism and elimination of commonly used compounds is accomplished through the three CYP families (CYP1, CYP2, and CYP3) in humans (Wilkinson 2005). Numbered investigations confirm that *CYP1A2*, *CYP2B6,* and *CYP3A4* genes are crucial in terms of OPs toxicity. Interestingly, all three isoforms were included in the conversion of various OPs, including azinphos-methyl, chlorpyrifos, diazinon, parathion (Buratti et al. 2003), and malathion to oxon forms (Buratti et al. 2005). The effects of isoforms on the metabolism of particular OP depend on many factors, including the structure of OP and mutual interactions directly related to inter-individual variability. For example, CYP2D6 is one of the most important CYPs involved in the metabolism of about 30% of the currently marketed xenobiotic (Zanger and Schwab 2013). The G1934A polymorphism of CYP2D6 leads to a disruption of the reading frame and a release of nonfunctional protein. Therefore, enzyme activity depends on genotypes. Thus, individuals with wild (GG) CYP2D6 genotype have good enzyme activity, while individuals with heterozygous (GA) and mutant homozygous (AA) CYP2D6 genotypes have poor and no enzyme activity, respectively (Hanioka et al. 1990).

Glutathione S-Transferases (GSTM1, GSTT1, and GSTP1)

Glutathione S-transferases (GSTs), a large family of Phase II detoxification enzymes, have important roles in regulating the conversion of toxic compounds to hydrophilic metabolites. Among the GSTs, alteration in the GSTM1, GSTT1, and GSTP1 genotypes may diminish enzyme activity and lead to the risk of cancer development (Allocati et al. 2018; Singh et al. 2011).

In a comprehensive study conducted by Singh et al. 2011., encompassing 230 subjects, including 115 workers occupationally exposed to OPs and an equal number of normal healthy controls, a significantly higher level of DNA damage was observed in workers with GSTM1 null genotype compared to GSTM1 positive genotype (Singh et al. 2011). A similar association between GSTM1 null genotype and DNA damage was found (Jiang et al. 2010). Also, deletions of the GSTM1/GSTT1 gene correlated with the increased level of DNA damage, as reported previously (Abhishek et al. 2010). However, contrary to results obtained by Abhishek et al. (Abhishek et al. 2010), no association between the null deletion of GSTT1 and DNA damage was found. That is in agreement with other studies performed on workers exposed to pesticides (Liu et al. 2006).

Regarding the GSTP1, it is known that Ile105Val polymorphism modifies the catalytic activity of the enzyme and arises as a consequence of a base substitution (A to G) in the GSTP1 binding site. Results obtained by Singh et al. 2011 (Singh et al. 2011) confirmed an increase in the level of DNA damage in workers with GSTP1 Ile–Ile genotype compared to GSTP1 Ile–Val and Val–Val genotypes, which is in agreement with previous observations (Liu et al. 2006; Wong et al. 2008).

According to the abovementioned findings, determination of GSTM1 and GSTP1 genotypes status could be helpful as an individual marker of susceptibility in the organophosphate-exposed population (Singh et al. 2011).

N-Acetyltransferase 2 (NAT2)

N-acetyltransferase 2 (NAT2) is a phase II drug metabolizing enzyme that has a primary role in the detoxification and biotransformation of many environmental toxicants and therapeutic drugs (Zahra et al. 2020).

Genetic polymorphisms in NAT2 have been associated with the production of proteins with variable enzyme activity or stability, leading to slow or rapid acetylation. Furthermore, the slow acetylator phenotype is associated with Parkinson's disease (Borlak and Reamon-Buettner 2006).

Although some studies suggested the relationship between OP poisoning and polymorphism rs1799931 (857G>A), the obtained results are not consistent and should be subjected to further research (Singh et al. 2012; Glass et al. 2017).

Paraoxonase-1 (PON1)

Paraoxonase-1 (PON1), an A-esterase, a key predictor of OP toxicity, is particularly important in this context. PON1 activity and its concentration showed significant inter-individual variability. Paraoxonases (POase), together with arylesterases (ARYase), are encoded by the paraoxonase-1 (*PON1*) gene located on chromosome 7 (q21.3) (Teodoro et al. 2019). In general, paraoxonases have several functions in biochemical processes, such as protection against oxidative damage and lipid peroxidation, immunity regulation, drug bioactivation, and regulation of cell proliferation/apoptosis (Ceron, Tecles, and Tvarijonaviciute 2014).

PON1 is synthesized in the liver, where it has the highest activity. In plasma, PON1 interacts with high-density lipoproteins (HDL) and protects lipoproteins, particularly low-density lipoprotein-cholesterol (LDL-c), from oxidation (Meneses et al. 2019). Moreover, PON is involved in the detoxification of several important OPs. PON1 hydrolyzes the active metabolites (oxons) of various organophosphorus insecticides like paraoxon (PO), the highly toxic metabolite of the insecticide parathion, chlorpyrifos oxon (CPO), the active metabolite of chlorpyrifos, diazoxon (DZO), the metabolite of diazinon, and also the organophosphate nerve agents such as sarin, soman, and VX (Morcos et al. 2015).

Human *PON1* has many single-nucleotide polymorphisms (SNPs), of which some induce changes in its level and activity that consequently may negatively influence the biotransformation and detoxification of pesticides (Shunmoogam, Naidoo, and Chilton 2018). Moreover, polymorphisms in the *PON1* gene can modulate the activity and functionality of the enzymes encoded and determine OP susceptibility among exposed individuals affecting the health outcomes. Only three common SNPs appear to have clinical relevance regarded OP toxicity. Two polymorphisms have been identified in the coding region (Q192R and L55M) and one polymorphism in the promoter (-108C/T) (Reichert, Levy, and Bydlowski 2020). The glutamine/arginine polymorphism rs662 (Q192R) influences enzymatic activity (catalytic efficiency), while the leucine/methionine polymorphism rs854560 (L55M) influences protein stability and change in PON1 serum concentrations. The polymorphism in position 192 seems to be more involved in the differences in PON1 activity between individuals, affecting catalytic efficiency in a substrate-dependent manner. It was established that R192 isoform hydrolyzes paraoxon, chlorpyrifos oxon and metabolizes oxidized LDL more effectively than the Q192 isoform, indicating increased susceptibility of $PON1_{Q192}$ homozygotes toward their toxicity. Controversial results have been reported for diazoxon. The initial studies indicated that the catalytic efficiency of the R isoform was more efficient than the Q isoform (Humbert et al. 1993). In contrast, further studies stated the opposite, demonstrating a higher hydrolytic rate of the Q isoform (Davies et al. 1996). In the following research conducted by Li et al. 2000., no significant difference between isoforms in their efficacy to hydrolyze the diazoxon was found (Li et al. 2000).

Another polymorphism identified in the coding region of PON1 that resulted in an amino acid substitution at codon 55 (L55M) has been associated with both variability in PON1 levels and PON1 activity in plasma (Costa et al. 2013). Results have shown higher serum enzymatic activity of L55 isoform compared to M isoform. In addition, some studies indicated a higher frequency of the M allele in patients with Parkinson's disease. Furthermore, the estimated relative risk was approximately two times higher when compared with LL homozygous individuals.

The SNP in the promoter region, rs705379 (-108C/T), has been reported to influence transcription rates, affect the regulation of the expression of the PON1 gene, and, accordingly, the PON1 enzyme levels (Costa et al. 2013).

An additional aspect of PON1 is its possibility of acting as a modulator of OP toxicity, potentially affecting the outcome of combined OP exposure. Thus, chlorpyrifos oxon, diazoxon, and paraoxon may potentiate the toxicity of malaoxon by inhibiting carboxylesterase (CarE) activity where the degree of potentiation depends on PON1 status (Jansen et al. 2009).

Taken together, the results of the studies pointed out that the PON1 gene plays an important role in the metabolism of OPs and the etiology of diseases related to pesticide exposure.

In addition, a recent systematic review indicates that prenatal pesticide exposure affects neurodevelopment by influencing the genotype of the PON1 gene, which subsequently alters POase and ARYase activity (Banhela, Naidoo, and Naidoo 2020).

Current Treatment Strategies to Reduce OPs Effects

Numerous *in vivo* and *in vitro* studies have shown that oxidative stress has a pivotal role in OPs induced cytotoxicity (Zheng et al. 2020; Guo et al. 2013; Soltaninejad and Abdollahi 2009; Toledo-Ibarra et al. 2016; Yan, Wang, and Zhu 2013). As mentioned before, OPs increase the generation of free radicals, which interact with cellular macromolecules, causing lipid peroxidation, dysfunctional enzymes, and genotoxicity. Since there is rising evidence that adequate supplementation may be effective against OP-induced cytotoxicity, various treatment strategies to prevent or mitigate cytotoxic effects and protect human health are suggested. In parallel, several strategies aimed to minimize the impact of pesticides on the environment have been proposed.

Antioxidant Therapy

As a lipid-soluble antioxidant that protects biological membranes from oxidative stress, vitamin E has an important role in maintaining cellular homeostasis by regulating the molecular signaling pathways (Azzi et al. 1995). It was reported that vitamin E has a beneficial impact on the modulation of the oxidant-antioxidant balance in the brain of animal models exposed to phosphamidon (Kosta et al. 2013). Also, selenium (Se), a cofactor of glutathione peroxidase, enhances the availability of glutathione, which effectively reduces lipid peroxidation and cell death. In rats exposed to fenthion, a significant reduction of oxidative stress parameters in the brain tissue was seen after treatment with selenium alone and with selenium combined with vitamin E (Cemek et al. 2010). In addition, similar effects were observed after adding vitamin E and selenium to the diet after dimethoate exposure (Ben Amara et al. 2011). According to the abovementioned study results, vitamin E and selenium attenuate this pesticide's toxicity by improving the antioxidants defense system. Thus, Se and vitamin E supplementation may protect against dimethoate-induced neurotoxicity (Ben Amara et al. 2011). However, it appears that vitamin E and/or selenium can only partially quench the deleterious effects of chronic toxicity of malathion (El-Desoky et al. 2012).

Some studies suggest that vitamin C can be effective against OP-induced neurotoxicity in animal models. For example, vitamin C was shown to mitigate chlorpyrifos-induced toxicity partly due to its antioxidant and AChE restoration properties (Ambali and Ayo 2011). Furthermore, the administration of vitamin C in both prophylactic and therapeutic groups of Wistar rats ameliorated the diazinon-induced adversity effect (Khazaie et al. 2019). Although the investigated parameters of oxidative stress did not return to the normal level, these findings suggest the need for further research of vitamin C, alone or in combination with another antioxidant(s), to mitigate damages induced by OPs. Accordingly, it was demonstrated that pre-treatment with Ceftriaxone (antimicrobial cephalosporin) combined with vitamin C provided near-complete protection against the induction of oxidative stress after diazinon exposure (Abdel-Daim 2016).

In rats exposed to malathion, zink (in the form $ZnCl_2$) reduced lipid peroxidation and increased GPx activity in the cerebral cortex as well as the glutathione reductase (GR) activity in the hippocampus. These findings suggest that zinc may be a useful agent in neuroprotection (Brocardo et al. 2007). Furthermore, the results of another study also showed that zinc attenuated chlorpyrifos-evoked alterations in hematological parameters due to its antioxidant properties (Ambali et al. 2010).

The N-acetylcysteine (NAC), as a precursor to glutathione and powerful antioxidant, has been investigated to alleviate OPs toxicity. N-acetylcysteine reduced neurotoxicity, immunotoxicity, and DNA oxidative damage after exposure to fenitrothion. Moreover, administration of NAC significantly reduced the release of inflammatory markers TNF-α and interleukin (IL)-2, prevented the secretion of pro-inflammatory mediators by reduction of their expression, and also reduced the activation of nuclear factor kappa B (NF-kB) (Alam et al. 2019). Besides, it was found that NAC enhanced the synthesis of GSH and glutathione S transferases (GST) activity and scavenged free radicals in the cerebrum and spleen tissues of male rats orally exposed to diazinon (Shadnia et al. 2007). Recently, it was reported that NAC reversed the chlorpyrifos-induced neurotoxicity by modulating oxidative stress, inflammation, and apoptosis parameters in rat brain tissue (Mahmoud et al. 2019).

Likewise, melatonin, a pineal hormone with a well-known antioxidant activity and modulatory effects on many behavioral processes, has been reported to neutralize free radicals and inhibit lipid peroxidation in cellular membranes (Garcia et al. 1997). Furthermore, it was indicated that melatonin administration decreased MDA and TNF-α levels and increased the activity of glutathione peroxidase in the brain tissue of rats treated with diazinon (Ahmed, Ahmed, and El-Morsy 2013). Research conducted by Buyukokuroglu et al. also supports these findings after fenthion exposure. They found that melatonin protected against oxidative damage diminished MDA level and elevated SOD, GPx, and CAT levels in the brain of rats (Buyukokuroglu et al. 2008).

Numbered studies have shown that curcumin, isolated from *Curcuma longa*, has strong antioxidant properties. It was reported that curcumin could decrease OP-induced neurotoxicity and prevent neurodegenerative diseases, including AD (Sarkar et al. 2017). In rats exposed to parathion after curcumin injection, ROS production was reduced, and antioxidant content in the hippocampus was enhanced (Canales-Aguirre et al. 2012).

Carnosic acid (isolated from rosemary) modulated the mitochondrial activities due to stimulating the PI3K/Akt/Nrf2 pathway. Also, the enhanced levels of GSH in mitochondria after chlorpyrifos exposure in Sh-Sy5y cells were observed (de Oliveira, Ferreira, and Schuck 2015). Additionally, pre-treatment with carnosic acid showed a neuroprotective effect against chlorpyrifos-induced oxidative stress, as evidenced by a dose-dependent increase in GSH, GPx, SOD, and CAT, as well as a decrease in the levels of MDA and NO in mice brains. Furthermore, the serum levels of pro-inflammatory cytokines, including interleukin-1β, IL-6, and tumor necrosis factor-α, were also decreased (AlKahtane et al. 2020). Similarly, it was observed that caffeic acid has the potential to neutralize chlorpyrifos-induced oxidative stress in rat brains (Ozkan et al. 2014).

As already mentioned, malathion exposure increased malondialdehyde and 8-OHdG levels, whereas it decreased glutathione levels. Moreover, AChE, SOD, and CAT activities were reduced in the blood, liver, kidney, heart, and brain tissues. The addition of crocin (isolated from saffron) after malathion exposure markedly ameliorated MDA and GSH levels in the striatum of rats. Also, it elevated the serum levels of AChE (Mohammadzadeh et al. 2018). Moreover, the aqueous extract of saffron (*Crocus sativus* L.) ameliorated diazinon-induced oxidative modification and inflammation in the brain of rats (Moallem et al. 2014).

Similarly, the antioxidant impact of *Nigella sativa* (Black Cumin) seen as enhanced TAC and GSH levels and diminished ROS production in neurons after dichlorvos exposure was observed in rat brains (Imam et al. 2018). In addition, thymoquinone, a phytochemical compound isolated from black cumin seeds and diallyl sulfide, an organosulfur compound isolated from *Allium sativum* L. have shown significant mitigation of the oxidative effects in malathion-intoxicated rats, probably by reducing the generation of reactive oxygen and nitrogen radicals, as well as by enhancing the antioxidant defense mechanisms (Abdel-Daim, Abushouk, Bungău, et al. 2020).

A study performed by Fereidouni et al. demonstrated that quercetin administration ameliorated the expression of Bax, Bcl-2 cytochrome c, caspase-8, and caspase-9 in the brain of animals exposed to chlorpyrifos, which resulted in a significant decrease in cell death (Fereidouni et al. 2019; Moallem et al. 2014). Furthermore, resveratrol (from red grapes) treatment could invert malathion-induced oxidative stress, lipid peroxidation, and activity of antioxidant enzymes (Akbel et al. 2018). Additionally, it was observed that trans-resveratrol reduced ROS generation and lipid peroxidation and increased GSH levels and mitochondrial membrane potential in PC12 cells exposed to monocrotophos (Kumar et al. 2013).

Abdel-Daim et al. 2016., revealed that supplementation by sesame oil and alpha-lipoic acid could alleviate the toxic effects of diazinon *via* their potent antioxidant and free radical-scavenging activities (Abdel-Daim et al. 2016).

The antioxidant and protective effects of microalga *Spirulina platensis* against diazinon-induced subacute toxicity in the blood, heart, liver, and kidneys of male Wistar albino rats were tested. The obtained results revealed that *Spirulina platensis* effectively alleviated the diazinon-induced hematologic alterations and organ injuries, probably through its antioxidant and anti-inflammatory activities (Abdel-Daim et al. 2018).

Effects of fucoidans, organic sulfated polysaccharides from *Laminaria japonicum* were investigated against diazinon-induced injuries in male Wistar rats. Results have shown that fucoidans ameliorated diazinon-induced oxidative stress, pro-inflammatory effects, and injuries to the cardiac, hepatic, and renal tissues. Interestingly, significant inhibition of diazinon-induced overproduction of IL-6 and TNF-α after fucoidans treatment was observed, indicating other mechanisms of action related to the inhibition of cyclooxygenase-II enzyme activity and downregulation of the transcription of activator protein-1 and inducible nitric oxide synthase (Abdel-Daim, Abushouk, Bahbah, et al. 2020).

Degradation Strategies for OPs

Nowadays, due to the increased use of OPs against different pest species, efforts are put to develop efficient methods for their degradation to non-toxic or less toxic forms to minimize their impact on the environment and human health.

Natural (Biotic) Degradation

Biodegradation, a process that involves the complete rupture of an organic compound to its inorganic constituents, is recognized as an effective tool for the remediation of pesticide contamination due to its low cost, easy application, high effectiveness, and lack of secondary contamination (Ning et al. 2012; Ramu and Seetharaman 2014; Ozdal et al. 2017). Microorganisms, mainly bacteria, play an important role in biodegradation due to their ability to metabolize OPs contaminants, in many cases using them as nutrient and/or energy sources. The microbial degradation of OPs involves hydrolysis, oxidation, alkylation, and dealkylation. The hydrolysis of P-O-alkyl and P-O-aryl bonds is crucial in detoxification (Singh and Walker 2006).

A variety of microorganisms have been reported to be effective in this process. For example, the species of genera *Arthrobacter*(Choi et al. 2009), *Enterobacter* (Singh et al. 2004), *Burkholderia* (Choi et al. 2009), *Pseudomonas* (Chanika et al. 2011), *Serratia* (Cycoń, Wójcik, and Piotrowska-Seget 2009), *Sphingobium* (Iyer, Iken, and Damania 2013) and *Bacillus*, *Flavobacterium*, *Micrococcus*, and *Plesiomonas* (Singh and Walker 2006), as well as fungi including *Penicillium oxalicum* ZHJ6 (Zhao, Bao, and Liu 2010), *Fusarium* sp. F1 (Zhao, Zhao, and Han 2009) and *Aspergillus sydowii* PA F-2 (Tian et al. 2016), among others.

In general, their action is related to their genes and associated enzymes that hydrolyze and detoxify OP pesticides (Singh and Walker 2006).Various functional genes have been reported to date, including opd (opdE, opdA, opdC), amp (ampA), oph (ophB) and mpd (Haque et al. 2018; Jiang et al. 2019). The opd gene product, parathion hydrolase, was the first enzyme identified for its ability to degrade organophosphorus compounds. Later the organophosphorus hydrolase (*oph*) gene was characterized. For example, chlorpyrifos was hydrolyzed with a gene (*oph*B) encoding a protein that was extracted from a strain of *Pseudomonas* sp. BF1–3 (Barman et al. 2014).

Other discovered genes, such as methyl parathion degradation (*mpd*), enable the degradation of chlorpyrifos, methyl paraoxon, and methyl parathion by following the different OP degrading pathways (Sidhu et al. 2019).

Future perspectives in this area imply that introducing all degrading genes into one organism will enable optimization of gene expression and elevate the degradation rate, consequently minimizing the cytotoxic effects.

Abiotic Degradation

Abiotic degradation is the approach that involves the direct chemical or mechanical breakdown of OPs into non-toxic forms utilizing non-living entities that are naturally present in the environment or by an adsorbent that binds and removes that pesticide from the environment (Kaushal, Khatri, and Arya 2021).

Many chemical oxidants were tested for this purpose, such as ferrate (VI), manganese dioxide, manganese dioxide with bisulfate, titanium-iron oxides, molybdates, and silver particles (Kaushal, Khatri, and Arya 2021). In addition, gamma radiations and UV-C light have also been used for the degradation and destruction of contaminants, including OPs, due to their efficacy in the decontamination of the surfaces of fruits and vegetables and various food preparations (Alsager, Alnajrani, and Alhazzaa 2018; Savic et al. 2019).

In recent years, this research field has intensified, and findings have been of tremendous relevance for reducing toxicity and impact on human health.

Conclusion

Multiple studies have shown that OPs can affect oxidant/antioxidant balance, which leads to oxidative stress. The appearance of lipid peroxidation in cell membranes, the modification of antioxidant capacity levels, DNA damage, and disruption of cellular homeostasis, in general, are the most prominent evidence of the role of oxidative stress in the cytotoxicity of OPs.

Moreover, the OPs–induced DNA damage and chromosomal aberrations are considered primary mechanisms that lead to chronic diseases, including cancer. Furthermore, pesticides are also known to induce epigenetic changes, heritable changes without any alteration in DNA sequences through DNA methylation, histone modifications, and expression of non-coding RNAs. Accordingly, there is an obvious need to investigate OPs further to achieve a complete knowledge of their genetic toxicology. In this context, biomonitoring studies are especially significant.

In the context of risk assessment, it is important to assess numerous events involved in OPs toxicity and estimate the degree of OPs impact on biological processes, which is crucial for predicting adverse effects. Further investigations are needed, especially of toxicity pathways modulation and detection of antioxidants that may act against biochemical alterations caused by OP to reduce their cytotoxicity.

The scientific data concludes that exposure to OPs is associated with a high risk of cytotoxicity and genotoxicity that may result in the development of various diseases. Thus, finally, world usage of OPs should be reduced and better controlled.

Acknowledgments

This research was funded by the Ministry of Education, Science and Technological Development of the Republic of Serbia, grant number 451-03-9/2021-14/200017.

References

Abass, K., P. Reponen, S. Mattila, and O. Pelkonen. 2010. "Metabolism of carbosulfan II. Human interindividual variability in its in vitro hepatic biotransformation and the identification of the cytochrome P450 isoforms involved." *Chem Biol Interact* 185 (3):163-73. doi: 10.1016/j.cbi.2010. 03.024.

Abdel-Daim, M. M. 2016. "Synergistic protective role of ceftriaxone and ascorbic acid against subacute diazinon-induced nephrotoxicity in rats." *Cytotechnology* 68 (2):279-289. doi: 10.1007/s10616-014-9779-z.

Abdel-Daim, M. M., A. I. Abushouk, M. I. Alkhalf, E. A. Toraih, M. S. Fawzy, H. Ijaz, L. Aleya, and S. G. Bungau. 2018. "Antagonistic effects of Spirulina platensis on diazinon-induced hemato-biochemical alterations and oxidative stress in rats." *Environ Sci Pollut Res Int* 25 (27):27463-27470. doi: 10.1007/s11356-018-2761-0.

Abdel-Daim, M. M., A. I. Abushouk, E. I. Bahbah, S. G. Bungau, M. S. Alyousif, L. Aleya, and S. Alkahtani. 2020. "Fucoidan protects against subacute diazinon-induced oxidative damage in cardiac, hepatic, and renal tissues." *Environ Sci Pollut Res Int* 27 (11):11554-11564. doi: 10.1007/ s11356-020-07711-w.

Abdel-Daim, M. M., A. I. Abushouk, S. G. Bungău, M. Bin-Jumah, A. F. El-kott, A. A. Shati, L. Aleya, and S. Alkahtani. 2020. "Protective effects of thymoquinone and diallyl sulphide against malathion-induced toxicity in rats." *Environmental Science and Pollution Research* 27 (10):10228-10235. doi: 10.1007/s11356-019-07580-y.

Abdel-Daim, M. M., R. Taha, E. W. Ghazy, and Y. S. El-Sayed. 2016. "Synergistic ameliorative effects of sesame oil and alpha-lipoic acid against subacute diazinon toxicity in rats: hematological, biochemical, and antioxidant studies." *Can J Physiol Pharmacol* 94 (1):81-8. doi: 10.1139/ cjpp-2015-0131.

Abhishek, S., N. Kaur, S. Kaur, M. Lata, J. K. Sharma, and A. Sharma. 2010. "Association of GSTM1 and GSTT1 gene deletions with susceptibility to DNA damage in the pesticide-exposed workers of Punjab." *Rejuvenation Res* 13 (2-3):281-4. doi: 10.1089/rej.2009.0931.

Ahmed, M. A., H. I. Ahmed, and E. M. El-Morsy. 2013. "Melatonin protects against diazinon-induced neurobehavioral changes in rats." *Neurochem Res* 38 (10):2227-36. doi: 10.1007/s11064-013-1134-9.

Akbel, E., D. Arslan-Acaroz, H. H. Demirel, I. Kucukkurt, and S. Ince. 2018. "The subchronic exposure to malathion, an organophosphate pesticide, causes lipid peroxidation, oxidative stress, and tissue damage in rats: the protective role of resveratrol." *Toxicology Research* 7 (3):503-512. doi: 10.1039/C8TX00030A.

Alam, R. T., T. S. Imam, A. M. A. Abo-Elmaaty, and A. H. Arisha. 2019. "Amelioration of fenitrothion induced oxidative DNA damage and inactivation of caspase-3 in the brain and spleen tissues of male rats by N-acetylcysteine." *Life Sci* 231:116534. doi: 10.1016/j.lfs.2019.06.009.

AlKahtane, A. A., E. Ghanem, S. G. Bungau, S. Alarifi, D. Ali, G. AlBasher, S. Alkahtani, L. Aleya, and M. M. Abdel-Daim. 2020. "Carnosic acid alleviates chlorpyrifos-induced oxidative stress and inflammation in mice cerebral and ocular tissues." *Environ Sci Pollut Res Int* 27 (11):11663-11670. doi: 10.1007/s11356-020-07736-1.

Allocati, N., M. Masulli, C. Di Ilio, and L. Federici. 2018. "Glutathione transferases: substrates, inihibitors and pro-drugs in cancer and neurodegenerative diseases." *Oncogenesis* 7 (1):8. doi: 10.1038/s41389-017-0025-3.

Alsager, O. A., M. N. Alnajrani, and O. Alhazzaa. 2018. "Decomposition of antibiotics by gamma irradiation: Kinetics, antimicrobial activity, and real application in food matrices." *Chemical Engineering Journal* 338:548-556. doi: 10.1016/j.cej.2018. 01.065.

Ambali, S., A. Abubakar, S. Muftau, L. Yaqub, S. Anafi, and A. Abdullahi. 2010. "Chlorpyrifos-Induced Alteration of Hematological Parameters in Wistar Rats: Ameliorative Effect of Zinc." *Research Journal of Environmental Toxicology* 4:55-66. doi: 10.3923/rjet.2010.55.66.

Ambali, S. F., and J. O. Ayo. 2011. "Sensorimotor performance deficits induced by chronic chlorpyrifos exposure in Wistar rats: mitigative effect of vitamin C." *Toxicological & Environmental Chemistry* 93 (6):1212-1226. doi: 10.1080/02772248.2011.585991.

Angerer, J., U. Ewers, and M. Wilhelm. 2007. "Human biomonitoring: State of the art." *International Journal of Hygiene and Environmental Health* 210 (3):201-228. doi: 10.1016/j.ijheh.2007.01.024.

Argentin, G., M. Divizia, and R. Cicchetti. 2015. "Oxidative Stress, Cytotoxicity, and Genotoxicity Induced by Methyl Parathion in Human Gingival Fibroblasts: Protective Role of Epigallocatechin-3-Gallate." *J Toxicol Environ Health A* 78 (19):1227-40. doi: 10.1080/15287394. 2015.1079582.

Ayed-Boussema, I., K. Rjiba, A. Moussa, N. Mnasri, and H. Bacha. 2012. "Genotoxicity associated with oxidative damage in the liver and kidney of mice exposed to dimethoate subchronic intoxication." *Environmental Science and Pollution Research* 19 (2):458-466. doi: 10.1007/s11356-011-0588-z.

Azzi, A., D. Boscoboinik, D. Marilley, N. K. Ozer, B. Stauble, and A. Tasinato. 1995. "Vitamin E: a sensor and an information transducer of the cell oxidation state." *Am J Clin Nutr* 62 (6 Suppl):1337S-1346S. doi: 10.1093/ ajcn/62.6.1337S.

Badr, A. M. 2020. "Organophosphate toxicity: updates of malathion potential toxic effects in mammals and potential treatments." *Environmental Science and Pollution Research* 27 (21):26036-26057. doi: 10.1007/s11356-020-08937-4.

Balaji, M., and K. Sasikala. 1993. "Cytogenetic effect of malathion in in vitro culture of human peripheral blood." *Mutat Res* 301 (1):13-7. doi: 10.1016/0165-7992(93)90050-6.

Banhela, N., P. Naidoo, and S. Naidoo. 2020. "Association between pesticide exposure and paraoxonase-1 (PON1) polymorphisms, and neurobehavioural outcomes in children: a systematic review." *Systematic Reviews* 9 (1):109. doi: 10.1186/s13643-020-01330-9.

Barman, D. N., M. A. Haque, S. M. Islam, H. D. Yun, and M. K. Kim. 2014. "Cloning and expression of ophB gene encoding organophosphorus hydrolase from endophytic Pseudomonas sp. BF1-3 degrades organophosphorus pesticide chlorpyrifos." *Ecotoxicol Environ Saf* 108:135-41. doi: 10.1016/j.ecoenv.2014.06.023.

Ben Amara, I., N. Soudani, A. Hakim, A. Troudi, K. M. Zeghal, T. Boudawara, and N. Zeghal. 2011. "Selenium and vitamin E, natural antioxidants, protect rat cerebral cortex against dimethoate-induced neurotoxicity." *Pesticide Biochemistry and Physiology* 101 (3):165-174. doi: 10.1016/j.pestbp. 2011.08.012.

Bernieri, T., M. F. Moraes, P. G. Ardenghi, and L. Basso da Silva. 2019. "Assessment of DNA damage and cholinesterase activity in soybean farmers in southern Brazil: High versus low pesticide exposure." *J Environ Sci Health B* 55 (4):355-360. doi: 10.1080/03601234.2019.1704608.

Binukumar, B. K., A. Bal, R. J. L. Kandimalla, and K. D. Gill. 2010. "Nigrostriatal neuronal death following chronic dichlorvos exposure: crosstalk between mitochondrial impairments, alpha synuclein aggregation, oxidative damage and behavioral changes." *Molecular Brain* 3. doi: 10.1186/1756-6606-3-35.

Blasiak, J., P. Jaloszynski, A. Trzeciak, and K. Szyfter. 1999. "In vitro studies on the genotoxicity of the organophosphorus insecticide malathion and its two analogues."

Mutation Research/Genetic Toxicology and Environmental Mutagenesis 445 (2):275-283. doi: 10.1016/S1383-5718(99)00132-1.

Bolognesi, C., A. Creus, P. Ostrosky-Wegman, and R. Marcos. 2011. "Micronuclei and pesticide exposure." *Mutagenesis* 26 (1):19-26. doi: 10.1093/mutage/geq070.

Bondzic, A. M., T. D. Lazarevic-Pasti, A. R. Leskovac, S. Z. Petrovis, M. B. Colovic, T. N. Parac-Vogt, and G. V. Janjic. 2020. "A new acetylcholinesterase allosteric site responsible for binding voluminous negatively charged molecules â€ "the role in the mechanism of AChE inhibition." *European Journal of Pharmaceutical Sciences* 151:105376. doi: 10.1016/j.ejps.2020.105376.

Borlak, J., and S. M. Reamon-Buettner. 2006. "N-acetyltransferase 2 (NAT2) gene polymorphisms in Parkinson's disease." *BMC Med Genet* 7:30. doi: 10.1186/1471-2350-7-30.

Boskabady, M., N. Marefati, T. Farkhondeh, F. Shakeri, A. Farshbaf, and M.H. Boskabady. 2018. "The effect of environmental lead exposure on human health and the contribution of inflammatory mechanisms, a review." *Environment International* 120:404-420. doi: 10.1016/j.envint.2018.08.013.

Brocardo, P. S., F. Assini, J. L. Franco, P. Pandolfo, Y. M. R. Muller, R. N. Takahashi, A. L. Dafre, and A. L. S. Rodrigues. 2007. "Zinc Attenuates Malathion-Induced Depressant-like Behavior and Confers Neuroprotection in the Rat Brain." *Toxicological Sciences* 97 (1):140-148. doi: 10.1093/ toxsci/kfm024.

Buratti, F. M., A. D'Aniello, M. T. Volpe, A. Meneguz, and E. Testai. 2005. "Malathion bioactivation in the human liver: the contribution of different cytochrome p450 isoforms." *Drug Metab Dispos* 33 (3):295-302. doi: 10.1124/dmd.104.001693.

Buratti, F. M., M. T. Volpe, A. Meneguz, L. Vittozzi, and E. Testai. 2003. "CYP-specific bioactivation of four organophosphorothioate pesticides by human liver microsomes." *Toxicol Appl Pharmacol* 186 (3):143-54. doi: 10.1016/s0041-008x(02)00027-3.

Buyukokuroglu, M. E., M. Cemek, Y. Yurumez, Y. Yavuz, and A. Aslan. 2008. "Antioxidative role of melatonin in organophosphate toxicity in rats." *Cell Biol Toxicol* 24 (2):151-8. doi: 10.1007/s10565-007-9024-z.

Canales-Aguirre, A. A., U. A. Gomez-Pinedo, S. Luquin, M.A. Ramirez-Herrera, M. L. Mendoza-Magana, and A. Feria-Velasco. 2012. "Curcumin protects against the oxidative damage induced by the pesticide parathion in the hippocampus of the rat brain." *Nutr Neurosci* 15 (2):62-9. doi: 10.1179/1476830511Y.0000000034.

Cemek, M., A. Buyukben, M. E. Buyukokuroglu, F. Aymelek, and L. Tur. 2010. "Protective roles of vitamin E (a-tocopherol), selenium and vitamin E plus selenium in organophosphate toxicity in vivo: A comparative study." *Pesticide Biochemistry and Physiology* 96 (3):113-118. doi: 10.1016/j. pestbp.2009.09.009.

Ceron, J. J., F. Tecles, and A. Tvarijonaviciute. 2014. "Serum paraoxonase 1 (PON1) measurement: an update." *BMC Vet Res* 10:74. doi: 10.1186/1746-6148-10-74.

Chanika, E., D. Georgiadou, E. Soueref, P. Karas, E. Karanasios, N. G. Tsiropoulos, E. A. Tzortzakakis, and D. G. Karpouzas. 2011. "Isolation of soil bacteria able to hydrolyze both organophosphate and carbamate pesticides." *Bioresource Technology* 102 (3):3184-3192. doi: 10.1016/j. biortech.2010.10.145.

Chen, T., J. Tan, Z. Wan, Y. Zou, H. K. Afewerky, Z. Zhang, and T. Zhang. 2017. "Effects of Commonly Used Pesticides in China on the Mitochondria and Ubiquitin-

Proteasome System in Parkinson's Disease." *International journal of molecular sciences* 18 (12):2507. doi: 10.3390/ijms18122507.

Choi, M. K., K. D. Kim, K. M. Ahn, D. H. Shin, J. H. Hwang, C. N. Seong, and J. O. Ka. 2009. "Genetic and phenotypic diversity of parathion-degrading bacteria isolated from rice paddy soils." *J Microbiol Biotechnol* 19 (12):1679-87. doi: 10.4014/jmb.0905.05057.

Colovic, M., D. Krstic, S. Petrovic, A. Leskovac, G. Joksic, J. Savic, M. Franko, P. Trebse, and V. Vasic. 2010. "Toxic effects of diazinon and its photodegradation products." *Toxicol Lett* 193 (1):9-18. doi: 10.1016/j.toxlet. 2009.11.022.

Cortes-Iza, S. C., and A.I. Rodriguez. 2018. "Oxidative stress and pesticide disease: A challenge for toxicology." *Revista de la Facultad de Medicina* 66:261-267. doi: 10.15446/revfacmed.v66n2.60783.

Costa, L. G., G. Giordano, T. B. Cole, J. Marsillach, and C. E. Furlong. 2013. "Paraoxonase 1 (PON1) as a genetic determinant of susceptibility to organophosphate toxicity." *Toxicology* 307:115-122. doi: 10.1016/j.tox. 2012.07.011.

Cycoń, M., M. Wójcik, and Z. Piotrowska-Seget. 2009. "Biodegradation of the organophosphorus insecticide diazinon by Serratia sp. and Pseudomonas sp. and their use in bioremediation of contaminated soil." *Chemosphere* 76 (4):494-501. doi: 10.1016/j.chemosphere.2009.03.023.

Davies, H. G., R. J. Richter, M. Keifer, C. A. Broomfield, J. Sowalla, and C. E. Furlong. 1996. "The effect of the human serum paraoxonase polymorphism is reversed with diazoxon, soman and sarin." *Nature Genetics* 14 (3):334-336. doi: doi 10.1038/ng1196-334.

de Oliveira, M. R., G. C. Ferreira, and P. F. Schuck. 2015. "Protective effect of carnosic acid against paraquat-induced redox impairment and mitochondrial dysfunction in SH-SY5Y cells: Role for PI3K/Akt/Nrf2 pathway." *Toxicology in vitro : an international journal published in association with BIBRA* 32:41-54. doi: 10.1016/j.tiv.2015.12.005.

Dhananjayan, V., B. Ravichandran, K. Panjakumar, K. Kalaiselvi, R. Kausic, A. Mala, G. Avinash, K. Shridhar, A. Manju, and W. Rajesh. 2019. "Assessment of genotoxicity and cholinesterase activity among women workers occupationally exposed to pesticides in tea garden." *Mutation Research/Genetic Toxicology and Environmental Mutagenesis* 841:1-7. doi: 10.1016/j.mrgentox.2019.03.002.

dos Santos, A. A., A. A. Naime, J. de Oliveira, D. Colle, D. B. dos Santos, M. A. Hort, E. L. G. Moreira, C. Sunol, A. F. de Bem, and M. Farina. 2016. "Long-term and low-dose malathion exposure causes cognitive impairment in adult mice: evidence of hippocampal mitochondrial dysfunction, astrogliosis and apoptotic events." *Archives of Toxicology* 90 (3):647-660. doi: 10.1007/s00204-015-1466-0.

Echeverri-Jaramillo, G., B. Jaramillo-Colorado, C. Sabater-Marco, and M. Á. Castillo-López. 2021. "Cytotoxic and estrogenic activity of chlorpyrifos and its metabolite 3,5,6-trichloro-2-pyridinol. Study of marine yeasts as potential toxicity indicators." *Ecotoxicology* 30 (1):104-117. doi: 10.1007/s10646-020-02315-z.

Eddleston, M., and F. R. Chowdhury. 2016. "Pharmacological treatment of organophosphorus insecticide poisoning: the old and the (possible) new." *Br J Clin Pharmacol* 81 (3):462-70. doi: 10.1111/bcp.12784.

Edwards, F. L., C. G. Yedjou, and P. B. Tchounwou. 2013. "Involvement of oxidative stress in methyl parathion and parathion-induced toxicity and genotoxicity to human liver carcinoma (HepG(2)) cells." *Environ Toxicol* 28 (6):342-8. doi: 10.1002/tox.20725.

El-Desoky, G., M. Abdelreheem, A. AL-Othman, Z. ALOthman, M. Mahmoud, and K. Yusuf. 2012. "Potential hepatoprotective effects of vitamin E and selenium on hepatotoxicity induced by malathion in rats." *African Journal of Pharmacy and Pharmacology* 6 (11):806-813. doi: 10.5897/AJPP11.628.

Fereidouni, S., R. R. Kumar, V. D. Chadha, and D. K. Dhawan. 2019. "Quercetin plays protective role in oxidative induced apoptotic events during chronic chlorpyrifos exposure to rats." *J Biochem Mol Toxicol* 33 (8):e22341. doi: 10.1002/jbt.22341.

Ferrer, A. 2003. "Pesticide poisoning." *An Sist Sanit Navar* 26 Suppl 1:155-71.

Fischer, R., and O. Maier. 2015. "Interrelation of Oxidative Stress and Inflammation in Neurodegenerative Disease: Role of TNF." *Oxidative Medicine and Cellular Longevity* 2015:610813. doi: 10.1155/2015/610813.

Garcia, J. J., R. J. Reiter, J. M. Guerrero, G. Escames, B. P. Yu, C. S. Oh, and A. Munoz-Hoyos. 1997. "Melatonin prevents changes in microsomal membrane fluidity during induced lipid peroxidation." *FEBS Lett* 408 (3):297-300. doi: 10.1016/s0014-5793(97)00447-x.

Glass, T., M. A. Dalvie, Z. Holtman, A. A. Vorster, R. S. Ramesar, and L. London. 2017. "DNA variants and organophosphate neurotoxicity among emerging farmers in the Western Cape of South Africa." *American Journal of Industrial Medicine* 61 (1):11-20. doi: 10.1002/ajim.22790.

Guo, C., L. Sun, X. Chen, and D. Zhang. 2013. "Oxidative stress, mitochondrial damage and neurodegenerative diseases." *Neural regeneration research* 8 (21):2003-2014. doi: 10.3969/j.issn.1673-5374.2013.21.009.

Hanioka, N., S. Kimura, U. A. Meyer, and F. J. Gonzalez. 1990. "The human CYP2D locus associated with a common genetic defect in drug oxidation: a G1934----A base change in intron 3 of a mutant CYP2D6 allele results in an aberrant 3' splice recognition site." *Am J Hum Genet* 47 (6):994-1001.

Haque, M. A., S. Y. Hong, C. E. Hwang, S. C. Kim, and K. M. Cho. 2018. "Cloning of an organophosphorus hydrolase (opdD) gene of Lactobacillus sakei WCP904 isolated from chlorpyrifos-impregnated kimchi and hydrolysis activities of its gene product for organophosphorus pesticides." *Applied Biological Chemistry* 61 (6):643-651. doi: 10.1007/s13765-018-0397-x.

Hernández, A. F., T. Parrón, A. M. Tsatsakis, M. Requena, R. Alarcón, and O. López-Guarnido. 2013. "Toxic effects of pesticide mixtures at a molecular level: their relevance to human health." *Toxicology* 307:136-45. doi: 10.1016/j.tox.2012.06.009.

Humbert, R., D.A. Adler, C. M. Disteche, C. Hassett, C. J. Omiecinski, and C. E. Furlong. 1993. "The molecular basis of the human serum paraoxonase activity polymorphism." *Nature Genetics* 3 (1):73-76. doi: 10.1038/ng0193-73.

Hung, J. H., C. Y. Chen, H. A. Omar, K. Y. Huang, C. C. Tsao, C. C. Chiu, Y. L. Chen, P. H. Chen, and Y. N. Teng. 2016. "Reactive oxygen species mediate Terbufos-induced apoptosis in mouse testicular cell lines via the modulation of cell cycle and pro-apoptotic proteins." *Environmental Toxicology* 31 (12):1888-1898. doi: 10.1002/tox.22190.

Imam, A., A. Ogunniyi, A. Ibrahim, W. I. Abdulmajeed, L. O. Aboyeji, A. H. Lawan, F. A. Sulaimon, M. Y. Adana, and S. M. Ajao. 2018. "Dichlorvos Induced Oxidative and Neuronal Responses in Rats: Mitigative Efficacy of Nigella sativa (Black Cumin)." *Niger J Physiol Sci* 33 (1):83-88.

Iyer, R., B. Iken, and A. Damania. 2013. "A comparison of organophosphate degradation genes and bioremediation applications." *Environ Microbiol Rep* 5 (6):787-98. doi: 10.1111/1758-2229.12095.

Jan, Y. H., J. R. Richardson, A. A. Baker, V. Mishin, D. E. Heck, D. L. Laskin, and J. D. Laskin. 2016. "Novel approaches to mitigating parathion toxicity: targeting cytochrome P450-mediated metabolism with menadione." *Annals of the New York Academy of Sciences* 1378 (1):80-86. doi: 10.1111/nyas.13156.

Jansen, K. L., T. B. Cole, S. S. Park, C. E. Furlong, and L. G. Costa. 2009. "Paraoxonase 1 (PON1) modulates the toxicity of mixed organophosphorus compounds." *Toxicol Appl Pharmacol* 236 (2):142-53. doi: 10.1016/j. taap.2009.02.001.

Jiang, B., N. Zhang, Y. Xing, L. Lian, Y. Chen, D. Zhang, G. Li, G. Sun, and Y. Song. 2019. "Microbial degradation of organophosphorus pesticides: novel degraders, kinetics, functional genes, and genotoxicity assessment." *Environ Sci Pollut Res Int* 26 (21):21668-21681. doi: 10.1007/s11356-019-05135-9.

Jiang, S., L. Yu, J. Cheng, S. Leng, Y. Dai, Y. Zhang, Y. Niu, H. Yan, W. Qu, C. Zhang, K. Zhang, R. Yang, L. Zhou, and Y. Zheng. 2010. "Genomic damages in peripheral blood lymphocytes and association with polymorphisms of three glutathione S-transferases in workers exposed to formaldehyde." *Mutation Research/Genetic Toxicology and Environmental Mutagenesis* 695 (1):9-15. doi: 10.1016/j.mrgentox. 2009.09.011.

Josse, R., A. Sharanek, C. C. Savary, and A. Guillouzo. 2014. "Impact of isomalathion on malathion cytotoxicity and genotoxicity in human HepaRG cells." *Chemico-Biological Interactions* 209:68-76. doi: 10.1016/j.cbi.2013. 12.002.

Jovicic, D., J. Pajic, B. Rakic, L. Radivojevic, M. Pajic, V. Janjic, and A. Milovanovic. 2013. "Cytogenetic biomonitoring in a Serbian population occupationally exposed to a complex mixture of pesticides." *Genetika* 45 (1):121-133. doi: 10.2298/ GENSR1301121J.

Kalyabina, V. P., E. N. Esimbekova, K. V. Kopylova, and V. A. Kratasyuk. 2021. "Pesticides: formulants, distribution pathways and effects on human health - a review." *Toxicol Rep* 8:1179-1192. doi: 10.1016/j.toxrep.2021.06. 004.

Kamel, F., and J. A. Hoppin. 2004. "Association of pesticide exposure with neurologic dysfunction and disease." *Environmental health perspectives* 112 (9):950-958. doi: 10.1289/ehp.7135.

Kapka-Skrzypczak, L., M. Cyranka, M. Skrzypczak, and M. Kruszewski. 2011. "Biomonitoring and biomarkers of organophosphate pesticides exposure - state of the art." *Ann Agric Environ Med* 18 (2):294-303.

Kaur, P., B. Radotra, R. W. Minz, and K. D. Gill. 2007. "Impaired mitochondrial energy metabolism and neuronal apoptotic cell death after chronic dichlorvos (OP) exposure in rat brain." *Neurotoxicology* 28 (6):1208-1219. doi: 10.1016/j.neuro.2007.08.001.

Kaushal, J., M. Khatri, and S. K. Arya. 2021. "A treatise on Organophosphate pesticide pollution: Current strategies and advancements in their environmental degradation and elimination." *Ecotoxicology and Environmental Safety* 207:111483. doi: 10.1016/j.ecoenv.2020.111483.

Khazaie, S., M. Jafari, J. Heydari, A. Asgari, K. Tahmasebi, M. Salehi, and M.S. Abedini. 2019. "Modulatory effects of vitamin C on biochemical and oxidative changes induced by acute exposure to diazinon in rat various tissues: Prophylactic and therapeutic roles." *Journal of Animal Physiology and Animal Nutrition* 103 (5):1619-1628. doi: 10.1111/jpn.13144.

Kim, K. H., E. Kabir, and S. A. Jahan. 2017. "Exposure to pesticides and the associated human health effects." *Science of The Total Environment* 575:525-535. doi: 10.1016/j.scitotenv.2016.09.009.

Kosta, P., A. K. Mehta, A. K. Sharma, N. Khanna, P. K. Mediratta, D. R. Mundhada, and S. Suke. 2013. "Effect of piracetam and vitamin E on phosphamidon-induced impairment of memory and oxidative stress in rats." *Drug Chem Toxicol* 36 (1):48-54. doi: 10.3109/01480545.2011.649093.

Kumar, V., V. K. Tripathi, A. K. Singh, M. Lohani, and M. Kuddus. 2013. "Trans-resveratrol restores the damages induced by organophosphate pesticide-monocrotophos in neuronal cells." *Toxicology international* 20 (1):48-55. doi: 10.4103/0971-6580.111571.

Lazarevic-Pasti, T., A. Leskovac, T. Momic, S. Petrovic, and V. Vasic. 2017. "Modulators of Acetylcholinesterase Activity: From Alzheimer's Disease to Anti-Cancer Drugs." *Curr Med Chem* 24 (30):3283-3309. doi: 10.2174/0929867324666170705123509.

Li, W. F., L. G. Costa, R. J. Richter, T. Hagen, D. M. Shih, A. Tward, A. J. Lusis, and C. E. Furlong. 2000. "Catalytic efficiency determines the in-vivo efficacy of PON1 for detoxifying organophosphorus compounds." *Pharmacogenetics and Genomics* 10 (9).

Lionetto, M. G., R. Caricato, A. Calisi, M. E. Giordano, and T. Schettino. 2013. "Acetylcholinesterase as a biomarker in environmental and occupational medicine: new insights and future perspectives." *Biomed Res Int* 2013:321213. doi: 10.1155/2013/321213.

Liu, Y. J., P. L. Huang, Y. F. Chang, Y. H. Chen, Y. H. Chiou, Z. L. Xu, and R. H. Wong. 2006. "GSTP1 Genetic Polymorphism Is Associated with a Higher Risk of DNA Damage in Pesticide-Exposed Fruit Growers." *Cancer Epidemiology Biomarkers* 15 (4):659-666. doi: 10.1158/1055-9965.epi-05-0617.

Lushchak, V. I., T. M. Matviishyn, V. V. Husak, J. M. Storey, and K. B. Storey. 2018. "Pesticide toxicity: a mechanistic approach." *EXCLI J* 17:1101-1136. doi: 10.17179/excli2018-1710.

Mahmoud, S. M., A. E. Abdel Moneim, M. M. Qayed, and N. A. El-Yamany. 2019. "Potential role of N-acetylcysteine on chlorpyrifos-induced neurotoxicity in rats." *Environ Sci Pollut Res Int* 26 (20):20731-20741. doi: 10.1007/s11356-019-05366-w.

Meneses, M. J., R. Silvestre, I. Sousa-Lima, and M. P. Macedo. 2019. "Paraoxonase-1 as a Regulator of Glucose and Lipid Homeostasis: Impact on the Onset and Progression of Metabolic Disorders." *International journal of molecular sciences* 20 (16):4049. doi: 10.3390/ijms20164049.

Mishra, V., and N. Srivastava. 2015. "Organophosphate pesticides-induced changes in the redox status of rat tissues and protective effects of antioxidant vitamins." *Environ Toxicol* 30 (4):472-82. doi: 10.1002/tox. 21924.

Moallem, S. A., A. T. Hariri, M. Mahmoudi, and H. Hosseinzadeh. 2014. "Effect of aqueous extract of Crocus sativus L. (saffron) stigma against subacute effect of diazinon on specific biomarkers in rats." *Toxicol Ind Health* 30 (2):141-6. doi: 10.1177/0748233712452609.

Mohammadzadeh, L., H. Hosseinzadeh, K. Abnous, and B. M. Razavi. 2018. "Neuroprotective potential of crocin against malathion-induced motor deficit and neurochemical alterations in rats." *Environ Sci Pollut Res Int* 25 (5):4904-4914. doi: 10.1007/s11356-017-0842-0.

Morcos, N. Y. S., M. K. El-Masry, M. M. Kamal, and S. Y. Galal. 2015. "Contrasting genetic influence of PON 1 coding gene polymorphisms L55M and Q192R on individuals' response to environmental agents." *Egyptian Journal of Medical Human Genetics* 16 (3):233-237. doi: 10.1016/j.ejmhg. 2015.02.007.

Moshou, H., A. Karakitsou, F. Yfanti, D. Hela, D. Vlastos, A. K. Paschalidou, P. Kassomenos, and I. Petrou. 2020. "Assessment of genetic effects and pesticide exposure of farmers in NW Greece." *Environmental Research* 186:109558. doi: 10.1016/j.envres.2020.109558.

Naravaneni, R., and K. Jamil. 2007. "Determination of AChE levels and genotoxic effects in farmers occupationally exposed to pesticides." *Hum Exp Toxicol* 26 (9):723-31. doi: 10.1177/0960327107083450.

Nebert, D. W., K. Wikvall, and W. L. Miller. 2013. "Human cytochromes P450 in health and disease." *Philosophical transactions of the Royal Society of London. Series B, Biological sciences* 368 (1612):20120431-20120431. doi: 10.1098/rstb.2012.0431.

Ning, J., G. Gang, Z. Bai, Q. Hu, H. Qi, A. Ma, X. Zhuan, and G. Zhuang. 2012. "In situ enhanced bioremediation of dichlorvos by a phyllosphere Flavobacterium strain." *Frontiers of Environmental Science & Engineering* 6 (2):231-237. doi: 10.1007/s11783-011-0316-4.

Nita, M., and A. Grzybowski. 2016. "The Role of the Reactive Oxygen Species and Oxidative Stress in the Pathomechanism of the Age-Related Ocular Diseases and Other Pathologies of the Anterior and Posterior Eye Segments in Adults." *Oxidative Medicine and Cellular Longevity* 2016:3164734-3164734. doi: 10.1155/2016/3164734.

Nwani, C. D., P. O. Somdare, E. O. Ogueji, J. C. Nwani, J. A. Ukonze, and A. O. Nwadinigwe. 2017. "Genotoxicity assessment and oxidative stress responses in freshwater African catfish *Clarias gariepinus* exposed to fenthion formulations." *Drug and Chemical Toxicology* 40 (3):273-280. doi: 10.1080/01480545.2016.1209772.

Ojha, A., and Y. K. Gupta. 2015. "Evaluation of genotoxic potential of commonly used organophosphate pesticides in peripheral blood lymphocytes of rats." *Hum Exp Toxicol* 34 (4):390-400. doi: 10.1177/ 0960327114537534.

Ojha, A., S. K. Yaduvanshi, S. C. Pant, V. Lomash, and N. Srivastava. 2013. "Evaluation of DNA damage and cytotoxicity induced by three commonly used organophosphate pesticides individually and in mixture, in rat tissues." *Environ Toxicol* 28 (10):543-52. doi: 10.1002/tox.20748.

Ozdal, M., O. G. Ozdal, O. F. Algur, and E. B. Kurbanoglu. 2017. "Biodegradation of α-endosulfan via hydrolysis pathway by Stenotrophomonas maltophilia OG2." *3 Biotech* 7 (2):113. doi: 10.1007/ s13205-017-0765-y.

Ozkan, D., D. Yuzbasioglu, F. Unal, S. Yilmaz, and H. Aksoy. 2009. "Evaluation of the cytogenetic damage induced by the organophosphorous insecticide acephate." *Cytotechnology* 59 (2):73-80. doi: 10.1007/s10616-009-9195-y.

Ozkan, U., A. Osun, K. Basarslan, S. Senol, I. Kaplan, and H. Alp. 2014. "Effects of intralipid and caffeic acid phenethyl ester on neurotoxicity, oxidative stress, and acetylcholinesterase activity in acute chlorpyriphos intoxication." *International journal of clinical and experimental medicine* 7 (4):837-846.

Proskocil, B. J., A. C. G. Grodzki, D. B. Jacoby, P. J. Lein, and A. D. Fryer. 2019. "Organophosphorus Pesticides Induce Cytokine Release from Differentiated Human THP1 Cells." *American journal of respiratory cell and molecular biology* 61 (5):620-630. doi: 10.1165/rcmb.2018-0257OC.

Ramu, S., and B. Seetharaman. 2014. "Biodegradation of acephate and methamidophos by a soil bacterium *Pseudomonas aeruginosa* strain Is-6." *Journal of Environmental Science and Health, Part B* 49 (1):23-34. doi: 10.1080/03601234.2013.836868.

Rani, M. V., and M. S. Rao. 1991. "In vitro effect of fenthion on human lymphocytes." *Bull Environ Contam Toxicol* 47 (2):316-20. doi: 10.1007/ BF01688658.

Reichert, C. O., D. Levy, and S. P. Bydlowski. 2020. "Paraoxonase Role in Human Neurodegenerative Diseases." *Antioxidants (Basel)* 10 (1). doi: 10.3390/antiox1001 0011.

Ren, N., M. Atyah, W. Y. Chen, and C. H. Zhou. 2017. "The various aspects of genetic and epigenetic toxicology: testing methods and clinical applications." *Journal of translational medicine* 15 (1):110-110. doi: 10.1186/s12967-017-1218-4.

Sams, C., H. J. Mason, and R. Rawbone. 2000. "Evidence for the activation of organophosphate pesticides by cytochromes P450 3A4 and 2D6 in human liver microsomes." *Toxicology Letters* 116 (3):217-221. doi: 10.1016/ S0378-4274(00) 00221-6.

Sarkar, B., M. Dhiman, S. Mittal, and A. K. Mantha. 2017. "Curcumin revitalizes Amyloid beta (25-35)-induced and organophosphate pesticides pestered neurotoxicity in SH-SY5Y and IMR-32 cells via activation of APE1 and Nrf2." *Metab Brain Dis* 32 (6):2045-2061. doi: 10.1007/s11011-017-0093-2.

Savic, J. Z., S. Z. Petroviic, A. R. Leskovac, T. D. Lazarevic Pasti, B. J. Nastasijevic, B. B. Tanovic, S. M. Gasic, and V. M. Vasic. 2019. "UV-C light irradiation enhances toxic effects of chlorpyrifos and its formulations." *Food Chemistry* 271:469-478. doi: 10.1016/j.foodchem.2018.07.207.

Shadnia, S., M. Dasgar, S. Taghikhani, A. Mohammadirad, R. Khorasani, and M. Abdollahi. 2007. "Protective Effects of a-Tocopherol and N-Acetyl-Cysteine on Diazinon-Induced Oxidative Stress and Acetylcholinesterase Inhibition in Rats." *Toxicology Mechanisms and Methods* 17 (2):109-115. doi: 10.1080/153765 10600860318.

Sharma, L. K., J. Lu, and Y. Bai. 2009. "Mitochondrial respiratory complex I: structure, function and implication in human diseases." *Current medicinal chemistry* 16 (10):1266-1277. doi: 10.2174/092986709787846578.

Shunmoogam, N., P. Naidoo, and R. Chilton. 2018. "Paraoxonase (PON)-1: a brief overview on genetics, structure, polymorphisms and clinical relevance." *Vasc Health Risk Manag* 14:137-143. doi: 10.2147/VHRM. S165173.

Sidhu, G. K., S. Singh, V. Kumar, D. S. Dhanjal, S. Datta, and J. Singh. 2019. "Toxicity, monitoring and biodegradation of organophosphate pesticides: A review." *Critical Reviews in Environmental Science and Technology* 49 (13):1135-1187. doi: 10.1080/10643389.2019.1565554.

Silins, I., and J. Högberg. 2011. "Combined Toxic Exposures and Human Health: Biomarkers of Exposure and Effect." *International Journal of Environmental Research and Public Health* 8 (3):629-647.

Singh, B. K., and A. Walker. 2006. "Microbial degradation of organophosphorus compounds." *FEMS Microbiology Reviews* 30 (3):428-471. doi: 10.1111/j. 1574-6976.2006.00018.x.

Singh, B. K., A. Walker, J. A. Morgan, and D. J. Wright. 2004. "Biodegradation of chlorpyrifos by enterobacter strain B-14 and its use in bioremediation of contaminated soils." *Appl Environ Microbiol* 70 (8):4855-63. doi: 10.1128/aem.70.8.4855-4863.2004.

Singh, S., V. Kumar, P. Singh, B. D. Banerjee, R. S. Rautela, S. S. Grover, D. S. Rawat, S. T. Pasha, S. K. Jain, and A. Rai. 2012. "Influence of CYP2C9, GSTM1, GSTT1 and NAT2 genetic polymorphisms on DNA damage in workers occupationally exposed to organophosphate pesticides." *Mutation Research/Genetic Toxicology and Environmental Mutagenesis* 741 (1):101-108. doi: 10.1016/j.mrgentox.2011.11.001.

Singh, S., V. Kumar, P. Singh, S. Thakur, B. D. Banerjee, R. S. Rautela, S. S. Grover, D. S. Rawat, S. T. Pasha, S. K. Jain, and A. Rai. 2011. "Genetic polymorphisms of GSTM1, GSTT1 and GSTP1 and susceptibility to DNA damage in workers occupationally exposed to organophosphate pesticides." *Mutation Research/Genetic Toxicology and Environmental Mutagenesis* 725 (1):36-42. doi: 10.1016/j.mrgentox. 2011.06.006.

Soltaninejad, K., and M. Abdollahi. 2009. "Current opinion on the science of organophosphate pesticides and toxic stress: a systematic review." *Med Sci Monit* 15 (3):RA75-90.

Suarez, B., F. Vela-Soria, F. Castiello, A. Olivas-Martinez, D. Acuna-Castroviejo, J. Gomez-Vida, N. Olea, M. F. Fernandez, and C. Freire. 2021. "Organophosphate pesticide exposure, hormone levels, and interaction with PON1 polymorphisms in male adolescents." *Science of the Total Environment* 769. doi: 10.1016/j.scitotenv. 2020.144563.

Tang, J. U. N., R. L. Rose, J. E. Chambers, and R. C. Gupta. 2006. " Metabolism of Organophosphorus and Carbamate Pesticides." In *Toxicology of Organophosphate & Carbamate Compounds*, 127-143. Burlington: Academic Press.

Teodoro, M., G. Briguglio, C. Fenga, and C. Costa. 2019. "Genetic polymorphisms as determinants of pesticide toxicity: Recent advances." *Toxicology reports* 6:564-570. doi: 10.1016/j.toxrep.2019.06.004.

Terry, A. V. 2012. "Functional consequences of repeated organophosphate exposure: Potential non-cholinergic mechanisms." *Pharmacology & Therapeutics* 134 (3):355-365. doi: 10.1016/j.pharmthera.2012.03.001.

Tian, J., Q. Dong, C. Yu, R. Zhao, J. Wang, and L. Chen. 2016. "Biodegradation of the Organophosphate Trichlorfon and Its Major Degradation Products by a Novel Aspergillus sydowii PA F-2." *J Agric Food Chem* 64 (21):4280-7. doi: 10.1021/acs.jafc.6b00909.

Toledo-Ibarra, G. A., K. J. G. Diaz Resendiz, G. H. Ventura-Ramon, F. Gonzalez-Jaime, A. Vega-Lopez, E. Becerril-Villanueva, L. Pavon, and M. I. Giron-Perez. 2016. "Oxidative damage in gills and liver in Nile tilapia (*Oreochromis niloticus*) exposed to diazinon." *Comparative Biochemistry and Physiology Part A: Molecular & Integrative Physiology* 200:3-8. doi: 10.1016/j.cbpa.2016.05.007.

Valencia-Quintana, R., R. M. Lopez-Duran, M. Milic, S. Bonassi, M. A. Ochoa-Ocana, M. O. Uriostegui-Acosta, G. A. Perez-Flores, J. L. Gomez-Olivares, and J. Sanchez-Alarcon. 2021. "Assessment of Cytogenetic Damage and Cholinesterasese Activity in Workers Occupationally Exposed to Pesticides in Zamora-Jacona, Michoacan, Mexico." *International Journal of Environmental Research and Public Health* 18 (12):6269.

Vinken, M., and B. J. Blaauboer. 2017. "In vitro testing of basal cytotoxicity: Establishment of an adverse outcome pathway from chemical insult to cell death." *Toxicology in vitro: an international journal published in association with BIBRA* 39:104-110. doi: 10.1016/j.tiv.2016.12.004.

Wahyuni, E. A., H. D. Lin, C. W. Lu, C. M. Kao, and S. C. Chen. 2021. "The cytotoxicity and genotoxicity of single and combined fenthion and terbufos treatments in human liver cells and zebrafish embryos." *Science of The Total Environment* 758:143597. doi: 10.1016/j.scitotenv.2020.143597.

WHO. 2016. *International code of conduct on pesticide management: guidelines on highly hazardous pesticides*. Geneva: World Health Organization & Food and Agriculture Organization of the United Nations.

Wilkinson, G. R. 2005. "Drug metabolism and variability among patients in drug response." *N Engl J Med* 352 (21):2211-21. doi: 10.1056/NEJMra032424.

Wong, R. H., S. Y. Chang, S. W. Ho, P. L. Huang, Y. J. Liu, Y. C. Chen, Y. H. Yeh, and H. S. Lee. 2008. "Polymorphisms in metabolic GSTP1 and DNA-repair XRCC1 genes with an increased risk of DNA damage in pesticide-exposed fruit growers." *Mutation Research/Genetic Toxicology and Environmental Mutagenesis* 654 (2):168-175. doi: 10.1016/j.mrgentox. 2008.06.005.

Wozniak, E., P. Sicinska, J. Michalowicz, K. Wozniak, E. Reszka, B. Huras, J. Zakrzewski, and B. Bukowska. 2018. "The mechanism of DNA damage induced by Roundup 360 PLUS, glyphosate and AMPA in human peripheral blood mononuclear cells - genotoxic risk assessment." *Food Chem Toxicol* 120:510-522. doi: 10.1016/j.fct.2018.07.035.

Wu, J. C., Y. C. Hseu, J. S. Tsai, L. C. Chen, S. M. Chye, C. H. Chen, and S. Ching Chen. 2011. "Fenthion and terbufos induce DNA damage, the expression of tumor-related genes, and apoptosis in HEPG2 cells." *Environmental and Molecular Mutagenesis* 52 (7):529-537. doi: 10.1002/ em.20652.

Yan, M. H., X. Wang, and X. Zhu. 2013. "Mitochondrial defects and oxidative stress in Alzheimer disease and Parkinson disease." *Free radical biology & medicine* 62:90-101. doi: 10.1016/j.freeradbiomed.2012.11.014.

Zahra, M. A., M. Kandeel, S. A. Aldossary, and A. Al-Taher. 2020. "Study on Genotyping Polymorphism and Sequencing of N-Acetyltransferase 2 (NAT2) among Al-Ahsa Population." *Biomed Res Int* 2020:8765347. doi: 10.1155/2020/8765347.

Zanger, U. M., and M. Schwab. 2013. "Cytochrome P450 enzymes in drug metabolism: Regulation of gene expression, enzyme activities, and impact of genetic variation." *Pharmacology & Therapeutics* 138 (1):103-141. doi: 10.1016/j.pharmthera.2012.12.007.

Zhao, J., D. Zhao, and J. Han. 2009. "Isolation and Characterization of Dimethoate Degrading Phytopathogen Fungus from Soil." *2009 3rd International Conference on Bioinformatics and Biomedical Engineering,* 11-13 June 2009.

Zhao, R. B., H. Y. Bao, and Y. X. Liu. 2010. "Isolation and Characterization of *Penicillium oxalicum* ZHJ6 for Biodegradation of Methamidophos." *Agricultural Sciences in China* 9 (5):695-703. doi: 10.1016/S1671-2927 (09)60145-0.

Zheng, F., F. M. Goncalves, Y. Abiko, H. Li, Y. Kumagai, and M. Aschner. 2020. "Redox toxicology of environmental chemicals causing oxidative stress." *Redox biology* 34:101475-101475. doi: 10.1016/j.redox.2020. 101475.

Chapter 3

Low-Level Chronic Organophosphate Exposure Contributes to the Development of Synaptic Vulnerability

Sanja Ivković*
Department of Molecular Biology and Endocrinology,
"Vinča" Institute for Nuclear Sciences,
National Institute of Republic of Serbia, University of Belgrade,
Belgrade, Serbia

Abstract

Organophosphates (OPs) are the main components of herbicides, insecticides, and pesticides used worldwide. In addition, OPs are used in the production of chemical weapons such as nerve gasses, sarin, and soman. The main mechanism of OP action in the case of lethal poisoning is the inhibition of acetylcholinesterase (AChE) activity resulting in excess of acetylcholine (ACh), and the induction of "cholinergic crisis," a clinical condition that often ends in coma and death. Thus, the levels of AChE in the blood are used as the primary biomarker for evaluating OP-induced poisoning. Normal levels of AChE in the blood have verified that low-level chronic OP exposure is relatively safe. However, recent findings associated a low-level, chronic OP exposure to neurological and cognitive impairments that can persist long after the normalization of AChE levels suggesting that mechanisms of action for acute and chronic OP poisoning may differ. One such mechanism is the contribution of chronic OP poisoning to the development of synaptic vulnerability that can further lead to neurodegenerative and neuropsychiatric pathologies such as Alzheimer's disease, Parkinson's disease, ALS, and ADHD, and

* Corresponding Author's Email: sivkovic@vin.bg.ac.rs.

In: Organophosphates
Editor: Tamara Lazarević-Pašti
ISBN: 978-1-68507-652-8

cognitive impairments in children. The potential mechanisms of action during low-level chronic OP exposure in various preclinical models and paradigms were discussed. In addition, the need for the development of new biomarkers in order to reevaluate systemic, neural, and neurodevelopmental toxicity, and thus the applicative doses, of various OPs that exist today is underscored. Finally, potential prophylactic measures that can be applied as the countermeasures for the increased synaptic vulnerability are suggested.

Keywords: organophosphates, chronic exposure, synaptic plasticity, neurodegeneration

Introduction

Organophosphates are a heterogeneous group of compounds that have wide application in our daily life in agriculture as insecticides and pesticides, industrial solvents, or in fuel production. They exert their function through the potent inhibition of acetylcholinesterase (AChE) activity. If used in lethal doses, as chemical warfare (in the form of soman or sarin nerve gas poisoning), or ingested in suicidal attempts, they induce "cholinergic crisis" inhibiting 80-90% of AChE enzymatic activity, thus leading to respiratory failure, paralysis, decreased blood pressure, parasympathetic discharge, and, finally, death. Even the moderate intoxication, as in Tokyo subway attack survivors, exhibited long-term effects, including deficiency in cognitive processing and even the brain structural alterations as revealed with neuroimaging (Sullivan and Blose, 1992; Singh and Sharma, 2000; Peter et al., 2014). However, many occupations today carry the risk of the low-level, chronic, OP exposure – including, but not limited to, agricultural work, aircraft, and military personnel, and workers involved in any of the steps in the production, transport, and application of billions of tons of pesticides yearly. Many cases of acute OP poisoning were documented among these working groups and together with more than 2 million OP-poisoning suicide attempts resulting in 200.000 deaths, recorded mainly in developing countries, designate OP exposure as a major global problem (Bertolote and Fleischman, 2015; Yurumez et al., 2007; Gunnell and Eddleston, 2003). The necessity to address this issue is reflected in the defined US-standard-of-care (SOC) medical countermeasures for acute OP exposure consisting of Pralidoxime (2-PAM) and Atropine Sulfate, and an anticonvulsant such as Midazolam or Diazepam.

The OPs' widespread usage led to their sustained presence in all the parts of our ecosystem – in water, soil, and air, and consequently in the living organisms, plants, and animals, including humans. The presence of OPs is so persistent in our daily life that each of us is exposed to chronic, low OP levels through food ingestion, inhalation of air, or just living near agricultural fields treated with OP pesticides/insecticides. One of the widely used OPs, whose usage was estimated to be more than 200.000 tons in 2015, is chlorpyrifos (CPF) (John and Shaike, 2015). However, chronic OP exposure can vary significantly between individuals, and studies regarding its effects in humans are challenging to synchronize. Several studies have shown that the cumulative exposure to low levels of pesticides over many years of agricultural work, for example, is affecting cognition, autonomic and motor function and vision, resulting in neurological impairment (Kamel et al., 2007; Rohlman et al., 2007) similar to the symptoms reported from the Gulf war veterans (Sullivan et al., 2018).

Although some studies reported the absence of positive correlations between OPs exposure and neurological impairment, the prevalence of findings indicating that chronic OPs exposure is associated with neuropsychiatric, neurodegenerative, and cognitive deficits is overpowering (for detailed review and references therein, see Voorhees et al., 2017). Furthermore, prenatal exposure to OPs is linked to the development of cognitive and motor abnormalities in children exposed to OP pesticides. Therefore, one of the major concerns regarding low-level OPs exposure (together with the long-term effects of the acute OPs poisoning in the survivors) is the subsequent development of neurotoxicity and related neuropathologies (Rohlman et al., 2011).

The Molecular Mechanisms of Acute OP Poisoning

The symptoms of acute OPs poisoning, as in the case of nerve gases soman and sarin, have been extensively studied, and the appropriate SOC procedure has been developed. The primary mechanism of OP action is the irreversible inhibition of AChE and the ensuing accumulation of ACh in synaptic clefts in both the peripheral and central nervous system resulting in the hyper-cholinergic crisis. It consists of neurological, motor, cardiovascular, and respiratory failures that end in coma and death in the majority of cases. In the case of severe OP poisoning, the increased levels of ACh in synapses disrupt the balanced GABAergic and glutamatergic activity. The increased glutamatergic activity can induce the avalanche effect creating a secondary neurotoxic cascade that can persist even months after the initial insult resulting in seizures that often progress to status epilepticus (Avanzini, 1993; Chen, 2012). The hyperstimulation of muscarinic cholinergic receptors results in the development of the initial seizures, which are then strengthened and maintained by glutamatergic hyperexcitation, which is mainly responsible for the further excitotoxic neuronal damage and death (Avanzini, 1993; Chen, 2012). Regarding the main role of glutamatergic signaling in the detrimental aftermath of the cholinergic crisis, additional treatments in the SOC procedure are now focusing on the administration of inhibitors of glutamatergic signaling. The additional administration of antagonists of AMPA/GluK1 (LY293558) and NMDA (caramiphen and ketamine) receptors (Lewine et al., 2018; de Araujo Furtado et al., 2020; Aroniadou-Aderjaska et al., 2020) in preclinical models of exposure to lethal and sub-lethal OPs doses resulted in the increased and sustained neuroprotection when compared to current treatments.

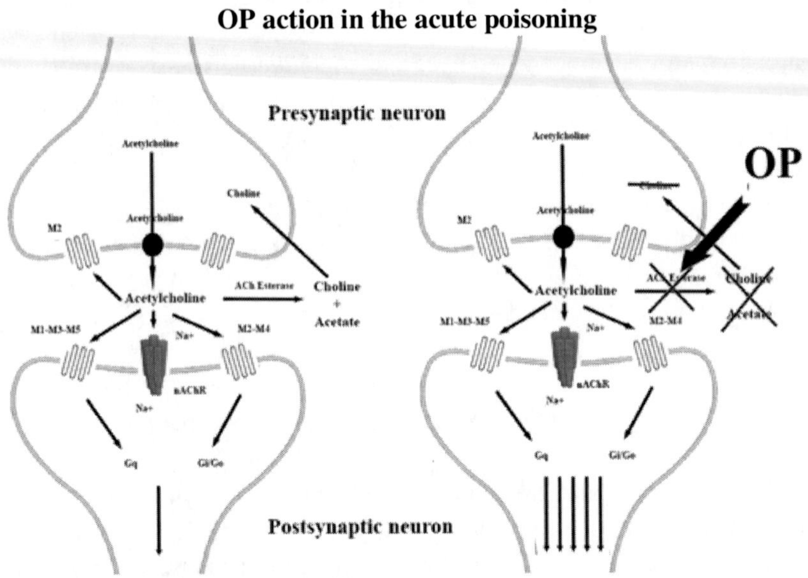

Figure 1. OP exposure impairs ACh signaling through the inhibition of AChE activity. Normally, ACh is secreted from the presynaptic neuron in the synaptic cleft, where it binds to its receptors (muscarinic, M1-5, or nicotinic, nAChR). The excess of unbound ACh is broken on acetate and choline through the action of ACh esterase (AChE), and choline is returned to the presynaptic neuron. The OP exposure inhibits AChE activity, leading to the accumulation of ACh at the synapse and elicits increased responses in the postsynaptic neuron.

The Low-Level OPs Exposure Association with Neurological Disorders

In contrast to the molecular mechanisms of acute OP poisoning, the effects of chronic, low-level OPs exposure are much less understood, and their connection with the subsequent neurodegenerative disorders is not so simple. Establishing the association between OP exposure and the subsequent development of neurodegeneration demands broad, well-organized cohorts of studies that should last for a substantial amount of time. Additionally, the levels of chronic exposure can vary between the different populations. The results obtained from the studies involving agricultural workers, for example, will differ from the results obtained from other populations that are not exposed to OPs to such an extent (for a detailed review, see Kamel and Hoppin, 2004). Finally, OPs based pesticides and insecticides are a heterogeneous group consisting of nearly 40 different chemical members registered by the US-EPA (www.epa.gov) with different toxicity and mode of action, making it difficult to obtain comparable analyses not only among human subjects but also in the preclinical models. The comparison of the results obtained in the preclinical studies is further complicated with different animal models, different doses of OPs used for the analyses, and many OPs currently in use in everyday life.

However, the evidence accumulated so far associates exposure to OPs with compromised health, and importantly, substantiated effects were found in the nervous system. Several parameters have been analyzed and measured, such as symptom prevalence, neurobehavioral performance, sensory and motor dysfunction, and direct measure of nerve function (Kamel and Hoppin, 2004). There is a considerable amount of data associating OP exposure with neurodegenerative diseases such as Parkinson's disease (PD), amyotrophic lateral sclerosis (ALS), and with the development of dementia and Alzheimer's disease (AD) (Kamel and Hoppin, 2004). In addition, occupational pesticide exposure was also associated with mild cognitive impairment (MCI) (Bosma, 2000), vascular dementia (Lindsay, 1997), and with risk of dementia among PD patients (Hubble et al., 1998). Notably, the association of OP exposure with the development and progression of AD is further complicated with the fact that the important part of AD therapy is the treatment with the inhibitors of acetylcholinesterase.

In regards to the nervous system impairments, OP pesticides can produce several distinct neurotoxic effects such as acute cholinergic toxicity, delayed ataxia known as organophosphorus ester-induced delayed neurotoxicity

(OPIDN), chronic neurotoxicity, and developmental neurotoxicity. Understandably, these effects can vary depending on the dose, frequency of exposure, and the type of OP. As OPs are often applied together with other classes of insecticides and pesticides, the study of their effects is further complicated with the possibility of synergistic effects that may ensue in such cases. Behavioral effects of OP exposure are primarily due to the inhibition of acetylcholine esterase or through the direct binding of OP to nicotinic receptors, and the most common clinical symptoms are – memory, concentration and learning deficits, anxiety, depression, psychotic symptoms, chronic fatigue, peripheral neuropathy, autonomic dysfunction and extrapyramidal symptoms such as dystonia, resting tremor, bradykinesia, postural instability and rigidity of face muscles (Mileson et al., 1998),

The assessment of the levels of AChE in blood serves as a primary biomarker in the evaluation of OP poisoning, either acute or mild. The normalization of AChE levels in the blood has led to the conclusion that OPs are, in general, relatively safe for use. However, various studies have confirmed that the effects of OP poisoning on neurobehavioral deficiencies can persist for a long time after the normalization of AChE levels suggesting that the systemic effects of the acute OP exposure can significantly differ from the chronic ones. More importantly, that would mean that chronic effects may not be alarming enough and can exist for a long time before being correctly diagnosed and treated. Such is the case with Persian Gulf Veterans' symptoms (Gulf War Illness, GWI) that include memory loss, concentration problems, and fatigue, suggesting the dysfunction in the CNS (Haley and Kurt, 1997; Jamal, 1998; Haley, 2003; Golomb, 2008). Similar findings were registered in the study analyzing 18.872 white male private pesticide applicators from 1993-1997, confirming the association between OP exposure and 23 neurological symptoms typically associated with pesticide intoxication (Kamel et al., 2005).

The Volume and Synaptic Transmission of ACh

In order to predict, understand, and anticipate the consequences of compromised ACh transmission, it is essential to understand the complex ACh actions in CNS. These actions are defined predominantly as neuromodulatory in contrast to the periphery, where ACh acts as a primary excitatory neurotransmitter (Picciotto et al., 2000, Wess et al., 2003). Neuromodulation implies any action that is neither directly excitatory nor inhibitory but can change the neuronal response to the later stimulation. ACh action can induce drastic behavioral transitions such as switching from sleep to wakefulness, distraction to attention, learning, and recall (Hasselmo and Sarter, 2011; Lee and Dan, 2012). Although the complete understanding of such transitions is lacking, the actions of neuromodulators, including ACh, are considered to regulate the shift from synchronized to desynchronized states of neuronal networks (Lee and Dan, 2012).

ACh is produced in several distinct regions in the brain that can be grouped in two main projections: the magnocellular basal forebrain and the brainstem cholinergic system (Mesulam et al., 1983; Everitt and Robbins, 1997) (Figure 2). The basal forebrain cholinergic system consists of the medial septal nucleus (MS), the vertical and horizontal limbs of the diagonal band of Broca (DB), and the nucleus basalis (nBM). The ACh projections from these sites innervate the neocortex, hippocampus, basolateral amygdala, olfactory bulb, and others (Mesulam et al., 1983).

Initially, cholinergic signaling was considered only a slow and diffuse process (through volume transmission) because of the lack of classical synapses on cholinergic terminals (Munoz and Rudy, 2014), with ACh executing its effects on distant locations. Although it was strongly suggested that the actions of ACh are executed mainly through the volume transmission (Umbriaco et al., 1994; Descarries and Mechawar, 2000; Sarter et al., 2009; Yamasaki et al., 2010), the new findings, starting from the early '90s were able to confirm the existence of actual synapses in the structures that were until then called cholinergic varicosities in the rat and macaque cingulate cortex (Umbriaco et al., 1994; Mrzijak et al., 1995). Finally, the use of optogenetic tools (Kalmbach et al., 2012) confirmed recently that ACh signaling occurs through direct fast point-to-point synaptic transmission.

The distribution pattern of cholinergic receptors in the CNS is essential for defining the specific modality through which ACh exerts its function. The muscarinic receptor M1 subtype is the most abundant ACh receptor in the cortex (Yamasaki, 2010). It is located predominantly extrasynaptically and not in any particular vicinity to the cholinergic varicosities, i.e., putative synaptic terminals, suggesting that the role of the M1 subtype is to exert the ACh effects at longer distances through the volume transmission. Thus, the precise mapping of the receptor subtypes subcellular localization can provide a quantitative method to predict the particular modality of ACh action (volume vs. synaptic transmission). Interestingly, the quantitative analysis of cholinergic synapses has shown striking differences among species and that synaptic transmission is the primary mechanism of ACh signaling in humans

(Smiley et al., 1997). However, the relationship between volume and synaptic transmission is still elusive. It is not well understood if one mode prevails over the other and if such equilibrium persists throughout the CNS. It is much more likely that these two modes of ACh actions are complementary, working together to modulate network activity in response to external stimuli through changing membrane excitability and synaptic activity. While ACh volume release is tonic and exerts its actions through extra-synaptic receptors, the point-to-point release of ACh in synaptic junctions will have a much more restricted effect on the smaller spatiotemporal scale (Munoz and Rudy, 2014).

ACh as a Neuromodulator

The effect of ACh release can be different depending on the specific cortical neuronal circuitry involved. As a neuromodulator, ACh can regulate the release of other neurotransmitters in both volume or synaptic transmission, and the latter depends on the combination of pre-and postsynaptic receptor components.

The specific execution of these functions is highly dependent on the neuronal subtypes, and CNS structures analyzed and are executed through two classes of receptors: metabotropic muscarinic receptors (mAChRs, M1 to M5) and ionotropic nicotinic receptors (nAChRs) (Figure 2). Although neuromodulation is canonically exerted through the metabotropic receptors, nicotinic ionotropic receptors appear to have the same function (Picciotto, 2003). Importantly, mAChRs are located pre-and postsynaptically and can produce opposite effects on brain activity (Gil et al., 1997; McCormick and Prince, 1985; Douglas et al., 2002; Higley et al., 2009).

Nicotinic receptors are comprised of a large family of alpha and beta subunits (α2-α7 and β2-β4, Picciotto et al., 2000) organized in the homomeric or heteromeric aggregates forming nonselective, excitatory cation channels (Changeaux et al., 1998; Picciotto et al., 2001) located throughout the neuronal surface. Combining different α and β subunits and cellular localization of nAChRs can modulate the release of different neurotransmitters, such as glutamate, GABA, DA, ACh, norepinephrine, and serotonin (McGehee et al., 1995; Wonnacott, 1997). In addition, ACh can modulate postsynaptic responses in the hippocampus and cortex through nAChRs (Bell et al., 2011; Gu and Yakel, 2011, Arroyo et al., 2012).

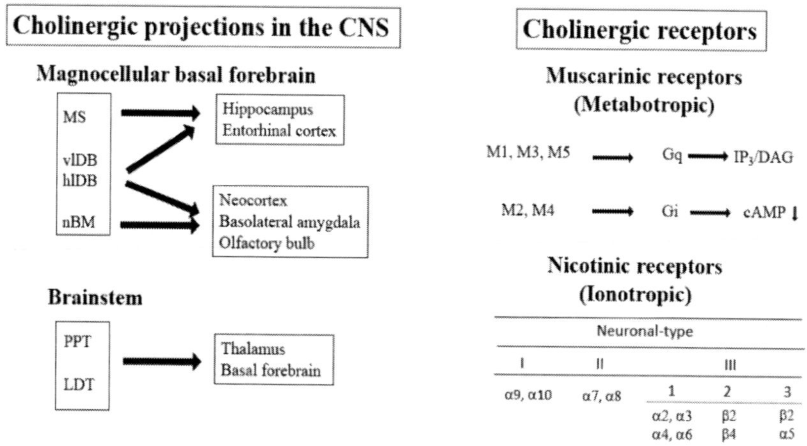

Figure 2. The schematic representation of cholinergic projections and receptors in the CNS. The main cholinergic projections in the CNS stem from the magnocellular basal forebrain and the brainstem. The cholinergic projections from the magnocellular basal forebrain originate in the three core structures: the medial septal nucleus (MS), the vertical and horizontal limbs of the diagonal band of Broca (vlDB and hlDB), and the nucleus basalis magnocellularis (nBM). The projections innervating the hippocampus and entorhinal cortices come from MS and the vertical limb of the DB. The horizontal limb of the DB and nBM project diffusely to the neocortex, basolateral amygdala, and olfactory bulb. The brainstem cholinergic projections are coming from pedunculopontine tegmental nucleus (PPT) and laterodorsal pontine tegmentum (LDT) and innervate mainly thalamus together with the basal forebrain regions. Cholinergic receptors are classified into two main groups – muscarinic (metabotropic) and nicotinic (ionotropic) receptors. Their organization in subgroups is represented schematically.

Synaptic transmission depends on the combination of pre-and postsynaptic receptor components. The neuromodulatory effects of specific neurotransmitters have an added layer of complexity as many neuronal subtypes can co-release multiple neurotransmitters – slow-acting neuro-modulators together with fast-acting neurotransmitters (Hnasko and Edwards, 2012). Most basal forebrain (BF) cholinergic neurons simultaneously release glutamate as a cotransmitter, and ACh secretion will, in this case, have a dual role. One would be to elicit a strong negative feedback auto-inhibition through the presynaptic M2 mAChRs (Dudar, 1975), and the other to inhibit the release of glutamate-mediated through the same type of mAChRs (Allen et al., 2006). In such a way, these neurotransmitters will act as antagonists if they are simultaneously released on the same axons. The recent findings that cholinergic neurons can elicit fast-acting GABA-ergic neurotransmission

(Henny and Jones, 2008; Granger et al., 2016) introduce another way of ACh-regulated synaptic activity. The co-release of GABA could potentiate the cholinergic disinhibition of cortical firing by inhibiting local interneurons. On the synaptic level, it can be regulated by nAChRs localized both pre-and postsynaptic (Saunders et al.,2015).

Although the precise role of such neurotransmitter co-releasing in cortical processing and synaptic plasticity is not entirely understood, its existence indicates added complexity in the regulation of brain activity and cognition and any dysregulation in these tightly organized processes, such as in the case of low-level, chronic, OP exposure, can have long-lasting effects on the CNS functioning.

OP Exposure Promotes Synaptic Vulnerability

The proper functioning of the nervous system is made possible through the plastic properties of synapses. Synaptic plasticity is the ability of synapses to change their strength depending on either their activity or as a result of the activity of surrounding synapses and is considered crucial for learning and memory (Kandel, 2001; Fusi et al., 2005). However, synapses are highly vulnerable to changes in neuronal homeostasis regardless of their source, and very often, neuronal death is preceded by the loss of synapses and their connections. Thus, the synapse might be the initial target of the insult, and the developing synaptopathies can trigger consequent neurodegeneration and neuronal loss. It was suggested that the survivors of organophosphate exposure exhibit delayed neuronal and behavioral impairments, together with cognitive disabilities (Menon et al.,2004; Toomey et al., 2007; Abou-Donia et al., 2017), which is the basis for the increased risk for the development of neurodegenerative and neuropsychiatric pathologies (Yokoyama et al., 1998; Bullman et al., 2005; Hayden et al., 2010; Jones 2010; Parron et al., 2011; Slotkin 2011).

Regarding the endangered neuronal homeostasis as a prequel to the induction of synaptic vulnerability, several studies have confirmed that low-level chronic OP exposure can cause distress in multiple neuronal compartments. Chronic dichlorvos exposure can cause a significant mitochondrial dysfunction. The most probable cause of these changes is the enhanced oxidative stress, decreased GSH (glutathione) levels, and decreased Mn-SOD (Manganese superoxide-dismutase) activity in the mitochondria in the brain after the OP exposure. The final result is the activation of caspase-3,

DNA fragmentation, and apoptotic neuronal degeneration (dos Santos et al., 201).

On the other hand, OPs can interact directly with M2 receptors (Verma et al., 2008), and some of the OPs (paraoxon, malaoxon, and chlorpyrifos oxon) can inhibit cAMP formation in a dose-dependent manner (Ward and Mundy, 1996). Considering that OP exposure unequivocally affects memory function, CREB protein cAMP-responsive element-binding protein) was analyzed as another molecular target for the OP action. CREB is a transcription factor heavily involved in regulating synaptic plasticity and transcription-dependent forms of memory. The low dose OP exposure reduced the phosphorylation of CREB (Schuh et al., 2002) together with the adenylyl cyclase-linked signaling pathway impairment (Verma et al., 2008), without any overt effect on the levels of AChE activity. In addition, impaired axonal transport (Terry et al., 2003, 2007) and alterations in gene expression (Ray et al., 2010) have been associated with the effects of CPF not directly linked to the AChE inhibition (Liu et al., 1999; Pope, 1999).

The Suggested Molecular Mechanisms Underlying Synaptic Vulnerability Induced by OP Exposure

Various preclinical studies have confirmed the OPs neurotoxic effect and revealed specific molecular mechanisms directly affected by OP exposure that may have long-lasting effects on the synaptic structure and function. However, the synaptotoxic and neurotoxic potentials of low-level, chronic OP exposure are not yet fully understood.

The study using repeated administration of subclinical doses of CPF has confirmed the extended effects of low-level OP exposure on synaptic transmission (Speed et al., 2012). The observed initial increase in synaptic transmission was in accordance with similar findings when the administration of AChE inhibitors resulted in improved hippocampal function and increased memory in AD patients (Birks, 2006). However, the analyses performed three months after the initial OP exposure showed a dramatic *de novo* decrease in synaptic density in both apical and basal dendrites in the CA1 hippocampal region, accompanied by a decrease in the basal hippocampal transmission. At the same time, no overt changes in the neuronal number and density in either CA1 or CA3 hippocampal regions, and no symptoms of cholinergic toxicity were observed (Speed et al., 2012).

One of the OPs, parathion, in use since 1944, is converted to paraoxon (Pxn) by oxidation in the soil and the liver (Spencer et al., 1980; Foxenberg et al., 2011) and exerts severe excitotoxic effects in the brain, causing changes in the behavior and cognition, triggering seizures and elongating epileptiform activity (Sanchez-Santed et al., 2004; Todorovic et al., 2012; Deshpande et al., 2014; Desphande et al., 2016). The infusion of Pxn into hippocampal explants was used as a model to mimic the action of OPs that may be essential for the development of the long-term symptoms of nerve agent exposure (Farizatto et al., 2019). On the other hand, the Pxn effects in such a model can also be interpreted as the mechanism of action induced during low-level OPs exposure. The single infusion of Pxn induced a significant loss of major constitutive synaptic proteins - Synapsin II, (SynII), Synaptophysin, and PSD95 (Post Synaptic Density protein 95) when measured at 24 h after the exposure. The hippocampal areas that were most affected, dendritic zones of dentate gyrus and CA1, exhibited a significant increase in GFAP (Glial Fibrillary Acidic Protein) expression, indicative of developing pathology. Notably, the substantial effects of Pxn exposure are evident shortly after the exposure (3h) and continue to increase with time, with more than 80% reduction in Synaptophysin and Syn IIb, and more than 90% reduction in the levels of PSD95 72h post-treatment. Remarkably, these robust changes in the synaptic markers were neither accompanied by changes in the neuronal density or morphology nor the increase in cell death at the time points analyzed, indicating that Pxn effects were primarily synaptotoxic.

In addition, Pxn exhibited a unique mode of action inducing the expression of integrin β. There are several possible explanations for this action. One of them is that the induction of integrin β serves as a neuroprotective response to the synaptotoxic effects of Pxn as was previously suggested (Bahr and Lynch, 1992; Clegg et al., 2003; Babayan et al., 2012). This increased production of integrinβ may be the product of activated astrocytes that increase the expression of extracellular matrix components, including integrin β (Hillen et al., 2018), upon the assault. The induced β1 integrin may also support neuronal resiliency through the stabilization of actin filaments via inactivation of actin severing protein cofilin, stabilizing the functional synapses (Kramar et al., 2006).

The exposure to another OP, soman, a deadly nerve agent (McDonough and Shih, 1997), triggers the release of glutamate, reinforcing the occurrence of seizures and thus, inducing excitotoxic neurodegeneration (Wade et al., 1987; Lallement et al., 1992; Deshpande et al., 1995; Raveh et al., 2003). The exposure to high-doses of soman in the model of hippocampal slices induces synaptotoxicity without the overall cell death and neuronal toxicity, indicating that synaptic vulnerability might be the trigger for the subsequent development of neurodegenerative changes. However, Munirathinam and Bahr (2004) have shown that the sub-lethal low dose exposure to soman does not induce synaptotoxicity but renders the hippocampus susceptible to further insults as early as one hour following soman exposure lasting up to 1 week or longer. Hippocampal slices exposed briefly to AMPA (α-amino-3-hydroxy-5-methyl-4-isoxazole propionic acid), a specific agonist for AMPA receptors, do not show, in general, any lasting effects on synaptic and cellular stability. However, repeated exposures to low levels of soman induce a significant hippocampal vulnerability towards previously non-effective AMPA exposure. The possible explanation is that the synergy between AMPA receptor activation and glutamate release after the accumulation of ACh post-soman exposure can cause the pathogenic effect on NMDA receptors (McDonough and Shih, 1997; Pavlovsky et al., 2003).

Remarkably, an enhanced cytotoxic effect also occurs when soman exposure takes place after the brief excitotoxic event. Therefore, the subtoxic soman exposure drastically decreases neuronal tolerance (Munirathinam and Bahr, 2004), making the CNS highly vulnerable to further insults. The authors proposed a molecular mechanism that can lead to the increased vulnerability. The repeated exposure to the sub-toxic soman levels leads to the induction of calpain levels and consequently the induction of caspases which promote cell death during excitotoxic insults (Wang, 2000; Blomgren et al., 2001). Interestingly, the levels of calpain after the exposure to the sub-lethal doses of OPs and the ones after the excitotoxic insult were similar to the ones after the acute exposure to soman alone, suggesting that the repeated low-dose soman exposure may decrease the threshold for the calpain activation, and enhance the following synaptic vulnerability.

These findings suggest that low-level OPs exposure can promote synaptic vulnerability through a molecular mechanism that is not including the inhibition of AChE activity. However, the exact mechanisms through which OPs exert their function and the consequences they elicit are not completely understood.

OP Exposure during Neurodevelopment

The epidemiological studies exploring the effects of the OPs exposure on neurodevelopment have revealed strong associations with neurological, cognitive, motor, and structural abnormalities among exposed children, such as the increased susceptibility to the development of ASD (Coyle, 2006; Shelton et al., 2012; Sealy et al., 2016), ADHD (Richardson et al., 2015; Yu et al., 2016), motor and learning impairment (Kofman et al., 2006), structural brain differences (Rauh et al., 2012), as well as neurodegenerative diseases such as PD (Bhatt et al., 1999; Ascherio et al., 2006; Hancock et al., 2008), and AD (Richardson et al., 2014). These associations were reported to be generally stronger if sex was used as a factor in statistical analyses. The lipophilic properties of OPs allow them to pass through the placenta and accumulate in the fatty organs such as the brain (Andersen et al., 2000). Furthermore, children can be exposed to OPs during postnatal development through breastfeeding. Again, it has been revealed that the effects of OPs cannot be fully explained only with the inhibition of acetylcholinesterase activity.

The critical component of establishing fully functional neuronal circuits is neuronal synapses that enable the communication between neurons. The formation of a fully functional synapse involves many components necessary for successful synaptogenesis. The complexity and vulnerability of synaptogenesis during development confirmed with many studies underscore the connection between the altered synaptic components and the subsequent development of specific neurodevelopmental pathologies (Wishart et al., 2006; Gilingwater and Wishart 2013).

The molecular mechanisms underlying the effects of OPs exposure during neurodevelopment have been studied in the preclinical models and confirmed their toxic effects on neuronal development (Todd et al., 2020). However, the studies on animal models or *in vitro* conditions are spread broadly, exploring the effects of different dosages, routes, and duration of exposure of different OPs. The complexity of preclinical studies is further increased with different experimental animal species used for these studies, which may differ significantly in their response to the specific OP. The variability of the OPs used and molecular targets measured adds to the increasing difficulty of comparing these results and assessing the exact molecular pathways of OP action. However, in many experimental paradigms, symptoms associated with CNS impairment persist a significant period after the levels of AChE activity have been restored in the blood and the brain (Gralewicz and Socko, 1997) as it was suggested that molecular mechanisms other than the inhibition of the catalytic activity of AChE are underlying the OP developmental neurotoxicity.

Keeping in mind that the levels of systemic OP toxicity may not correspond to their neurodevelopmental toxicity, the necessity to develop a different approach in order to determine the safe applicative doses is clear. Chlorpyrifos, for example, was shown to have detrimental effects on neuronal maturation and synaptic activity that are not the result of AChE inhibition (Pope 1999; Barone et al. 2000; Qiao et al. 2002; Yanai et al. 2002; Qiao et al. 2003; Casida and Quistad 2004; Gupta 2004). Furthermore, different OPs that may have similar systemic toxicity may differ in their toxic effects on neurodevelopment. Comparison between three widely used OPs, chlorpyrifos, diazinon, and parathion, showed that chlorpyrifos and diazinon are far more toxic to the developing brain than parathion (Slotkin et al., 2006). All of these OPs exert neurotoxic effects even when used in doses below the maximum tolerated official dose. On the other hand, parathion exerts neurotoxicity only in doses that exceed the levels determined to induce overt toxicity.

The Suggested Molecular Mechanisms Underlying OP Exposure Toxicity during Neurodevelopment

Molecular mechanisms other than the inhibition of the catalytic activity of AChE are also underlying the OP developmental neurotoxicity. For example, chronic low-level OP exposure is associated with the reduction in the presynaptic choline transporter (ChAT) (Qiao et al., 2002), together with the changes in vesicular acetylcholine transporter (vAChT) and muscarinic receptors (Richardson and Chambers, 2004). These molecular changes correlate with subsequent behavioral changes in the offspring as they exhibited learning and memory deficit and increased locomotor activity (Levin et al., 2002). Interestingly, the activity of presynaptic ChAT in mice exposed to chlorpyrifos *in utero* shifts during different developmental windows. It is decreased neonatally, unaltered during weaning, and reappears in adolescence and adulthood (Qiao et al., 2003). Thus, the clarification of the developmental neurotoxic OPs effects is further complicated because neuronal development continues throughout adolescence and young adulthood. The disturbance of these neurotransmitter systems in specific developmental periods can have different outcomes. In such a way, OP-induced changes in choline and dopamine systems early on can have specific, OP-associated neurologic outcomes that differ from the pathologies specific for each of these neurotransmitters (Hatcher et al., 2008a,b). In addition, serotonergic neurotransmission appears to be directly affected. Exposure to chlorpyrifos is associated with the altered expression of presynaptic 5HT transporter, postsynaptic 5 HT receptors (5HT1a, 5HT2), and downstream 5HT signal transduction (Aldridge et al., 2003; Aldridge et al., 2004).

Synaptic disturbances were also reported as a consequence of the OPs exposure during postnatal development. Neonatal chlorpyrifos exposure is associated with a decreased level of CAMKII, a molecule that is important for synaptic plasticity maintenance and has a role in learning and memory. A similar decrease in synaptophysin levels, a CAMKII substrate, important for synaptic vesicle recycling and, thus, impaired learning was reported in the cortex of the same mice. These molecular changes were accompanied by a decreased habituation in behavioral studies (Lee et al., 2015).

Another system that is important for neuronal development and synaptogenesis and not directly a molecular target of AChE is the endocannabinoid system (eCB), and the direct binding of several OPs to the cannabinoid receptor 1 (CB1R) was confirmed (Quistad et al., 2002). In addition, OPs were able to inhibit the activity of enzymes important for the proper functioning of the eCB system, such as fatty acid amide hydrolase (FAAH) and mono-acylglycerol lipase (MAGL) (Quistad et al., 2001, 2002, 2006; Alexander et al., 2016;). Differential activation of CB1R and CB2Rs has different outcomes during neural differentiation. Long-term activation favors differentiation into neurons (Jiang, 2005; Soltys et al., 2010; Compagnucci et al., 2013) and short-term exposure into glial cells (Soltys et al., 2010). Based on their findings Todd and colleagues (2020), suggesting the potential molecular mechanisms of OPs action through the regulation of the eCB system during neurodevelopment depending on the endogenous levels of eCB activity. In the case of the low tonic eCB activity, OP (chlorpyrifos) exposure can directly (receptor activation or binding) or indirectly (regulation of FAAH levels) increase CB1R activity in NPCs and increase the neuronal differentiation. When the endogenous tonic eCB activity is high and maintained by 2-AG (2-arachidonyglycerol), chlorpyrifos can reduce CB1R activity by either acting as partial CB1R agonist or through the induction of AEA, inhibiting CB1R activation via 2-AG. In this case, the differentiation into mature neurons is diminished. In addition, eCB can regulate axonal elongation and, consequently, the postsynaptic target selection (Berghuis et al., 2007). Therefore, the effects of OP exposure on the eCB during neurodevelopment can have detrimental consequences, altering the formation and organization of neuronal circuits and leading to dysfunctional synaptic connections. Interestingly, although sexual dimorphism in the eCB system is present in adulthood, this difference also exists during neurodevelopment

(Fattore and Fratta, 2010) and can be partially responsible for the sex-biased effects of chlorpyrifos.

Finally, OP exposure can affect the production of different neurotrophins important for proper neuronal development and synaptogenesis. The exposure to CPS induced a decrease in NGF mRNA and protein levels in postnatal rats, together with the altered expression markers for the developing neural cell types, reelin for developing neurons, GFAP (glial fibrillary acidic protein) for astrocytes, and MAG (myelin-associated glycoprotein) for oligodendrocytes (Betancourt et al., 2006). In addition, the transient upregulation of BDNF, an important regulator of synaptic plasticity, has been reported in young rats treated with OPs (Betancourt et al., 2007) and in primary hippocampal neurons in vitro (Tan et al., 2009). However, the long-term effects of such changes in the expression profile of regulatory neurotransmitters are not entirely understood.

The Eligibility of Applicative OP Doses That Are Considered Safe for Neurodevelopment

These findings raise the concern about the eligibility of applicative OP doses that are considered safe for use and warrant further studies that will determine the systemic and neuronal toxicity of each OPs in wide use today. Furthermore, it is clear that the systemic levels of cholinesterase inhibition are not enough to determine all of the OPs toxic effects and that new biomarkers for the different toxic influences are needed.

In order to validate the developmental and overall neurotoxicity of different pesticides, an in vitro system was developed involving the studies of the inhibition of neurite outgrowth in rat pheochromocytoma cell line PC12 (Christen, 2017). Different pesticides, including OPs, inhibit neurite outgrowth and the expression of several genes that may be associated with developmental neurotoxicity, such as Ca2+/calmodulin-dependent protein kinase class of enzymes - specifically camk2α, camk2β, and gap-43 (growth-associated protein 43). camk2α and camk2β regulate axonal and dendritic arborization and synaptogenesis (Redondo et al., 2010; Mockett et al., 2011). GAP-43 is highly expressed in axonal growth cones (Garcia et al.,2003), during neurite outgrowth (Skene, 1989, Kanazir et al., 1996), and during axonal remodeling and regeneration (Grasselli and Strata, 2013). The induction of GAP-43 may represent the adaptive brain response (Hamers et al., 1996; Reagan et al., 2000; Alm et al., 2010) in response to toxic stimulants.

The strong correlation between the inhibition of the neurite outgrowth and the induction of expression of gap-43 was observed when PC12 cells were exposed to different OPs (Christen et al., 2017), and the analyses of expressional changes in gap-43 were proposed to serve as a marker for assessing the levels of developmental neurotoxicity for different pesticides including OPs.

When discussing the toxic effects of OPs, it is important to consider that they can be metabolized in the exposed organisms. One of the enzymes involved in metabolizing pesticides is paraoxonase 1 (PON1), and its expression levels differ during specific developmental periods (Cole et al., 2003). In addition, low maternal PON1 expression during pregnancy was shown to strengthen the association between OP gestational exposure and subsequent impairment in cognitive development (Eskenazi et al., 2014). These findings are of particular interest, indicating that differences in individual neurodevelopmental susceptibility to OPs intoxication may in part depend on the endogenous, genetically determined PON1 levels. However, a direct connection between the genetic polymorphism of the PON1 protein and differential susceptibility to OP exposure has not been fully confirmed either in the adult population or during gestational development.

Conclusion

These findings can be used to additionally explain the proposed associations between the OP exposure and neurological impairments through the altered synaptic plasticity on the molecular level during both neuronal development and adulthood. The studies done so far on the effects of OPs on synaptic structure and function strongly suggest the contribution to the development of other pathophysiologies resulting in neurodevelopmental, neurodegenera-tive, and neuropsychiatric disorders. However, the low-level OP toxicity is difficult to recognize therefore making it impossible to apply any countermeasures. The effect on the synaptic function and plasticity is particularly hard to identify and measure before the onset of pathology as there are no biomarkers in wide use that can diagnose, with a high level of certainty, the ongoing synaptic impairment. That leaves many individuals exposed to chronic low levels of OPs vulnerable to additional insults. Thus, the further evaluation of proposed in vitro essays, such as the use of PC12 cells to evaluate the toxic effects of OPs on the neurite outgrowth in correlation with gap-43 expression

(Christen et al., 2017), together with the development of novel peripheral biomarkers for synaptic vulnerability, is warranted.

The current treatment for acute poisoning with OPs consists of atropine to reduce ACh levels, together with diazepam or, recently considered replacement, midazalom. Diazepam and midazolam inhibit NMDA receptors and reduce seizure activity after OP exposure that occurs due to the increased glutamatergic signaling. Ketamine, an NMDA receptor antagonist, was also suggested in addition to the current SOC measures as it provided clinically relevant protection against sarin even when its administration was significantly delayed (Lewine et al., 2018). Recently, LY293558 (Tezampanel), an inhibitor of AMPA and GluK1 receptors, was proposed to be very effective against seizures and neurotoxicity, even when used for an extended period after the exposure (Aroniadou-Anderjaska et al., 2020). However, the low-level OP exposure gives a more complicated clinical picture, and no particular treatment was devised in order to countermeasure such poisoning.

Until new biomarkers and treatments for the low-levels OP exposure are developed, the question arises of what could be the potential prophylactic measures that can countermeasure the synaptotoxic effects of OPs. So far, the primary focus has been on developing prophylactic measures that can establish the normal levels of AChE after acute OP exposure. PON-1 (human paraoxonase-1) was shown to have the characteristics of a catalytic bioscavenger (Masson et al., 1998; Cherny et al., 2013), and elevated levels of PON-1 in genetically engineered mice can provide protection against OP exposure in rats (Costa et al., 1990).

However, considering the danger of low-level, chronic OP poisoning, the development of prophylactic measures addressing the effects of OPs not directly connected to the changes in AChE is necessary. One of such suggested treatments is the application of fatty acid amide hydrolase (FAAH) inhibitor AM5206 that acts as an inhibitor of endocannabinoid deactivation (Farizattoet al., 2017). The application of this inhibitor 30 min before administering the high dose of Pxn exhibited the protective effects resulting in the reduction of seizure severity (by 86%) and improved balance and coordination. Notably, the AM5206 application prevented the Pxn induced loss of synIIb *in vivo* and hippocampal slices (Farizatto et al., 2017).

One of the effective prophylactic treatments that are gaining in importance as the potential regulators of brain plasticity is dietary intervention, including caloric restrictions (CR) and intermittent fasting (IF)(for a detailed review, see Murphy et al., 2014). The slow, inevitable reduction in synaptic plasticity, possibly similar to the one induced with chronic exposure to OPs, is one of the hallmarks of aging (Burke and Barns, 2006). The environmental treatments have a significant effect on the reversal of age-related changes. The lifelong CR can enhance synaptic remodeling and reverse the age-related decrease in synaptic proteins (Adams et al., 2008; Mladenovic-Djordjevic et al., 2010), particularly in the CA3 region of the hippocampus (Adams et al., 2008). However, besides the lifelong CR, the one started in midlife (Means et al., 1993) or short exposures at a late age (Kaur et al., 2008) can elicit beneficial effects. The IF can similarly improve the age-related synaptic decline (Mladenovic-Djordjevic 2010, Singh et al., 2012). Interestingly, even the short-term IF substantially restored the expression of synaptophysin and calcineurin, suggesting the recovery of synaptic density loss and an increase in neurotransmission (Singh et al., 2012), possibly through the regulation of calcium homeostasis (Singh et al., 2012). A similar reversal of the age-related impairment in the neuronal plasticity marker neural cell adhesion molecule was observed upon the IF regimen (Kaur et al., 2008). Nevertheless, IF and CR can have different outcomes on specific signaling pathways and energy homeostasis in different brain regions (Smiljanic et al., 2018). Further investigations regarding the effect of different dietary interventions on impaired synaptic plasticity following OP exposure are required to evaluate their therapeutic effect. Paradoxically, the application of any dietary change comes with restrictions towards the food that is already OP compromised.

Acknowledgments

This work was supported by the Ministry of Education, Science and Technological Development of the Republic of Serbia, Grant 451-03-68/2020-14/200017.

References

Abou-Donia MB, Conboy LA, Kokkotou E, Jacobson E, Elmasry EM, ELkafrawy P, Neely M. et al. (2017). "Screening for novel central nervous system biomarkers in veterans with Gulf War Illness." *Neurotoxicology and Teratology* 61: 36–46.

Adams MM, Shi L, Linville MC, et al. (2008). "Caloric restriction and age affect synaptic proteins in hippocampal CA+ and spatial learning ability." *Experimental Neurology* 211: 141-149.

Aldridge JE, Seidler FJ, Meyer A, Thillai I, Slotkin TA. (2003). "Serotonergic systems targeted by developmental exposure to chlorpyrifos: Effects during different critical periods." *Environmental Health Perspective* 111: 1736–1743.

Aldridge JE, Seidler FJ, Slotkin TA. (2004). "Developmental exposure to chlorpyrifos elicits sex-selective alterations of serotonergic synaptic function in adulthood: Critical periods and regional selectivity for effects on the serotonin transporter, receptor subtypes, and cell signaling." *Environmental Health Perspective* 112: 148–155.

Alexander S, Carter W, Clark J, Lo T, Garle M. (2006). "A novel fluorescence-based assay for FAAH activity." In: *16th Annual Symposium on the Cannabinoids.* International Cannabinoid Research Society, Burlington, Vermont, p. 121.

Allen TG, Abogadia FC, Brown DA. (2006). "Simultaneous release of glutamate and acetylcholine from single magnocellular "cholinergic" basal forebrain neurons." *Journal of Neuroscience* 26: 1588-1595.

Alm H, Scholz B, Kultima K, Nilsson A, Andrén PE, Savitski MM, Bergman A, Stigson M, Fex-Svenningsen A, Dencker L. (2010). "In vitro neurotoxicity of PBDE- 99: immediate and concentration-dependent effects on protein expression in cere- bral cortex cells." *Journal of Proteome Research* 9: 1226–1235.

Andersen HR, Nielsen JB, Grandjean P. (2000). "Toxicologic evidence of developmental neurotoxicity of environmental chemicals." *Toxicology* 144: 121–127.

Aroniadou-Anderjaska V, Figueiredo TH, Apland JP, Braga MF. (2020). "Targeting the glutamatergic system to counteract organophosphate poisoning: a novel therapeutic strategy." *Neurobiology of Disease* 133: 104406.

Arroyo S, Bennett C, Aziz D, Brown SP, Hestrin S. (2012). "Prolonged disynaptic inhibition in the cortex mediated by slow, non-alpha7 nicotinic excitation of a specific subset of cortical interneurons." *Journal of Neuroscience* 32: 3859-3864.

Ascherio A, Chen H, Weisskopf MG, O'Reilly E, McCullough ML, Calle EE, Schwarzschild MA, Thun MJ. (2006). "Pesticide exposure and risk for parkinson's disease." *Annals of Neuro*logy 60: 197–203.

Avanzini G. Fariello R, Heinemann, Mutani R (Eds) (1993). "*Epileptogenic and Excitotoxic Mechanisms*." John Libbey, London.

Babayan AH, Kramar EA, Brett RM, Jafari M, Haettig J, Chen LY, Rex CS, et al. (2012). "Integrin dynamics produce a delayed stage of long-term potentiation and memory consolidation." *Journal of Neuroscience* 32: 12854–12861.

Bahr BA, Lynch G. (1992). "Purification of an Arg-Gly-Asp selective matrix receptor from brain synaptic plasma membranes." *Biochemical Journal* 281: 137–142.

Barone S, Das KP, Lassiter TL, White LD. (2000). "Vulnerable processes of nervous system development: a review of markers and methods." *Neurotoxicology* 21: 15–36.

Bell KA, Shim H, Chen CK, McQuiston AR. (2011). "Nicotinic excitatory postsynaptic potentials in hippocampal CA1 interneurons are predominantly mediated by nicotinic receptors that contain alpha4 and beta2 subunits." *Neuropharmacology* 61: 1379-1388.

Berghuis P, Rajnicek AM, Morozov YM, Ross RA, Mulder J, Urban GM, Monory K, et al. (2007). "Hardwiring the brain: endocannabinoids shape neuronal connectivity." *Science* 316: 1212–1216.

Bertolote J, Fleischmann A. (2015). "Suicide by self-poisoning with pesticides: the need for action." *Suicidology* 13 (1).

Betancourt AM, Burgess SC, Carr RL. (2006). "Effect of developmental exposure to chlorpyrifos on the expression of neurotrophin growth factors and cell-specific markers in neonatal rat brain." *Toxicological Science* 92: 500–506.

Betancourt AM, Burgess SC, Carr RL. (2007). "Effect of developmental exposure to chlorpyrifos on the expression of neurotrophin growth factors and cell-specific markers in neonatal rat brain." *Toxicological Science* 92: 500-506.

Bhatt MH, Elias MA, Mankodi AK. (1999). "Acute and reversible parkinsonism due to organophosphate pesticide intoxication: Five cases." *Neurology* 52: 1467–1471.

Birks JS. (2006). "Cholinesterase inhibitors for Alzheimer's disease." *Cochrane Database Syst Rev* (1): CD005593.

Blomgren K, Zhu C, Wang X, Karlsson J-O, Andersson A-L, Bahr BA, Mallard C, et al. (2001). "Synergistic activation of caspase-3 by m-calpain after neonatal hypoxia-ischemia—a mechanism of "pathological apoptosis"?" *Journal of Biological Chemistry* 276: 10191–10198.

Bosma H, van Boxtel MP, Ponds RW, Houx PJ, Jolles J. (2000). "Pesticide exposure and risk of mild cognitive dysfunction." *Lancet* 356: 912–913.

Bullman TA, Mahan CM, Kang HK, Page WF. (2005). "Mortality in US Army Gulf War veterans exposed to 1991 Khamisiyah chemical munitions destruction." *American Journal of Public Health* 95: 1382–1388.

Burke SN, Barnes CA. (2006). "Neural plasticity in the ageing brain." *Nature Reviews Neuroscience* 7: 30-40.

Casida JE, Quistad GB. (2004). "Organophosphate toxicology: safety aspects of nonacetylcholinesterase secondary targets." *Chemical Research in Toxicology* 17: 983–998.

Changeux J, Bertrand D, Corringer PJ, Dehaene S, Edelstein S, Lena, C, Le Novere N, et al. (1998). "Brain nicotinic receptors: structure and regulation, role in learning and reinforcement." *Brain Research Reviews* 26: 198–216.

Chen Y. (2012). "Organophosphate-induced brain damage: mechanisms, neuropsychiatric and neurological consequences, and potential therapeutic strategies." *Neurotoxicology* 33: 391–400.

Cherny I, Greisen P Jr, Ashani Y, Khare SD, Oberdorfer G, Leader H, et al. (2013). "Engineering V-type nerve agents detoxifying enzymes using computationally focused libraries." *ACS Chemical Biology* 8: 2394–403.

Christen V, Rusconi M, Crettaz P, Fent K (2017). "Developmental neurotoxicity of different pesticides in PC-12 cells in vitro." *Toxicology and Applied Pharmacology* 325: 25-36.

Clegg DO, Wingerd KL, Hikkita ST, Tolhurst EC. (2003). "Integrins in the development, function and dysfunction of the nervous system." *Frontiers in Bioscience* 8: d723–d750.

Cole TB, Jampsa RL, Walter BJ, Arndt TL, Richter RJ, Shih DM, Tward A, et al. (2003). "Expression of human paraoxonase (Pon1) during development." *Pharmacogenetics* 13: 357–364.

Compagnucci C, Di Siena S, Bustamante MB, Di Giacomo D, Di Tommaso M, Maccarrone M, Grimaldi P, et al. (2013). "Type-1 (CB1) cannabinoid receptor promotes neuronal differentiation and maturation of neural stem cells." *PLoS One* 8: e54271.

Costa LG, McDonald BE, Murphy SD, Omenn GS, Richter RJ, Motulsky AG, et al. (1990). "Serum paraoxonase and its influence on paraoxon and chlorpyrifos-oxon toxicity in rats." *Toxicology and Applied Pharmacology*103: 66–76.

Coyle, J.T. (2006). "Glutamate and schizophrenia: Beyond the dopamine hypothesis." *Cellular and Molecular Neurobiology* 26: 365–384.

De Araujo Furtado M, Aroniadou-Anderjaska V, Figueiredo TH, Apland JP, Braga MFM. (2020). "Electroencephalographic analysis in soman-exposed 21-day-old rats and the effects of midazolam or LY293558 with caramiphen. "*Annals of New York Academy of Scienc*e 1479: 122-133.

Descarries L, Mechawar N. (2000). "Ultrastructural evidence for diffuse transmission by monoamine and acetylcholine neurons of the central nervous system," in *Progress in Brain Research* eds L. F. Agnati, K. Fuxe, C. Nicholson and Syková E (Amsterdam: Elsevier), 27–47.

Deshpande LS, Phillips K, Huang B, Delorenzo RJ. (2014). "Chronic behavioral and cognitive deficits in a rat survival model of paraoxon toxicity." *Neurotoxicology* 44: 352–357.

Deshpande LS, Blair RE, Phillips KF, DeLorenzo RJ. (2016). "Role of the calcium plateau in neuronal injury and behavioral morbidities following organophosphate intoxication." *Annals of New York Academy of Science* 1374: 176–183.

Deshpande SS, Smith, CD, Filbert MG. (1995). "Assessment of primary neuronal culture as a model for soman-induced neurotoxicity and effec- tiveness of memantine as a neuroprotective drug." *Archives of Toxicology* 69: 384– 390.

dos Santos AA, Naime AA, de Oliveira J, Colle D, Bonfanti dos Santos D, Hort MA, Moreira ELG, et al. (2016). "Long-term and low-dose malathion exposure causes impairment in adult mice: evidence of hippocampal mitochondrial dysfunction, astrogliosis and apoptotic event." *Archives of Toxicology* 90: 647-660.

Douglas CL, Baghdoyan HA, Lydic R. (2002). "Postsynaptic muscarinic M1 receptors activate prefrontal cortical EEG of C57BL/6J mouse." *Journal of Neurophysiology* 88: 3003–3009.

Dudar JD. (1975). "The effect of septal nuclei stimulation on the release of acetylcholine from the rabbit hippocampus." *Brain Research*83: 123–33.

Eskenazi B, Kogut K, Huen K, Harley KG, Bouchard M, Bradman A, Boyd-Barr D, et al. (2014). "Organophosphate pesticide exposure, PON1, and neurodevelopment in school-age children from the CHAMACOS study." *Environmental Research*134: 149-157.

Farizatto KLG, McEwan SA, Naidoo V, Nikas SP, Shukla VG, Almeida MF, Byrd A, et al. (2017). "Inhibitor of endocannabinoid deactivation protects against in vitro and in vivo neurotoxic effects of paraoxon." *Journal of Molecular Neuroscience* 63: 115-122.

Farizatto KLG, Almeida MF, Long RT, Bahr BA (2019). "Early synaptic alterations and selective adhesion signaling in hippocampal dendritic zones following orhanophosphate exposure. "*Scientific Reports* 9: 6532.

Fattore L, Fratta W. (2010). "How important are sex differences in cannabinoid action?"*British Journal of Pharmacology*160: 544–548.

Foxenberg RJ, Ellison CA, Knaak JB, Ma C, Olson JR. (2011). "Cytochrome P450-specific human PBPK/PD models for the organophosphorus pesticides: chlorpyrifos and parathion." *Toxicology* 285: 57–66.

Fusi S, Drew PJ, Abbott LF. (2005). "Cascade models of synaptically stored memories." *Neuron* 45: 599–611.

Garcia SJ, Seidler FJ, Slotkin TA. (2003). "Developmental neurotoxicity elicited by prenatal or postnatal chlorpyrifos exposure: effects on neurospecific proteins indicate changing vulnerabilities." *Environmental Health Perspective* 111: 294–303.

Gil Z, Connors BW, Amitai Y. (1997). "Differential regulation of neocortical synapses by neuromodulators and activity." *Neuron* 19: 679–686.

Gilingwater TH, Wishart TM (2013). "Mechanisms underlying synaptic vulnerability and degeneration in neurodegenerative disease." *Neuropathology and Applied Neurobiology* 39: 320-344.

Golomb BA. (2008). "Acetylcholinesterase inhibitors and Gulf War illnesses." *Proceedings of the National Academy of Science USA* 105: 4295–300.

Gralewicz S, Socko R. (1997). "Persisting behavioural and electroencephalogra-phic effects of exposure to chlorphenvinphos, an organophosphorous pesticide, in laboratory animals." *International Journal of Occupational Medicine and Environmental Health* 10: 375-94.

Granger AJ, Mulder N, Saunders A, Sabatini BL. (2016). "Cotransmission of acetylcholine and GABA." *Neuropharmacology* 100, 40–46.

Grasselli G, Strata P. (2013). "Structural plasticity of climbing fibers and the growth-associated protein GAP-43." *Frontiers in Neural Circuits* 7: 25.

Gu Z, Yakel JL. (2011). "Timing-dependent septal cholinergic induction of dynamic hippocampal synaptic plasticity." *Neuron* 71: 155-165.

Gunnell D, Eddleston M. (2003). "Suicide by intentional ingestion of pesticides: a continuing tragedy in developing countries." *International Journal of Epidemiology* 32: 902–909.

Gupta RC. (2004). "Brain regional heterogeneity and toxicological mechanisms of organophosphates and carbamates." *Toxicology Mechanisms and Methods* 14: 103–143.

Haley RW, Kurt TL. (1997). "Self-reported exposure to neurotoxic chemical combinations in the Gulf War. A cross-sectional epidemiological study." *JAMA* 277: 231-237.

Haley RW. (2003). "Gulf war syndrome: narrowing the possibilities." *Lancet Neurology* 2: 272-273.

Hamers FP, Plantinga LC, Verhaagen J, Neijt JP, Gispen WH. (1996). "Upregulation of B50/GAP-43 protein mRNA in rat dorsal root ganglia during cisplatin intoxication." *Journal of Neuroscience Research* 44: 142–148.

Hancock DB, Martin ER, Mayhew GM, Stajich JM, Jewett R, Stacy MA, Scott BL, et al. (2008). "Pesticide exposure and risk of parkinson's disease: A family-based case-control study." *BMC Neurology* 8: 6.

Hatcher JH, Delea KC, Richardson JR, Pennell KD, Miller GW. (2008a). "Disruption of dopamine transport by DDT and its metabolites." *Neurotoxicology* 29: 682-90.

Hatcher JM, Pennell KD, Miller GW. (2008b). "Parkinson's disease and pesticides: A toxicological perspective." *Trends in Pharmacological Science*29: 322–329.

Hasselmo ME, Sarter M. (2011). "Modes and models of forebrain cholin- ergic neuromodulation of cognition." *Neuropsychopharmacology* 36: 52–73.

Hayden, K. M. et al. (2010). "Occupational exposure to pesticides increases the risk of incident AD: the Cache County study." *Neurology* 74: 1524–1530.

Henny P, Jones BE (2008). "Projections from basal forebrain to prefrontal cortex comprise cholinergic, GABAergic and glutamatergic inputs to pyramidal cells or interneurons." *European Journal of Neuroscience* 27: 654–670.

Higley MJ, Soler-Llavina GJ, Sabatini BL. (2009). "Cholinergic modulation of multivesicular release regulates striatal synaptic potency and integration." *Nature Neuroscience* 12: 1121–1128.

Hillen AEJ, Burbach JPH, Hol EM. (2018). "Cell adhesion and matricellular support by astrocytes of the tripartite synapse." *Progress in Neurobiology* 165-167: 66–86.

Hnasko TS, Edwards RH. (2012). "Neurotransmitter corelease: mechanism and physiological role." *Annual Review of Physiology* 74:225–243.

Hubble JP, Kurth JH, Glatt SL, Kurth MC, Schellenberg GD, Hassanein RE, et al. (1998). "Gene-toxin interaction as a puta- tive risk factor for Parkinson's disease with dementia." *Neuroepidemiology* 17: 96-104.

Jamal GA. (1997). "Neurological syndromes of organophosphorus compounds." *Adverse Drug Reactions and Toxicological Reviews* 16: 133–170.

Jiang W. (2005). "Cannabinoids promote embryonic and adult hippocampus neurogenesis and produce anxiolytic- and antidepressant-like effects." *Journal of Clinical Investigations* 115: 3104–3116.

John, E. M.; Shaike, J. M. (2015). "Chlorpyrifos: pollution and remediation." *Environmental Chemistry Letters* 13, 269−291.

Jones, N. (2010). "Alzheimer disease: Risk of dementia and Alzheimer disease increases with occupational pesticide exposure." *Nature Reviews Neurology* 6: 353.

Kalmbach A, Hedrick T, Waters J. (2012). "Selective optogenetic stimulation of cholinergic axons in neocortex." *Journal of Neurophysiology* 107: 2008-2019.

Kamel F, Hoppin JA. (2004). "Association of pesticide exposure with neurologic dysfunction and disease." *Environmental Health Perspective* 112: 950-958.

Kamel F, Engel LS, Gladen BC, Hoppin JA, Alavanja MCR, Sandler DP. (2005). "Neurologic Symptoms in Licensed Private Pesticide Applicators in the Agricultural Health Study." *Environmental Health Perspective* 113: 877–882.

Kamel F, Engel LS, Gladen BC, Hoppin JA, Alavanja MCR, Sandler DP. (2007). "Neurologic symptoms in licensed pesticide applicators in the Agricultural Health Study." *Human Experimental Toxicology* 26: 243-50.

Kanazir S, Ruzdijic S, Vukosavic S, Ivkovic S, Milosevic A, Zecevic N, Rakic L (1996). "GAP-43 mRNA expression in early development of human nervous system." *Brain Research Molecualr Brain* 38: 145-55.

Kandel ER. (2001). "The molecular biology of memory storage: a dialogue between genes and synapses." *Science* 294: 1030–1038.

Kaur M, Sharma S, Kaur G. (2008). "Age-related impairments in neuronal plasticity markers and astrocytic GFAP and their reversal by late-onset short term dietary restriction." *Biogerontology* 6: 441-454.

Kofman O, Berger A, Massarwa A, Friedman A, Jaffar AA. (2006). " Motor inhibition and learning impairments in school-aged children following exposure to organophosphate pesticides in infancy." *Pediatric Research* 60: 88–92.

Kramár EA, Lin B, Rex CS, Gall CM, Lynch G. (2006). "Integrin-driven actin polymerization consolidates long-term potentiation." *Proceedings of the National Academy of Science USA*103: 5579–5584.

Lallement G, Denoyer M, Collet A, Pernot-Marino I, Baubichon D, Monmaur P, Blanchet G. (1992). "Changes in hippocampal acetylcholine and glutamate extracellular levels during soman-induced seizures: influence of septal cholinoceptive cells." *Neuroscience Letters* 139:104–109.

Lee I, Eriksson P, Fredriksson A, Buratovic S, Viberg H. (2015). "Develop-mental neurotoxic effects of two pesticides: Behavior and biomolecular studies on chlorpyrifos and Carbaryl." *Toxicology and Applied Pharmacology* 288: 429–438.

Lee S-H, Dan Y. (2012). "Neuromodulation of brain states." *Neuron* 76: 209–222.

Levin ED, Addy N, Baruah A, Elias A, Christopher NC, Seidler FJ, Slotkin TA. (2002). "Prenatal chlorpyrifos exposure in rats causes persistent behavioral alterations." *Neurotoxicology and Teratology* 24: 733–741.

Lewine JD, Weber W, Gigliotti A, McDonald JD, Doyle-Eisele M, Bangera N, Paulson K, et al. (2018). "Addition of ketamine to standard-of-care counter-measures for acute organophosphate poisoning improves neurobiological outcomes."*Neurotoxicology* 69: 37-46.

Lindsay J, Hebert R, Rockwood K. (1997). "The Canadian Study of Health and Aging: risk factors for vascular dementia." *Stroke* 28: 526–530.

Liu J, Olivier K, Pope CN. (1999). "Comparative neurochemical effects of repeated methyl parathion or chlorpyrifos exposures in neonatal and adult rats." *Toxicology and Applied Pharmacology* 158: 186-196.

Masson P, Josse D, Lockridge O, Viguié N, Taupin C, Buhler C. (1998). "Enzymes hydrolyzing organophosphates as potential catalytic scavengers against organo-phosphate poisoning." *Journal of Physiology-Paris* 92: 357–62.

McCormick DA, Prince DA. (1985). "Two types of muscarinic response to acetylcholine in mammalian cortical neurons." *Proceedings of the National Academy of Science USA* 82: 6344–6348.

McDonough J Jr, Shih T. (1997). "Neuropharmacological mechanisms of nerve agent-induced seizure and neuropathology." *Neuroscience and Biobehavio-ral Reviews* 21:559–579.

McGehee DS, Heath MJ, Gelber S, Devay P, Role LW. (1995). "Nicotine enhancement of fast excitatory synaptic transmission in CNS by presynaptic receptors." *Science* 269: 1692–1696.

Means LW, Higgins JL, Fernandez TJ. (1993). "Mid-life onset of dietary restriction extends life and prolongs cognitive functioning." *Physiology and Behavior* 54: 503-508.

Menon PM, Nasrallah HA, Reeves RR, Ali JA. (2004). "Hippocampal dysfunction in Gulf War syndrome a proton MR spectroscopy study." *Brain Research* 1009: 189–194.

Mesulam MM, Mufson EJ, Levey AI, Wainer BH. (1983b). "Cholinergic innervation of cortex by the basal forebrain: cytochemistry and cortical connections of the septal area, diagonal band nuclei, nucleus basalis (substantia innominata), and hypothalamus in the rhesus monkey." *Journal of Comparative Neurology* 214: 170–197.

Mileson BE, Chambers, JE, Chen WL, Dettbarn W, Ehrich M, Eldefrawi AT, Gaylor DW, et al. (1998). "Common mechanism of toxicity: A case study of organophosphorus pesticides." *Toxicological Science* 41: 8–20.

Mladenovic Djordjevic A, Perovic M, Tesic V, et al. (2010). "Long- term dietary restriction modulates the level of presynaptic proteins in the cortex and hippocampus of the aging rat." *Neurochemistry International* 56: 250-255.

Mockett BG, Guévremont D, Wutte M, Hulme SR, Williams JM, Abraham WC. (2011). "Calcium/calmodulin-dependent protein kinase II mediates group I metabotropic glutamate receptor-dependent protein synthesis and long-term depression in rat hippocampus." *Journal of Neuroscience* 31: 7380–7391.

Mrzijak, L, Pappy M, Leranth, C, Goldman-Rakic PS. (1995). "Cholinergic synaptic circuitry in the macaque prefrontal cortex." *Journal of Comparative Neurology* 357: 603–617.

Munirathinam S, Bahr BA. (2004). "Repeated contact with subtoxic soman leads to synaptic vulnerability in hippocampus. J Neurosci Res 77: 739–746Ning L, et al. (2013) Interactions between ICAM-5 and β1 integrins regulate neuronal synapse formation." *Journal of Cell Science* 126: 77–89.

Muñoz W, Rudy B. (2014). "Spatiotemporal specificity in cholinergic control of neocortical function." *Current Opinion in Neurobiology* 26: 149–160.

Murphy T, Dias GP, Thuret S. (2014). "Effects of diet on brain plasticity in animal and human studies: mind the gap." *Neural Plasticity* ID 563160.

Parron T, Requena M, Hernández AF, Alarcón R. (2011). "Association between environmental exposure to pesticides and neurodegenerative diseases." *Toxicology and Applied Pharmacology* 256: 379–385.

Pavlovsky L, Browne RO, Friedman A. (2003). "Pyridostigmine enhances glutamatergic transmission in hippocampal CA1 neurons." *Experimental Neurology* 179: 181–187.

Peter JV, Sudarsan TI, Moran JL. (2014). "Clinical features of organophosphate poisoning: a review of different classification systems and approaches." *Indian Journal of Critical Care in Medicine* 18: 735–745.

Picciotto MR, Caldarone BJ, King SL, Zachariou V. (2000). "Nicotinic receptors in the brain. Links between molecular biology and behavior." *Neuropsycho-pharmacology* 22: 451–465.

Picciotto MR, Caldarone BJ, Brunzell DH, Zachariou V, Stevens TR, King SL. (2001). "Neuronal nicotinic acetylcholine receptor subunit knockout mice: physiological and behavioral phenotypes and possible clinical implications." *Pharmacological Therapeutics* 92: 89–108.

Picciotto MR. (2003). "Nicotine as a modulator of behavior: beyond the inverted U." *Trends in Pharmacological Science* 24: 493–499.

Pope CN. (1999). "Organophosphorus pesticides: do they all have the same mechanism of toxicity?" *Journal of Toxicology and Environmental Health* 2: 161–181.

Qiao D, Seidler FJ, Padilla S, Slotkin TA. (2002). "Developmental neurotoxicity of chlorpyrifos: What is the vulnerable period?" *Environmental Health Perspective* 110: 1097–1103.

Qiao D, Seidler FJ, Tate CA, Cousins MM, Slotkin TA. (2003). "Fetal chlorpyrifos exposure: Adverse effects on brain cell development and cholinergic biomarkers emerge postnatally and continue into adolescence and adulthood." *Environmental Health Perspective* 111: 536–544.

Quistad GB, Sparks SE, Casida JE. (2001). "Fatty acid amide hydrolase inhibition by neurotoxic organophosphorus pesticides." *Toxicology and Applied Pharmacology* 173: 48–55.

Quistad GB, Nomura DK, Sparks SE, Segall Y, Casida JE. (2002). "Cannabinoid CB1 receptor as a target for chlorpyrifos oxon and other organophosphorus pesticides." *Toxicology Letters* 135: 89–93.

Quistad GB, Klintenberg R, Caboni P, Liang SN, Casida JE. (2006). "Monoacylglycerol lipase inhibition by organophosphorus compounds leads to elevation of brain 2-arachidonoylglycerol and the associated hypomotility in mice." *Toxicology and Applied Pharmacology* 211: 78–83.

Rauh VA, Perera FP, Horton MK, Whyatt RM, Bansal R, Hao X, Liu J, et al. (2012). "Brain anomalies in children exposed prenatally to a common organophosphate pesticide." *Proceedings of the National Acadamy of Science USA* 109: 7871–7876.

Raveh L, Brandeis R, Gilat E, Cohen G, Alkalay D, Rabinovitz I, Sonego H, et al. (2003). "Anticholinergic and antiglutamatergic agents protect against soman-induced brain damage and cognitive dysfunction." *Toxicological Science* 75: 108–116.

Ray A, Liu J, Ayobi P, Pope C. (2010). "Dose-related gene expression changes in forebrain following acute, low-level chlorpyrifos exposure in neonatal rats." *Toxicology and Applied Pharmacology* 248: 144-155.

Reagan KE, Jensen KF, Friedman MA, Abou-Donia MB. (2000). "Patterns of GAP-43 (B-50) expression and accumulation in subchronic acrylamide neurotoxicity." *Neuroscience Research Communication* 27: 199–212.

Redondo RL, Okuno H, Spooner PA, Frenguelli BG, Bito H, Morris RG. (2010). "Synaptic tagging and capture: differential role of distinct calcium/ calmodulin kinases in protein synthesis-dependent long-term potentiation." *Journal of Neuroscience* 30: 4981–4989.

Richardson JR, Chambers JE. (2004). "Neurochemical effects of repeated gestational exposure to chlorpyrifos in developing rats." *Toxicological Science* 77: 83–90.

Richardson JR, Roy A, Shalat SL, von Stein RT, Hossain MM, Buckley B, Gearing M, et al. (2014). "Elevated serum pesticide levels and risk for Alzheimer disease." *JAMA Neurology* 71: 284–290.

Richardson JR, Taylor MM, Shalat SL, Guillot TS 3rd, Caudle WM, Hossain MM, Mathews TA, et al. (2015). "Developmental pesticide exposure reproduces features of attention deficit hyperactivity disorder." *FASEB Journal* 29: 1960–1972.

Rohlman DS, Lasarev M, Anger WK, Scherer J,Stupfel J, McCauley L. (2007). "Neurobehavioral performance of adult and adolescent agricultural workers." *Neurotoxicology* 28: 374-80.

Rohlman DS, Anger WK, Lein PJ. (2011)." Correlating neurobehavioral performance with biomarkers of organophosphorus pesticide exposure." *Neurotoxicology* 32: 268-276.

Sánchez-Santed F, Cañadas F, Flores P, López-Grancha M, Cadona D. (2004). "Long-term functional neurotoxicity of paraoxon and chlorpyrifos: behavioral and pharmacological evidence." *Neurotoxicology and Teratology* 26: 305–317.

Sarter M, Parikh V, Howe WM. (2009). "Phasic acetylcholine release and the volume transmission hypothesis: time to move on." *Nature Review Neuroscience* 10: 383–390.

Saunders A,Granger AJ, Sabatini BL. (2015). "Corelease of acetylcholine and GABA from cholinergic forebrain neurons." *eLife* 4: e06412.

Schuh RA, Lein PJ, Beckles RA, Jett DA. (2002). "Noncholinesterase mechanisms of chlorpyrifos neurotoxicity: altered phosphorylation of Ca2+/cAMP response element binding protein in cultured neurons." *Toxicological and Applied Pharmacology* 182: 176–185.

Sealey LA, Hughes BW, Sriskanda AN, Guest JR, Gibson AD, Johnson-Williams L, Pace DG, et al. (2016). "Environmental factors in the development of autism spectrum disorders." *Environmental International* 88: 288–298.

Shelton JF, Hertz-Picciotto I, Pessah IN. (2012). "Tipping the balance of autism risk: Potential mechanisms linking pesticides and autism." *Environmental Health Perspective* 120: 944–951.

Singh R, Lakhanpal D, Kumaretal S. (2012). "Late-onset intermittent fasting dietary restriction as a potential intervention to retard age-associated brain function impairments in male rats." *Age* 917-933.

Singh S, Sharma N. (2000). "Neurological syndromes following organopho-sphate poisoning." *Neurology India* 48: 308–313.

Skene JH. (1989). "Axonal growth-associated proteins." *Annual Review Neuroscience* 12: 127–156.

Slotkin TA, Levin ED, Seidler FJ. (2006). "Comparative developmental neuroto-xicity of organophosphate insecticide: effects on brain development are separable from systemic toxicity." *Environmental Health Perspective* 114: 746-751.

Slotkin, TA. (2011). "Does early-life exposure to organophosphate insecticides lead to prediabetes and obesity?" *Reproductive Toxicology* 31: 297–301.

Smiley JF, Morrell F, Mesulam MM. (1997). "Cholinergic synapses in human cerebral cortex: an ultrastructural study in serial sections." *Experimental Neurology* 144: 361–368.

Smiljanic K, Todorovic S, Mladenovic Djordjevic A, Vanmierlo T, Lutjohann D, Ivkovic S, Kanazir S. (2018). "Limited daily feeding and intermittent feeding have different effects on regional brain energy homeostasis during aging." *Biogerontology* 19: 121-132.

Soltys J, Yushak M, Mao-Draayer Y. (2010). "Regulation of neural progenitor cell fate by anandamide." *Biochemistry and Biophysics Research Communication* 400: 21–26.

Speed HE, Blaiss CA, Kim A, Haws ME, Melvin NR, Jennings M, et al. (2012). "Delayed reduction of hippocampal synaptic transmission and spines following exposure to repeated subclinical doses of organophosphorus pesticide in adult mice." *Toxicological Science* 125: 196–208.

Spencer WF, Shoup TD, Spear RC. (1980). "Conversion of parathion to paraoxon on soil dusts as related to atmospheric oxidants at three California locations." *Journal of Agriculture and Food Chemistry* 28: 1295–1300.

Sullivan JB Jr, Blose J. (1992). "*Organophosphate and Carbamate Insecticides*." Baltimore, MD: William and Wilkins.

Sullivan K, Krengel M, Bradford W, Stone C, Thompson TA, Heeren T, White RF. (2018). "Neuropsychological functioning in military pesticide applicators from the Gulf War: Effects on information processing speed, attention and visual memory." *Neurotoxicolology and Teratology* 65: 1–13.

Tan DH, Peng SQ, Wu YL, Wang YM, Lu CF, Ding W, Wang QX, et al. (2009). "Chlorpyrifos induces delayed cytotoxicity after withdrawal in primary hippocampal neurons through extracellular signal-regulated kinase inhibition." *Biological and Pharmaceutical Bulletin* 32: 1649–1655.

Terry AV Jr, Stone JD, Buccafusco JJ, Sickles DW, Sood A, Prendergast MA. (2003). "Repeated exposures to subthreshold doses of chlorpyrifos in rats: hippocampal damage, impaired axonal transport, and deficits in spatial learning." *Journal of Pharmacology and Experimental Therapeutics* 305:375-84.

Terry AV Jr, Gearhart DA, Beck WD Jr, Truan JN, Middlemore ML, Williamson LN, et al. (2007). "Chronic, intermittent exposure to chlorpyrifos in rats: protracted effects on axonal transport, neurotrophin receptors, cholinergic markers, and information processing." *Journal of Pharmacology and Experimental Therapeutics* 322: 1117-28.

Todd SW, Lumsden EW, Aracava Y, Mamczarz J, Albiequerque EX, Pereira EFR (2020). "Gestational exposures to organophosphorus insecticides: from acute poisoning to developmental neurotoxicity." *Neuropharmacology* 180:108271.

Todorovic M, Cowan M, Balint C, Sun C, Kapur J. (2012). "Characterization of status epilepticus induced by two organophosphates in rats." *Epilepsy Research* 101: 268–276.

Toomey R, Kang HK, Karlinsky J, Baker DG, Vasterling JJ, Alpern R, Reda DJ, et al. (2007). "Mental health of US Gulf War veterans 10 years after the war." *British Journal of Psychiatry* 190: 385–393.

Umbriaco D, Watkins KC, Descarries L, Cozzari C, Hartman BK. (1994). "Ultrastructural and morphometric features of the acetylcholine innervation in adult rat parietal cortex: an electron microscopic study in serial sections." *Journal of Comparative Neurology* 348: 351–373.

Verma SK, Raheja G, Gill KD. (2009). "Role of muscarinic signal transduction and CREB phosphorylation in dichlorvos-induced memory deficits in rats: an acetylcholine independent mechanism." *Toxicology* 256: 175-182.

Voorhees JR, Rohlman DS, Lein PJ, Pieper AA. (2017). "Neurotoxicity in preclinical models of occupational e0xposure to organophosphorus compo-unds." *Frontiers in Neuroscience*10: 590.

Wade JV, Samson FE, Nelson SR, Pazdernik TL. (1987). "Changes in extracellular amino acids during soman- and kainic acid-induced seizures." *Journal of Neurochemistry* 49:645–650.

Wang KK. (2000). "Calpain and caspase: can you tell the difference?" *Trends in Neuroscience* 23: 20–26.

Ward TR, Mundy WR. (1996). "Organophosphorus compounds preferentially affect second messenger systems coupled to M2/M4 receptors in rat frontal cortex." *Brain Research Bulletin* 39: 49-55.

Wess J. (2003). "Novel insights into muscarinic acetylcholine receptor func- tion using gene targeting technology." *Trends in Pharmacological Science* 24: 414–420.

Wishart TM, Parson SH, Gilingwater TH. (2006). "Synaptic vulnerability in neurodegenerative disease." *Journal of Neuropathology and Experimental Neurology* 65: 733-739.

Wonnacott S. (1997). "Presynaptic nicotinic ACh receptors." *Trends in Neuroscience* 20: 92–98.

Yamasaki M, Matsui M, Watanabe M. (2010). "Preferential localization of muscarinic M1 receptor on dendritic shaft and spine of cortical pyramidal cells and its anatomical evidence for volume transmission." *Journal of Neuroscience* 30: 4408–4418.

Yanai J, Vatury O, Slotkin TA. (2002). "Cell signaling as a target and underlying mechanism for neurobehavioral teratogenesis." *Annals of New York Academy of Science* 965: 473–478.

Yokoyama K, Araki S, Murata K, Nishikitani M, Okomura T, Ishimatsu S, Takasu N, et al. (1998). "Chronic neurobehavioral effects of Tokyo subway sarin poisoning in relation to posttraumatic stress disorder." *Archives of Environmental Health* 53: 249–256.

Yu CJ, Du JC, Chiou HC, Chung MY, Yang W, Chen YS Fuh MR, et al. (2016). "Increased risk of attention-deficit/hyperactivity disorder associated with exposure to organophosphate pesticide in Taiwanese children." *Andrology* 4: 695-705.

Yurumez Y, Durukan P, Yavuz Y, Ikizceli I, Avsarogullari L, Ozkan S, et al. (2007). "Acute organophosphate poisoning in university hospital emergency room patients." *Internal Medicine* 46: 965–969.

Chapter 4

Organophosphates and Depression

Tamara Tasić[*] and Tamara Lazarević-Pašti[†]
Department of Physical Chemistry, "Vinča" Institute of Nuclear Sciences-National Institute of the Republic of Serbia, University of Belgrade, Belgrade, Serbia

Abstract

Organophosphates are the most widely used group of pesticides globally and are considered one of the most hazardous to vertebrates. The neurotoxic effects of high-level acute poisoning are well established and involve changes in peripheral, autonomic and central nervous system function. These events result in a constellation of physical, cognitive and psychiatric symptoms. The impact of long-term low-level exposure to organophosphates on human health is less known. Organophosphates disrupt many other neurotransmitters, which are involved in mood regulation, such as serotonin and acetylcholine. Depression is a common mental health condition affecting millions of people globally, and its etiology is connected with disturbances in neurotransmitter functioning and genetic factors. Epidemiological studies have reported higher incidences of depression in particular groups of individuals with chronic health conditions and individuals in specific occupations, such as farming, fishing and forestry. The reason behind the high risk of mood disorder in farming populations is indistinguishable. Research showed that individuals suffering from depression have elevated levels of the neurotransmitter acetylcholine. Since organophosphates are potent inhibitors of acetylcholinesterase, the enzyme responsible for acetylcholine's hydrolysis, there are strong indications that acetylcholine

---

[*] Corresponding Author's Email: tamara@vin.bg.ac.rs.
[†] Corresponding Author's Email: lazarevictlj@yahoo.com.

In: Organophosphates
Editor: Tamara Lazarević-Pašti
ISBN: 978-1-68507-652-8

level elevation could be the crucial cause of the symptoms of clinical depression in exposed individuals.

Keywords: organophosphate, neurological disorder, mental health, depression, acetylcholine, acetylcholinesterase

Introduction

Depression is a common mental health condition affecting millions of people around the world. Globally, more than 300 million people of all ages suffer from depression (Anićijević and Lazarević-Pašti 2020, WHO 2017). Several factors have been implicated in the disease's onset and course. It includes various biological, psychological, and social factors. The rate of depression seen in women from western industrialized nations is twice that of men (Patel 2005). Additionally, unemployed individuals of low socioeconomic status or who have experienced traumatic life events are thought to be at increased risk of developing depression (Brown and Harris 1978). Disturbances in neurotransmitter functioning and genetic factors have also been connected with the etiology of depression (Marchand, Dilda and Jensen 2005). Epidemiological studies have also reported higher incidences of depression in particular groups of individuals such as those with chronic health conditions (National Collaborating Centre for Mental Health 2005) and individuals in specific occupations, such as farming, fishing and forestry (Sanne et al. 2003). The reason behind the higher risk of mood disorder in farming populations is unclear. Some researchers have reported a link between exposure to pesticides, mood disorder and suicidal behavior (Harrison and Mackenzie Ross 2016). Organophosphate pesticides (OP), in particular, are associated with an elevated risk of neuropsychiatric disorder (Harrison and Mackenzie Ross 2016). They are the most widely used group of pesticides globally (Lazarevic-Pasti et al. 2011, Lazarevic-Pasti et al. 2010, Lazarevic-Pasti, Nastasijevic and Vasic 2011). They are considered by the World Health Organisation (WHO 2009) to be one of the most hazardous pesticides to vertebrate animals, responsible for many cases of poisoning worldwide, particularly in developing countries where protective measures are lacking (Harrison and Mackenzie Ross 2016). The neurotoxic effects of high-level acute poisoning are well established and involve changes in peripheral, autonomic and central nervous system function, resulting in a constellation of physical, cognitive and psychiatric symptoms. However, OPs disrupt many other neurotransmitters involved in mood regulation, such as serotonin (Harrison and Mackenzie Ross 2016). It could explain the link between pesticide exposure and mood disorders observed in former studies. This connection appears strongest in individuals who report previous acute poisonings (Harrison and Mackenzie Ross 2016). Nevertheless, the impact of long-term exposure to OPs in doses lower than that causing acute toxicity on human health is less clear. Some studies have found evidence of ill health, mood disorder, and cognitive impairment following low-level exposure to OPs, while others have not (Aničijević and Lazarević-Pašti 2020, Harrison and Mackenzie Ross 2016).

Organophosphates' Effects on Acetylcholinesterase in the Human Organism

The OPs exert their main toxicological effects through non-reversible phosphorylation of esterases in the central nervous system (Colovic et al. 2013). The acute toxic effects are related to the irreversible inactivation of acetylcholinesterase (AChE) (Colovic et al. 2013). OPs are substrate analogs to acetylcholine (ACh). Just like the natural substrate ACh, OPs enter the active site covalently binding to the serine –OH group. As in acetylation, OP is split, and the enzyme is phosphorylated. The acyl-enzyme is quickly hydrolyzed to regenerate the free enzyme, but dephosphorylation is very slow and takes several days. During that time, the phosphorylated enzyme cannot hydrolyze the neurotransmitter (Boublik et al. 2002). The inhibition of AChE leads to the accumulation of ACh in the synaptic cleft. This event results in over-stimulation of nicotinic and muscarinic ACh receptors and delayed neurotransmission. Acute poisoning with OPs is characterized by agitation, muscle weakness, muscle fasciculations, miosis, hypersalivation, and sweating. On the other hand, severe poisonings may cause respiratory failure, unconsciousness, confusion, convulsions and death (Colovic et al. 2013, Costa 2006, Hayden et al. 2010).

The Role of Acetylcholine in Depression

Exposure to chronic stress is another factor that can trigger ACh release in the forebrain and regulate its function on the hypothalamic-pituitary-adrenal (HPA) axis. This primary body system is essential for maintaining homeostasis and modulation of various psychological and emotional responses (Janowsky, el-Yousef and Davis 1974b, Mitic and Lazarevic-Pasti 2021, Newman et al. 2001). Accordingly, the stress response impairment leads to a dysfunctional ACh system known to contribute to psychiatric and neurological illnesses, including depression (Dagyte, Den Boer and Trentani 2011b, Mitic and Lazarevic-Pasti 2021). One of the symptoms frequently observed in depression, next to low mood and anhedonia, is cognitive problems that could be attributed to dysfunction in the cholinergic system, mainly if the duration of the disease is longer and therapy not efficient enough. This idea originates back in 1950 from clinical studies that suggested that elevation in central ACh levels could lead to depression (Dulawa and Janowsky 2019, Janowsky, el-Yousef and Davis 1974b, Mitic and Lazarevic-Pasti 2021, Rowntree, Nevin and Wilson 1950). Later was supported in the cholinergic-adrenergic hypothesis of mania and depression. According to this theory, mood perturbances may result from a complex disbalance between adrenergic and cholinergic transmission occurring in brain areas, favoring the predominance of the central cholinergic system. Later it was reformulated by including the dopamine role in regulating mood disorders (Dulawa and Janowsky 2019, Mitic and Lazarevic-Pasti 2021, van Enkhuizen et al. 2015). This hypothesis was a direct consequence of studies in which an increase in central ACh levels resulted from AChE inhibitors (AChEIs) application and its correlation with depressive phenotype. Later, there were reports of fruit pickers who developed depression symptoms after being chronically exposed to organophosphorus insecticides, which belong to the class of AChEIs (Gershon and Shaw 1961, Mitic and Lazarevic-Pasti 2021).

Over the years, several clinical and preclinical studies, in vivo, in vitro, and post-mortem, have been performed. Their results were remarkably consistent across species suggesting a negative implication of increased cholinergic activity and brain ACh levels in depression (Janowsky, el-Yousef and Davis 1974b, Mitic and Lazarevic-Pasti 2021). Besides elevated ACh levels, some researchers reported increased ACh precursor levels, choline, the brains of depressed patients, and its reversal to control values after treatment (Charles et al. 1994, Mitic and Lazarevic-Pasti 2021, Steingard et al. 2000). They were further substantiated with many studies that reported that various AChEIs and agonists induce depressive symptomatology in humans (Gershon and Shaw 1961, Janowsky, el-Yousef and Davis 1974b, Mitic and Lazarevic-Pasti 2021, Rowntree, Nevin and Wilson 1950). For example, the exposure or application of AChEIs to healthy people induced depression, while at the same time in depressed patients, it worsened their symptoms or caused relapse with mild cognitive impairment (Dulawa and Janowsky 2019, Katona 2009, Mitic and Lazarevic-Pasti 2021).

Furthermore, it has been shown that AChEIs can modulate the HPA axis activity and cause changes in sleep architecture (Dulawa and Janowsky 2019, Mitic and Lazarevic-Pasti 2021, Modell and Lauer 2007). AChEIs treatment increased neuroendocrine molecules, like serum adrenocorticotropic hormone, cortisol, epinephrine, and β-endorphine serum levels. The levels of these molecules are raised in endogenous depression (Janowsky et al. 1986, Mitic and Lazarevic-Pasti 2021). Results from clinical studies were corroborated with significant discoveries from rodent experiments. Stress decreased AChE activity and consequently increased ACh levels in CNS, resulting in a shift towards more depression-like behavior (Dagyte, Den Boer and Trentani 2011b, Das et al. 2000, Mitic and Lazarevic-Pasti 2021). Indeed, pharmacological manipulation confirmed this idea. Both systemic and specific intracerebral drug treatment with AChEIs and other cholinergic agonists and brain region-specific viral knockdown of AChE induced depression-like behavior in animal models (Dulawa and Janowsky 2019, Mineur et al. 2013a, Mitic and Lazarevic-Pasti 2021).

In contrast, treatments with antidepressants succeed in relieving and neutralize depressive symptoms by reducing cholinergic activity, mainly through suppression of activation of muscarinic-1 acetylcholine receptors (M1-AChRs) in cholinergic interneurons in the nucleus accumbens (NAc) and produce overall antidepressant-like effects (Chau et al. 2011, Mitic and Lazarevic-Pasti 2021, Rada et al. 2006). Interestingly, activation of M2-AChRs and M4-AChR autoreceptors decreased ACh release from cholinergic terminals and exhibited antidepressant-like effects (Dulawa and Janowsky 2019, Mitic and Lazarevic-Pasti 2021). Besides muscarinic receptors, human and rodent studies suggest that nicotinic receptors also contribute to the development of mood disorders. Most of these studies reported that nicotine withdrawal in both humans and rodents induced anhedonia and depressive behavior (Covey, Glassman and Stetner 1997, Kenny and Markou 2001, Mitic and Lazarevic-Pasti 2021). At the same time, nicotine administration to rodents and treatment with antagonists, particularly β2* or α7 nAChRs, produced an antidepressant-like effect (Mineur et al. 2018, Mitic and Lazarevic-Pasti 2021, Philip et al. 2010). Also, genetic studies showed that the viral knockdown of genes for specific nAChR subtypes in particular brain structures induced antidepressant-like effects (Mineur et al. 2018, Mitic and Lazarevic-Pasti 2021). Besides the antidepressant effect, applying nicotinic receptor antagonists in stressed animals showed that blocking the nicotinic receptors decreased the HPA axis activity and produced an anxiolytic effect (Mitic and Lazarevic-Pasti 2021, Newman et al. 2001).

Nevertheless, there are some inconsistencies when nAChRs agonists and antagonists are given (Dagyte, Den Boer and Trentani 2011b, Mitic and Lazarevic-Pasti 2021, Shytle et al. 2002). Some studies report that activating nAChRs with agonists can also induce an antidepressant effect (Dagyte, Den Boer and Trentani 2011b, Gatto et al. 2004, Mitic and Lazarevic-Pasti 2021). More research will be required to investigate which nAChR subtypes will be the most promising targets for treating mood disorders. Besides various components of complex depressive phenotype in rodents, different behavioral tests also showed a positive reaction when assessed anxious-like behavior (Dulawa and Janowsky 2019, Mineur et al. 2018, Mitic and Lazarevic-Pasti 2021, Picciotto et al. 2015). Because of its nature and multiple experimental advantages, animal models provided more information regarding Ach's anatomical, molecular, and genetic signature in stress response and depression. These experiments allowed us to map the specific brain regions and neural circuits involved, triggered, and affected in disease and may provide novel therapeutics for mood disorders. These studies demonstrate that depression and chronic stress exposure cause atrophy of neurons in cortical and limbic brain regions implicated in depression. Brain imaging studies show altered connectivity and network function in the brains of depressed patients. Secondly, cholinergic depletion has been found to impair experience-dependent plasticity in the cortex and hippocampus, and similar mechanisms are involved in mediating recovery from injury. Thirdly, acetylcholine modulates the expression of neurotrophic factors, such as brain-derived neurotrophic factor (BDNF), that play a significant role in neuronal survival and plasticity in adulthood. Fourth, recent work shows a fascinating interaction between acetylcholine and estrogen supporting hippocampal plasticity in aging females (Craig, Hong and McDonald 2011, Mitic and Lazarevic-Pasti 2021).

Both nicotinic and muscarinic receptors also regulate other neurotransmitter systems implicated in mood regulation, including noradrenergic, serotonergic, dopaminergic, GABAergic, glutamatergic and cannabinoid (Dulawa and Janowsky 2019, Mitic and Lazarevic-Pasti 2021) neurotransmitter systems. The dysfunction of glutamatergic signalization, particularly its excessive transmission and overload, is considered a core feature of stress-related mental illnesses (Dagyte, Den Boer and Trentani 2011a, Mitic and Lazarevic-Pasti 2021). Multiple studies reported significant stress-induced glutamate release in limbic and cortical areas responsible for emotional homeostasis, particularly in the hippocampus, prefrontal cortex, amygdala and basal ganglia (Dagyte, Den Boer and Trentani 2011a, Mitic and Lazarevic-Pasti 2021). Literature data also show that glutamatergic and cholinergic systems act synergistically. Changes in one transmission may trigger changes in another, causing altered network connectivity and function and perturbations in various physiological processes learning and memory and linking them with neurodegeneration and AD (Dagyte, Den Boer and Trentani 2011a, Mitic and Lazarevic-Pasti 2021). Even though the molecular mechanisms responsible for these alterations are not identified, there are strong indications that synergistic action of increased glucocorticoids and specific proinflammatory cytokines together with stress-induced excitotoxicity represent and disruption of glutamate-excitatory projection neurons represent the key factors in the etiology of stress-related illnesses (Duman et al. 2016, Mitic and Lazarevic-Pasti 2021).

Measuring the Exposure to Organophosphates

The contradictions in the results of various studies may probably result from methodological differences between the different studies in the field. For example, studies investigating this issue have used multiple methods to explore the role of exposure. Some of them are an investigation of diverse occupational groups with different levels of contact with OPs, routes and nature of exposure, use of protective clothing, demographic and geographic profiles, and over a wide variety of time frames. In addition, they tend to use different definitions and measures of exposure, which is further complicated since there is no agreement in the literature over what constitutes acute or chronic exposure. Finally, objective exposure measures are rarely available, making it difficult to establish clear dose-response relationships(Harrison and Mackenzie Ross 2016).

Measuring the Mood

Besides problematic organophosphates exposure measurement, measuring mood is also a challenge. Methods used to evaluate the incidence of mood disorder in OP exposed populations vary from merely asking participants whether or not they have ever been diagnosed with depression/anxiety by a physician to asking participants to complete standardized symptom questionnaires with differing degrees of sensitivity and specificity (Harrison and Mackenzie Ross 2016). Self-report measures may be useful for screening in large populations. Still, it is not always obvious whether findings are analogous with what would be resulting from controlled clinical interviews undertaken by mental health professionals using internationally agreed diagnostic criteria for mental disorders. Indeed, previous research with other patient cohorts has shown that self-report measures may not be good indicators of psychiatric disorder (Golden, Conroy and O'Dwyer 2007, Mitchell, Meader and Symonds 2010, Poole and Morgan 2006, Watson et al. 2014). A recent study implies this may be particularly true for research with OP exposed groups (Harrison and Mackenzie Ross 2016). On the other hand, structured interviews standardize the coverage of specific psychiatric issues (thus reducing the possibility of missed diagnoses) and entail the systematic assessment of relevant symptoms to minimize misdiagnosis. Thus, clinical interviews are considered the 'gold standard for diagnosis' (Harrison and Mackenzie Ross 2016).

Review of the Connection of Organophosphate Exposure and Depression

In the last two decades, two published studies have reported evaluations of OP exposed participants using strict diagnostic criteria. First, Amr et al. (Amr, Halim and Moussa 1997) examined 208 Egyptian pesticide formulators, 172 pesticide applicators and 223 control subjects. All subjects underwent a full psychiatric interview. The diagnoses were made following the Diagnostic and Statistical Manual of Psychiatric Disorders (DSM-III-R). Psychiatric disorders were found to be more widespread in exposed subjects, particularly depression and dysthymic disorder. Nevertheless, the problem was the authors did not collect in-depth information about participants' exposure history. Having that in mind, it is impossible to determine whether the effect is related to long-term low-level exposure since the risk of acute poisonings cannot be ruled out (Aníćijević and Lazarević-Pašti 2020).

Second, Salvi et al. (Salvi et al. 2003) evaluated 37 tobacco workers from Brazil exposed to OP pesticides. They evaluated them shortly after exposure and then for a second time, following three months without exposure. Diagnoses were made following the Diagnostic and Statistical Manual of Psychiatric Disorders (DSM-IV). Almost half of the sample suffered from a psychiatric disorder (mostly anxiety and depression) when first interviewed.

Still, after three months of not using pesticides, the rate of mental illness had decreased by almost 50%. It suggests a strong association between mental health and pesticide exposure (Aničijević and Lazarević-Pašti 2020).

A problem spotted in earlier work has been the failure to consider possible confounding variables involved in the etiology of psychiatric symptoms when assessing exposed cohorts. For example, farming is a stressful and physically demanding occupation. Some researchers disagree that socioeconomic and psychosocial factors are responsible for the prominent rates of psychological distress observed in this occupational group (Beard et al. 2011, Harrison and Mackenzie Ross 2016). Prior research reporting associations between exposure to pesticides and psychiatric illness have often failed to consider these issues. In addition, they lack detailed evaluations of participants' psychiatric history and background. These shortcomings make it difficult to determine whether psychosocial or toxicological factors are involved in the etiology of farmers' mental health complaints. Moreover, the effect of participants' physical health is hardly ever considered, even though poor health is a risk factor for mood disorder results. Numerous studies simply compare rates of anxiety and depression in exposed and unexposed groups (Ames et al. 1995, Bazylewicz-Walczak, Majczakowa and Szymczak 1999, Farahat et al. 2003, Harrison and Mackenzie Ross 2016, Jamal, Hansen and Julu 2002, Reidy et al. 1992, Roldan-Tapia et al. 2006, Steenland et al. 2000, Stephens et al. 1995).

An exception is the work of Stallones and Beseler and colleagues (Beseler et al. 2006, Beseler and Stallones 2008, Beseler et al. 2008, Stallones and Beseler 2002a, b). Within their work, the influence of demographic factors such as age, gender, marital status, alcohol use, health status and financial difficulties (known risk factors for depression) were investigated additionally to pesticide exposure. The higher rates of depression in individuals with a history of pesticide poisoning were found, even once corrections were made for other risk factors. Also, pesticide poisoning was more obviously connected with depression than lower levels of exposure. Nevertheless, they found a relationship between high cumulative exposure and depression in a subgroup of individuals who did not have a history of acute poisoning (Harrison and Mackenzie Ross 2016).

In 2010 Mackenzie Ross et al. published the findings of a cross-sectional study of 127 UK sheep farmers, which were required to determine the neurotoxicity of low-level exposure to OPs. They examined sheep farmers who had been repeatedly exposed to low levels of OPs during the process of sheep dipping (Mackenzie Ross et al. 2010). The neuropsychological assessment found exposed subjects performed significantly worse than matched controls and standardization samples on tests of memory, response speed, fine motor control, mental flexibility and strategy making, even after controlling for covariates. Besides, more than 40% of study participants reported via a survey that they were suffering from considerable levels of anxiety and depression compared to less than 23% of controls. According to strict diagnostic criteria, the true prevalence rate for these conditions required further investigation (Harrison and Mackenzie Ross 2016).

In 2016, Mackenzie Ross and Harrison reported findings from both self-report measures of mental health and the results of comprehensive clinical examinations to obtain a more accurate picture of the prevalence of psychological distress and psychiatric disorders amongst their cohort of sheep farmers investigated in 2010. This study aimed to examine whether farmworkers with a history of low-level exposure to OPs report higher levels of psychological distress on subjective symptom questionnaires than unexposed controls. The goal was to determine do they meet internationally agreed diagnostic criteria for a psychiatric disorder more often than unexposed controls, even after controlling for variables known to affect mood (namely demographic and psychosocial factors and physical health ratings) (Harrison and Mackenzie Ross 2016). Comparisons between groups based on self-report measures revealed long-term low-level exposure to OPs might be responsible for the increased levels of anxiety and depression detected in the farming cohort. Major effect within the exposed group was established even after age, and psychosocial risk factors such as ill-health, pain and stressful life events were considered. Demographic factors are not likely to illuminate the increased levels of distress seen in the cohort of farmworkers as they were equivalent in both exposed and unexposed cohorts (except in the case of age; it was included as a covariate). However, this findings pattern was only true for anxiety when mood disorder was diagnosed using in-depth structured clinical interviews (Harrison and Mackenzie Ross 2016). Age, exposure and pain were good predictors of anxiety diagnoses. On the other hand, depression was only foreseen by subjective health ratings. Therefore, although participants in the exposed cohort may report more symptoms commonly associated with depression than controls, they were not more likely to meet strict diagnostic criteria for a diagnosis of Major Depression (Harrison and Mackenzie Ross 2016).

There are only two studies from the USA reporting associations between low-level exposure and formally diagnosed psychiatric disorder in occupationally exposed cohorts. Beseler et al. (Beseler and Stallones 2008) evaluated the relationship between depression and pesticide exposure in 17,585 pesticide applicators. They found both acute high intensity and lower level exposures (in the absence of a history of acute poisoning) were associated with depression. More recently, Beard et al. (Beard et al. 2014) estimated links between mood disorder diagnosis and exposure to ten different classes of pesticides and fifty specific pesticides in 21,208 pesticide applicators. The results implied that depression was associated with several specific pesticides, including OPs, even after controlling other risk factors. However, either study did not explore anxiety (Harrison and Mackenzie Ross 2016).

Since depression is frequently reported in acutely exposed individuals (Freire and Koifman 2016), it suggests that only exposures to higher levels of OPs are associated with clinically significant depressive symptoms. A notion that gains some support from Beseler and colleagues' work found pesticide poisoning is more strongly associated with depression than lower levels of exposure (Beseler et al. 2006, Beseler et al. 2008). In contrast, anxiety may be associated with low-level exposure to OPs. It is certainly something that deserves further study. Unfortunately, not enough researchers provide adequate information about exposure history to allow a more detailed analysis of thresholds beyond which psychiatric problems may develop (Harrison and Mackenzie Ross 2016).

Even if it is the case that apparent ill health is the main predictor of depression, that does not rule out OP exposure as a potential cause. It may just indicate a relationship more complicated than dose-response at hand (Harrison and Mackenzie Ross 2016). Several researchers have found a possible biological mechanism through which OP pesticides may cause ill health in some individuals. For example, paraoxonase (PON1) is a liver and plasma enzyme that contributes significantly to the detoxification of OPs. Two studies have found PON1 polymorphisms and low PON1 activity levels more common in people reporting ill-health (Cherry et al. 2002, Mackness et al. 2003). In other words, people who reported ill health were less efficient at metabolizing OP pesticides. As such, it is possible that those who are less able to metabolize OPs are more likely to develop ill health and are more likely to become depressed due to that, suggesting that the relationship between exposure and symptomatology is more complex than simple dose-response (Harrison and Mackenzie Ross 2016).

It is known that cholinesterase inhibitors, especially organophosphates, are endocrine disruptors. A few existing studies link self-reports of exposure with increased depression and anxiety (Beseler et al. 2006, Beseler and

Stallones 2008, Beseler et al. 2008, Harrison and Mackenzie Ross 2016, Suarez-Lopez et al. 2021). Some evidence suggests that associations may be stronger in women. Still, the mechanism of this gender difference is unclear. In 2020. Suarez-Lopez et al. assessed whether acetylcholinesterase inhibition between 2-time points (reflecting greater cholinesterase inhibitor exposure) during different agricultural seasons in the year was associated with anxiety/depression symptoms (Suarez-Lopez et al. 2021). In their interesting and significant contribution, the authors examined 300 adolescents (ages 11-17 years, 51% female) living near agricultural settings in Ecuador (ESPINA study) twice in 2016: April and July-October. They assessed AChE activity, estradiol, testosterone, dehydroepiandrosterone, cortisol (saliva) and anxiety and depression scales (CDI-2 and MASC-2). They observed that exposures to cholinesterase inhibitors, characterized as inhibition of AChE, between two agricultural seasons in the same year was associated with increased depression symptoms in adolescents. Inhibition of AChE activity seemed to have a stronger association with depression scores among females compared to men and older compared to younger adolescents. However, they observed consistent positive associations in all age groups and genders (greater inhibition associated with greater depression scores), indicating internal consistency of the findings across these groups. Additionally, they did not observe associations with anxiety symptoms. These findings parallel earlier cross-sectional analyses in the full sample of our cohort study (n = 529), in which lower AChE activity was also associated with increased depression scores (but not anxiety), with stronger associations among girls compared to boys (Suarez-Lopez et al. 2019). The study form 2020. provides stronger evidence that cholinesterase inhibitors can affect mood regulation, as the degree of cholinesterase inhibition between two agricultural seasons was strongly related to depression symptoms. The findings of this study also suggest that cholinesterase inhibition can result in subacute alterations in the mood of adolescents, adding to data from earlier studies in children that have reported transient alterations in neurobehavioral performance associated with pesticide spray seasons (Khan et al. 2014, Rohlman et al. 2016, Suarez-Lopez et al. 2017). Earlier, the authors found that children examined closer to the end of the Mother's Day flower harvest, marking the end of a peak pesticide spray season, had lower neurobehavioral performance in 3 of 5 domains assessed compared to children examined later during a period of lower floricultural production and pesticide use (Suarez-Lopez et al. 2017). Research assessing the short- and medium-term effects of pesticide exposures on children's health is complex, and the present study contributes to this understudied field of research. Beyond the transient impact on mental health processes, it is plausible that repeated exposures throughout childhood may lead to long-lasting developmental alterations.

In the study from 2020, the authors also observed stronger associations between AChE inhibition and depression symptoms among girls compared to boys, and these findings are in agreement with the cross-sectional results in the full cohort of participants assessed in July-October 2016 ($n = 529$) (Suarez-Lopez et al. 2019). While the biological mechanism explaining why AChE inhibition affected girls more than boys is unclear, it is known that adolescent girls have higher rates of depression compared to boys (Whiteford et al. 2013) and this increased susceptibility to internalizing behaviors may be a contributing factor to the stronger associations observed in the study from 2020. The authors hypothesized that an important explanation of this effect modification by gender was that girls had greater AChE activity inhibition than boys. Nevertheless, this was not the case; in the sensitivity analysis using the frequency-matched sample across the four categories of AChE inhibition between the genders, they observed that the associations were substantially stronger in girls than boys. Furthermore, the difference between genders was strengthened in this subgroup. Modification for cortisol, testosterone and dehydroepiandrosterone decreased gender differences by 18-62%. No links were observed with anxiety. Inhibition of AChE activity at 2 points in time during different pesticide spray periods was connected with larger depression symptoms, affecting girls more than boys. Gender differences may be partly explained by endocrine disruption. These findings suggest that AChE inhibition may transiently affect the mood of adolescents (Suarez-Lopez et al. 2021).

Reduced acetylcholine hydrolysis due to cholinesterase inhibitors, such as carbamate and organophosphate pesticides, results in overstimulation of nicotinic and muscarinic receptors (Colovic et al. 2013, Lazarevic-Pasti et al. 2017, Mitic and Lazarevic-Pasti 2021). The cholinergic system has a key role in the central nervous system and modulates mood and behavior. Both clinical and animal studies have demonstrated that blocking the activation of nicotinic and muscarinic receptors results in antidepressant signs and symptoms (Furey and Drevets 2006, Mineur et al. 2013b). In mice, exposure to physostigmine, a cholinesterase inhibitor, led to increased depression and anxiety behaviors reversed once nicotinic or muscarinic acetylcholine receptor antagonists were administered (Mineur et al. 2013b). Likewise, physostigmine was found to induce depressive symptoms and exacerbate mood disorders in normal and manic subjects (Janowsky, El-Yousef and Davis 1974a, Steinberg et al. 1997).

Another study similarly found that rats who received the organophosphate insecticide chlorpyrifos had increased depressive behaviors and that such behaviors subsided upon receiving atropine, an anticholinergic agent (Siqueira et al. 2019). In addition, rats exposed to cholinesterase inhibitor pesticides had a reduction in hippocampal, striatal and prefrontal cortical AChE activity and plasma butyrylcholinesterase activity (Siqueira et al. 2019). These studies corroborate past reports of human exposure to cholinesterase inhibitors that resulted in the cholinergic balance hypothesis of mania and depression (Janowsky, El-Yousef and Davis 1974a, Suarez-Lopez et al. 2021). The hypothesis was based upon reports and case studies of both the emergence of depressive symptoms among psychotic patients who were administered a cholinergic inhibitor and by reports of cholinesterase inhibitor insecticide poisoning leading to symptoms of depression and parasympathetic toxicity (Dagyte, Den Boer and Trentani 2011b, Suarez-Lopez et al. 2021). Considering the role of the cholinergic system on mood modulation, anticholinergic drugs have been investigated as potential treatments for depression, but inconsistent antidepressant effects have been found (Shytle et al. 2002, Suarez-Lopez et al. 2021). Parallel to the cholinergic system, serotonin is a neurotransmitter directly linked with mood regulation. There is evidence that organophosphate pesticides can affect serotonergic mechanisms and serotonin-related emotional behaviors in animal models (Chen et al. 2011, Suarez-Lopez et al. 2021). The associations observed in these experimental and observational studies, including our own, may be partly explained by the interplay of the cholinergic and serotonergic systems in regulating behavior and neurobehavioral processes (Jeltsch-David, Koenig and Cassel 2008, Sparks et al. 2018, Steckler and Sahgal 1995). The cholinergic system is one of many factors involved in mood regulation (Suarez-Lopez et al. 2021).

In 2019. Siqueira et al. (Siqueira et al. 2019) investigated behavioral and biochemical consequences of acute exposure to the OP chlorpyrifos in male rats and the effectiveness of the antidotes atropine and pralidoxime on reversing these changes. A sub-lethal dose of commercial chlorpyrifos (20 mg/kg, i.p.) elicited signs of acute toxicosis during the first hours after its injection in rats. Twenty-four hours after treatment, this single dose of chlorpyrifos induced depressive-like behavior in the rat forced swimming test without impairing locomotor activity. At this time (24 h), in rats, chlorpyrifos decreased plasma butyrylcholinesterase (BuChE) activity and hippocampal, striatal and prefrontal cortical AChE activity. The behavioral and biochemical consequences of acute chlorpyrifos poisoning do not seem to be long lasting, since 30 days later, they were absent. The authors evaluated whether these behavioral and biochemical consequences of acute chlorpyrifos treatment would be reversed by the antidotes atropine (10 mg/kg i.p.) and pralidoxime (40 mg/kg; i.p.) given 1 h after poisoning. Pralidoxime partially reactivated the AChE activity in the prefrontal cortex but not in the hippocampus and striatum. Atropine attenuated the depressive-like behavior induced by chlorpyrifos in rats. The results suggested that acute chlorpyrifos poisoning causes a transient depressive-like behavior possible related to hippocampal AChE inhibition. Also, the treatment with atropine and pralidoxime seems insufficient to counteract all the effects of OP acute poisoning, at least in rats (Siqueira et al. 2019).

The study of Liu et al. from 2021. (Liu, Hester and Pope 2021) was aiming to evaluate the persistence of changes in the forced swim test out to 4 months. They also compared responses in a sucrose preference test as a measure of anhedonia. Adult male rats were treated with vehicle (peanut oil, 1 mL/kg, sc) or diisopropylfluorophosphate (DFP) (2.0, 2.25 or 2.5 mg/kg). It was followed by sacrifice 4 h later for measurement of OP-sensitive serine hydrolases in the hippocampus. Additional rats were treated similarly and assessed for signs of acute toxicity from 30 min to 6 days. Next, motor activity forced swim behavior and sucrose preference were evaluated at approximately one week, one month, and four months after dosing. All dosages of DFP elicited serine hydrolase inhibition. Bodyweight was decreased in all DFP-treated groups during the first two weeks. The lethality was prominent with the higher dosages. Involuntary movements were elicited in all DFP treatment groups during the first week, but both times of onset and rate of recovery were dose-related. There was a significant reduction in ambulation at one week after the highest dosage (2.5 mg/kg), but no other significant locomotor changes were noted. Immobility was increased, and swimming was decreased in the forced swim test at all three time-points by 2.25 mg/kg DFP and at 2 of 3 time-points by the other dosages. While the length of water deprivation and time after DFP dosing affected sucrose preference, DFP treatment had no main effect. The authors concluded that the forced swim test (a measure of behavioral despair/coping mechanism for inescapable stress) is a robust and persistent neurobehavioral consequence of acute DFP intoxication. In contrast, sucrose preference, a measure of anhedonia and a common symptom of major clinical depression, is not affected (Liu, Hester and Pope 2021).

Conclusion

Organophosphates are the most often used group of pesticides globally. They are potent inhibitors of acetylcholinesterase. There are strong indications that increased acetylcholine levels could be a key cause of clinical depression symptoms development in individuals exposed to organophosphates. However, there is not enough data on the research on the organophosphate and depression development connection. A small number of available studies predominantly contain the data collected inadequately. Since humans are extremely exposed to organophosphates in everyday life, it is necessary to change the approach and gather relevant data regarding this topic. It is of high importance for human society in the future, having in mind increasing evidence that long-term low-level exposure to organophosphates influences mammalian organisms in a much more negative manner than we thought. Furthermore, long-term exposure to organophosphates leads to numerous diseases where cancer development and depression are most threatening. With all this in mind, relevant research on these points and careful data collection are necessary.

Acknowledgments

This research was funded by the Ministry of Education, Science and Technological Development of the Republic of Serbia, grant number 451-03-9/2021-14/200017.

References

Ames, R. G., Steenland, K., Jenkins, B., Chrislip, D. and Russo, J. 1995. Chronic Neurologic Sequelae to Cholinesterase Inhibition among Agricultural Pesticide Applicators. *Archives of Environmental Health: An International Journal* 50(6):440-444.

Amr, M. M., Halim, Z. S. and Moussa, S. S. 1997. Psychiatric Disorders among Egyptian Pesticide Applicators and Formulators. *Environmental Research* 73(1):193-199.

Aničijević, V. and Lazarević-Pašti, T. 2020. Organophosphate Pesticides. Marquis, F. (ed), pp. 1-42, Nova Science Publishers, New York.

Bazylewicz-Walczak, B., Majczakowa, W. and Szymczak, M. 1999. Behavioral effects of occupational exposure to organophosphorous pesticides in female greenhouse planting workers. *NeuroToxicology* 20(5):819-826.

Beard, J. D., Umbach, D. M., Hoppin, J. A., Richards, M., Alavanja, M. C. R., Blair, A., Sandler, D. P. and Kamel, F. 2011. Suicide and Pesticide Use among Pesticide Applicators and Their Spouses in the Agricultural Health Study. *Environmental Health Perspectives* 119(11):1610-1615.

Beard, J. D., Umbach, D. M., Hoppin, J. A., Richards, M., Alavanja, M. C. R., Blair, A., Sandler, D. P. and Kamel, F. 2014. Pesticide exposure and depression among male private pesticide applicators in the agricultural health study. *Environmental Health Perspectives* 122(9):984-991.

Beseler, C., Stallones, L., Hoppin, J. A., Alavanja, M. C. R., Blair, A., Keefe, T. and Kamel, F. 2006. Depression and Pesticide Exposures in Female Spouses of Licensed Pesticide Applicators in the Agricultural Health Study Cohort. *Journal of Occupational and Environmental Medicine* 48(10).

Beseler, C. L. and Stallones, L. 2008. A Cohort Study of Pesticide Poisoning and Depression in Colorado Farm Residents. *Annals of Epidemiology* 18(10):768-774.

Beseler, C. L., Stallones, L., Hoppin, J. A., Alavanja, M. C. R., Blair, A., Keefe, T. and Kamel, F. 2008. Depression and Pesticide Exposures among Private Pesticide Applicators Enrolled in the Agricultural Health Study. *Environmental Health Perspectives* 116(12):1713-1719.

Boublik, Y., Saint-Aguet, P., Lougarre, A., Arnaud, M., Villatte, F., Estrada-Mondaca, S. and Fournier, D. 2002. Acetylcholinesterase engineering for detection of insecticide residues. *Protein Eng.* 15(1):43-50.

Brown, G. W. and Harris, T. O. (1978) *Social Origins of Depression: A Study of Psychiatric Disorder in Women*, Tavistock Publications, London.

Charles, H. C., Lazeyras, F., Krishnan, K. R., Boyko, O. B., Payne, M. and Moore, D. 1994. Brain choline in depression: *in vivo* detection of potential pharmacodynamic effects of antidepressant therapy using hydrogen localized spectroscopy. *Progress in Neuro-Psychopharmacology & Biological Psychiatry* 18(7):1121-1127.

Chau, D. T., Rada, P. V., Kim, K., Kosloff, R. A. and Hoebel, B. G. 2011. Fluoxetine Alleviates Behavioral Depression while Decreasing Acetylcholine Release in the Nucleus Accumbens Shell. *Neuropsychopharmacology* 36(8):1729-1737.

Chen, W. Q., Yuan, L., Xue, R., Li, Y. F., Su, R. B., Zhang, Y. Z. and Li, J. 2011. Repeated exposure to chlorpyrifos alters the performance of adolescent male rats in animal models of depression and anxiety. *NeuroToxicology* 32(4):355-361.

Cherry, N., Mackness, M., Durrington, P., Povery, A., Dippnall, M., Smith, T. and Mackness, B. 2002. Paraoxonase (PON1) polymorphisms in farmers attributing ill health to sheep dip. *The Lancet* 359(9308):763-764.

Colovic, M. B., Krstic, D. Z., Lazarevic-Pasti, T. D., Bondzic, A. M. and Vasic, V. M. 2013. Acetylcholinesterase inhibitors: pharmacology and toxicology. *Current Neuropharmacology* 11(3):315-335.

Costa, L. G. 2006. Current issues in organophosphate toxicology. *Clinica Chimica Acta* 366(1):1-13.

Covey, L., Glassman, A. and Stetner, F. 1997. Major depression followingsmoking cessation. *Am. J. Psychiatry* 154:263-265.

Craig, L. A., Hong, N. S. and McDonald, R. J. 2011. Revisiting the cholinergic hypothesis in the development of Alzheimer's disease. *Neurosci. Biobehav. Rev.* 35(6):1397-1409.

Dagyte, G., Den Boer, J. A. and Trentani, A. 2011. The cholinergic system and depression. *Behav. Brain Res.* 221(2):574-582.

Das, A., Kapoor, K., Sayeepriyadarshini, A. T., Dikshit, M., Palit, G. and Nath, C. 2000. Immobilization stress-induced changes in brain acetylcholinesterase activity and cognitive function in mice. *Pharmacol. Res.* 42(3):213-217.

Dulawa, S. C. and Janowsky, D. S. 2019. Cholinergic regulation of mood: from basic and clinical studies to emerging therapeutics. *Molecular Psychiatry* 24(5):694-709.

Duman, R. S., Aghajanian, G. K., Sanacora, G. and Krystal, J. H. 2016. Synaptic plasticity and depression: new insights from stress and rapid-acting antidepressants. *Nature Medicine* 22(3):238-249.

Farahat, T. M., Abdelrasoul, G. M., Amr, M. M., Shebl, M. M., Farahat, F. M. and Anger, W. K. 2003. Neurobehavioural effects among workers occupationally exposed to organophosphorous pesticides. *Occupational and Environmental Medicine* 60(4):279-286.

Freire, C. and Koifman, S. 2016. Pesticides, depression and suicide: A systematic review of the epidemiological evidence. *International Journal of Hygiene and Environmental Health* 216(4):445-460.

Furey, M. L. and Drevets, W. C. 2006. Antidepressant Efficacy of the Antimuscarinic Drug Scopolamine: A Randomized, Placebo-Controlled Clinical Trial. *Archives of General Psychiatry* 63(10):1121-1129.

Gatto, G. J., Bohme, G. A., Caldwell, W. S., Letchworth, S. R., Traina, V. M., Obinu, M. C., Laville, M., Reibaud, M., Pradier, L., Dunbar, G. and Bencherif, M. 2004. TC-1734: an orally active neuronal nicotinic acetylcholine receptor modulator with antidepressant, neuroprotective and long-lasting cognitive effects. *CNS Drug Reviews* 10(2):147-166.

Gershon, S. and Shaw, F. H. 1961. Psychiatric sequelae of chronic exposure to organophosphorus insecticides. *Lancet* 1(7191):1371-1374.

Golden, J., Conroy, R. M. and O'Dwyer, A. M. 2007. Reliability and validity of the Hospital Anxiety and Depression Scale and the Beck Depression Inventory (Full and FastScreen scales) in detecting depression in persons with hepatitis C. *Journal of Affective Disorders* 100(1):265-269.

Harrison, V. and Mackenzie Ross, S. 2016. Anxiety and depression following cumulative low-level exposure to organophosphate pesticides. *Environ. Res.* 151:528-536.

Hayden, K. M., Norton, M. C., Darcey, D., Ostbye, T., Zandi, P. P., Breitner, J. C. and Welsh-Bohmer, K. A. 2010. Occupational exposure to pesticides increases the risk of incident AD: the Cache County study. *Neurology* 74(19):1524-1530.

Jamal, G. A., Hansen, S. and Julu, P. O. O. 2002. Low level exposures to organophosphorus esters may cause neurotoxicity. *Toxicology* 181-182:23-33.

Janowsky, D. S., El-Yousef, K. M. and Davis, J. M. 1974. Acetylcholine and Depression. *Psychosomatic Medicine* 36(3):248-257.

Janowsky, D. S., Risch, S. C., Kennedy, B., Ziegler, M. and Huey, L. 1986. Central muscarinic effects of physostigmine on mood, cardiovascular function, pituitary and adrenal neuroendocrine release. *Psychopharmacology* 89(2):150-154.

Jeltsch-David, H., Koenig, J. and Cassel, J. C. 2008. Modulation of cholinergic functions by serotonin and possible implications in memory: General data and focus on 5-HT1A receptors of the medial septum. *Behavioural Brain Research* 195(1):86-97.

Katona, C. 2009. Dementia: Does depression predict donepezil response in MCI? *Nat. Rev. Neurol.* 5(11):585-586.

Kenny, P. J. and Markou, A. 2001. Neurobiology of the nicotine withdrawal syndrome. *Pharmacol. Biochem. Behav.* 70(4):531-549.

Khan, K., Ismail, A. A., Abdel Rasoul, G., Bonner, M. R., Lasarev, M. R., Hendy, O., Al-Batanony, M., Crane, A. L., Singleton, S. T., Olson, J. R. and Rohlman, D. S. 2014. Longitudinal assessment of chlorpyrifos exposure and self-reported neurological symptoms in adolescent pesticide applicators. *BMJ Open* 4(3):e004177.

Lazarevic-Pasti, T., Colovic, M., Savic, J., Momic, T. and Vasic, V. 2011. Oxidation of diazinon and malathion by myeloperoxidase. *Pesticide Biochemistry and Physiology* 100(2):140-144.

Lazarevic-Pasti, T., Leskovac, A., Momic, T., Petrovic, S. and Vasic, V. 2017. Modulators of Acetylcholinesterase Activity: From Alzheimer's Disease to Anti-Cancer Drugs. *Current Medicinal Chemistry* 24(30):3283-3309.

Lazarevic-Pasti, T., Momic, T., Onjia, A., Vujisic, L. and Vasic, V. 2010. Myeloperoxidase-mediated oxidation of organophosphorus pesticides as a pre-step in their determination by AChE based bioanalytical methods. *Microchimica Acta* 170(3):289-297.

Lazarevic-Pasti, T., Nastasijevic, B. and Vasic, V. 2011. Oxidation of chlorpyrifos, azinphos-methyl and phorate by myeloperoxidase. *Pesticide Biochemistry and Physiology* 101(3):220-226.

Liu, J., Hester, K. and Pope, C. 2021. Dose- and time-related effects of acute diisopropylfluorophosphate intoxication on forced swim behavior and sucrose preference in rats. *NeuroToxicology* 82:82-88.

Mackenzie Ross, S. J., Brewin, C. R., Curran, H. V., Furlong, C. E., Abraham-Smith, K. M. and Harrison, V. 2010. Neuropsychological and psychiatric functioning in sheep farmers exposed to low levels of organophosphate pesticides. *Neurotoxicology and Teratology* 32(4):452-459.

Mackness, B., Durrington, P., Povey, A., Thomson, S., Dippnall, M., Mackness, M., Smith, T. and Cherry, N. 2003. Paraoxonase and susceptibility to organophosphorus poisoning in farmers dipping sheep. *Pharmacogenetics* 13(2):81-88.

Marchand, W. R., Dilda, D. V. and Jensen, C. R. 2005. Neurobiology of mood disorders: a clinical review article. *Hosp. Physician* 43:17-26.

Mineur, Y. S., Mose, T. N., Blakeman, S. and Picciotto, M. R. 2018. Hippocampal a7 nicotinic ACh receptors contribute to modulation of depression-like behaviour in C57BL/6J mice. *Br. J. Pharmacol.* 175(11):1903-1914.

Mineur, Y. S., Obayemi, A., Wigestrand, M. B., Fote, G. M., Calarco, C. A., Li, A. M. and Picciotto, M. R. 2013. Cholinergic signaling in the hippocampus regulates social stress resilience and anxiety- and depression-like behavior. *Proc. Natl. Acad. Sci. U S A* 110(9):3573-3578.

Mitchell, A. J., Meader, N. and Symonds, P. 2010. Diagnostic validity of the Hospital Anxiety and Depression Scale (HADS) in cancer and palliative settings: A meta-analysis. *Journal of Affective Disorders* 126(3):335-348.

Mitic, M. and Lazarevic-Pasti, T. 2021. Does the application of acetylcholinesterase inhibitors in the treatment of Alzheimer's disease lead to depression? *Expert Opinion on Drug Metabolism and Toxicology* 17(7):841-856.

Modell, S. and Lauer, C. J. 2007. Rapid eye movement (REM) sleep: an endophenotype for depression. *Curr. Psychiatry Rep.* 9(6):480-485.

National Collaborating Centre for Mental Health (2005) *Depression in Adults with a Chronic Physical Health Problem: Treatment and Management*, British Psychological Society, Leicester (UK).

Newman, M. B., Nazian, S. J., Sanberg, P. R., Diamond, D. M. and Shytle, R.D. 2001. Corticosterone-attenuating and anxiolytic properties of mecamylamine in the rat. *Prog. Neuropsychopharmacol. Biol. Psychiatry* 25(3):609-620.

Patel, V. (2005) *Gender in mental health research*, WHO, Geneva.

Philip, N. S., Carpenter, L. L., Tyrka, A. R. and Price, L. H. 2010. Nicotinic acetylcholine receptors and depression: a review of the preclinical and clinical literature. *Psychopharmacology* 212(1):1-12.

Picciotto, M. R., Lewis, A. S., van Schalkwyk, G. I. and Mineur, Y. S. 2015. Mood and anxiety regulation by nicotinic acetylcholine receptors: A potential pathway to modulate aggression and related behavioral states. *Neuropharmacology* 96(Pt B):235-243.

Poole, N. A. and Morgan, J. F. 2006. Validity and reliability of the Hospital Anxiety and Depression Scale in a hypertrophic cardiomyopathy clinic: the HADS in a cardiomyopathy population. *General Hospital Psychiatry* 28(1):55-58.

Rada, P., Colasante, C., Skirzewski, M., Hernandez, L. and Hoebel, B. 2006. Behavioral depression in the swim test causes a biphasic, long-lasting change in accumbens acetylcholine release, with partial compensation by acetylcholinesterase and muscarinic-1 receptors. *Neuroscience* 141(1):67-76.

Reidy, T. J., Bowler, R. M., Rauch, S. S. and Pedroza, G. I. 1992. Pesticide exposure and neuropsychological impairment in migrant farm workers. *Archives of Clinical Neuropsychology* 7(1):85-95.

Rohlman, D. S., Ismail, A. A., Rasoul, G. A., Bonner, M. R., Hendy, O., Mara, K., Wang, K. and Olson, J. R. 2016. A 10-month prospective study of organophosphorus pesticide exposure and neurobehavioral performance among adolescents in Egypt. *Cortex* 74:383-395.

Roldan-Tapia, L., Nieto-Escamez, F. A., del Aguila, E. M., Laynez, F., Parron, T. and Sanchez-Santed, F. 2006. Neuropsychological sequelae from acute poisoning and long-term exposure to carbamate and organophosphate pesticides. *Neurotoxicology and Teratology* 28(6):694-703.

Rowntree, D. W., Nevin, S. and Wilson, A. 1950. The effects of diisopropyl-fluorophosphonate in schizophrenia and manic depressive psychosis. *J Neurol Neurosurg Psychiatry* 13(1):47-62.

Salvi, R. M., Lara, D. R., Ghisolfi, E. S., Portela, L. V., Dias, R. D. and Souza, D. O. 2003. Neuropsychiatric evaluation in subjects chronically exposed to organophosphate pesticides. *Toxicological Sciences* 72(2):267-271.

Sanne, B., Mykletun, A., Dahl, A. A., Moen, B. E. and Tell, G. S. 2003. Occupational differences in levels of anxiety and depression: The Hordaland Health Study. *Journal of Occupational and Environmental Medicine* 45(6):628-638.

Shytle, R. D., Silver, A. A., Lukas, R. J., Newman, M. B., Sheehan, D. V. and Sanberg, P. R. 2002. Nicotinic acetylcholine receptors as targets for antidepressants. *Molecular Psychiatry* 7(6):525-535.

Siqueira, A. A., Cunha, A. F., Marques, G. L. M., Felippe, I. S. A., Minassa, V. S., Gramelich, T. C. D. S., Cicilini, M. A., Alarcon, T. A., Pires, R. G. W., Sampaio, K. N. and Beijamini, V. 2019. Atropine counteracts the depressive-like behaviour elicited by acute exposure to commercial chlorpyrifos in rats. *Neurotoxicology and Teratology* 71:6-15.

Sparks, D. W., Tian, M. K., Sargin, D., Venkatesan, S., Intson, K. and Lambe, E. K. 2018. Opposing Cholinergic and Serotonergic Modulation of Layer 6 in Prefrontal Cortex. *Frontiers in Neural Circuits* 11(107).

Stallones, L. and Beseler, C. 2002a. Pesticide Illness, Farm Practices, and Neurological Symptoms among Farm Residents in Colorado. *Environmental Research* 90(2):89-97.

Stallones, L. and Beseler, C. 2002b. Pesticide Poisoning and Depressive Symptoms among Farm Residents. *Annals of Epidemiology* 12(6):389-394.

Steckler, T. and Sahgal, A. 1995. The role of serotonergic-cholinergic interactions in the mediation of cognitive behaviour. *Behavioural Brain Research* 67(2):165-199.

Steenland, K., Dick, R. B., Howell, R. J., Chrislip, D. W., Hines, C. J., Reid, T. M., Lehman, E., Laber, P., Krieg Jr, E.F. and Knott, C. 2000. Neurologic function among termiticide applicators exposed to chlorpyrifos. *Environmental Health Perspectives* 108(4):293-300.

Steinberg, B. J., Trestman, R., Mitropoulou, V., Serby, M., Silverman, J., Coccaro, E., Weston, S., deVegvar, M. and Siever, L. J. 1997. Depressive Response to Physostigmine Challenge in Borderline Personality Disorder Patients. *Neuropsychopharmacology* 17(4):264-273.

Steingard, R. J., Yurgelun-Todd, D. A., Hennen, J., Moore, J. C., Moore, C. M., Vakili, K., Young, A. D., Katic, A., Beardslee, W. R. and Renshaw, P. F. 2000. Increased orbitofrontal cortex levels of choline in depressed adolescents as detected by in vivo proton magnetic resonance spectroscopy. *Biol. Psychiatry* 48(11):1053-1061.

Stephens, R., Spurgeon, A., Calvert, I. A., Beach, J., Levy, L. S., Harrington, J. M. and Berry, H. 1995. Neuropsychological effects of long-term exposure to organophosphates in sheep dip. *The Lancet* 345(8958):1135-1139.

Suarez-Lopez, J. R., Checkoway, H., Jacobs, D. R., Al-Delaimy, W. K. and Gahagan, S. 2017. Potential short-term neurobehavioral alterations in children associated with a peak pesticide spray season: The Motherâ€™s Day flower harvest in Ecuador. *NeuroToxicology* 60:125-133.

Suarez-Lopez, J. R., Hood, N., Suarez-Torres, J., Gahagan, S., Gunnar, M. R. and Lopez-Paredes, D. 2019. Associations of acetylcholinesterase activity with depression and anxiety symptoms among adolescents growing up near pesticide spray sites. *International Journal of Hygiene and Environmental Health* 222(7):981-990.

Suarez-Lopez, J. R., Nguyen, A., Klas, J., Gahagan, S., Checkoway, H., Lopez-Paredes, D. and Noble, M. 2021. Associations of acetylcholinesterase inhibition between pesticide spray seasons with depression and anxiety symptoms in adolescents, and the role of sex and adrenal hormones on gender moderation. *Expo Health* 13(1):51-64.

van Enkhuizen, J., Janowsky, D. S., Olivier, B., Minassian, A., Perry, W., Young, J. W. and Geyer, M. A. 2015. The catecholaminergic-cholinergic balance hypothesis of bipolar disorder revisited. *Eur. J. Pharmacol.* 753:114-126.

Watson, T. M., Ford, E., Worthington, E. and Lincoln, N. B. 2014. Validation of Mood Measures for People with Multiple Sclerosis. *International Journal of MS Care* 16(2):105-109.

Whiteford, H. A., Degenhardt, L., Rehm, J., Baxter, A. J., Ferrari, A. J., Erskine, H. E., Charlson, F. J., Norman, R. E., Flaxman, A. D., Johns, N., Burstein, R., Murray, C. J. L. and Vos, T. 2013. Global burden of disease attributable to mental and substance use disorders: findings from the Global Burden of Disease Study 2010. *The Lancet* 382(9904):1575-1586.

WHO. 2009. *The WHO Recommended Classification of Pesticides by Hazard and Guidelines to Classification* 2009.

WHO (2017) *Depression and Other Common Mental Disorders*. WHO (ed), WHO.

Chapter 5

Exposure to Organophosphates as a Risk Factor for Cancer

Lidija Todorović*, PhD

Laboratory for Radiobiology and Molecular Genetics,
"Vinča" Institute of Nuclear Sciences, National Institute of the Republic of Serbia,
University of Belgrade, Belgrade, Serbia

Abstract

Organophosphorous compounds have been widely used as pesticides in agricultural, commercial and residential settings. A number of studies have reported that people exposed to these compounds in occupational and environmental settings are more prone, compared with the general population, to develop cancer. However, a direct link in terms of causality between organophosphate exposure and carcinogenesis in humans has not been established, mainly because the exact biologic mechanisms underlying these associations are still unclear. This chapter aims to link epidemiologic studies to available evidence from basic science research to provide a better understanding of the role organophosphate exposures play in the molecular pathogenesis of different types of cancer.

Keywords: organophosphates, cancer, cell signaling, gene expression, molecular pathways

---

* Corresponding Author's Email: lidijat@vin.bg.ac.rs.

In: Organophosphates
Editor: Tamara Lazarević-Pašti
ISBN: 978-1-68507-652-8

Introduction

The global burden of cancer incidence and mortality is high and continues to increase. This is the result of aging and the growth of the population, but at the same time, it reflects changes in the prevalence and distribution of the main risk factors for cancer, many of which are associated with socioeconomic development. With current trends in demographics and exposure, the cancer burden has been shifting from developed to developing countries (Sung et al. 2021; WHO 2020).

The etiology of cancer is very complex and multifactorial. Epidemiologic studies have shown that both environmental (exogenous) and non-environmental, individual factors, including genetic predisposition, contribute to the development of cancer. Most often, it is the interaction between external causes and individual genetic susceptibility that ultimately results in the development of cancer. Environmental exposures, particularly chemical exposures, have been repeatedly implicated as risk factors across many cancer types (Lewandowska et al. 2019).

Organophosphate (OP) compounds have been extensively used for decades in agricultural, commercial and residential settings as the main components of herbicides, pesticides, and insecticides because of their effectiveness and relatively low toxicity in mammals. They are also widely used in the production of solvents and plastics (Pierre 2021). The widespread use of OPs has led to severe environmental pollution since residual amounts of OPs have been detected in the soil, groundwater, drinking water, house dust, vegetables, grains and other food, causing environmental and public health concerns (Bhanti and Taneja 2007; Poet et al. 2004; Yu et al. 2017; Gill et al. 2020; Kapeleka et al. 2020; Li et al. 2019). According to a number of reports, OP metabolites have been repeatedly detected in the urine of the majority of the US and European populations (Barr et al. 2004; Barr et al. 2005; Kavvalakis and Tsatsakis 2012), adding evidence to the ubiquitous exposure of the general population. In developing countries, OP exposure is even greater due to high availability, widespread use and poor safety regulation (Jaga and Dharmani 2005; Ali et al. 2017).

Organophosphates are broadly known to exert toxic effects in humans following high-dose acute exposure. On the other hand, knowledge about chronic low-dose effects of specific OPs is more limited and represents an area of growing scientific interest.

During the last decades, a number of epidemiological as well as laboratory studies have shown that exposure to some organophosphate compounds may be associated with the development of cancer. According to the International Agency for Research on Cancer (IARC), some of the widely used organophosphorous pesticides such as malathion, diazinon and glyphosate have been classified as probably carcinogenic to humans (Group 2A), while parathion and tetrachlorvinphos have been classified as possibly carcinogenic to humans (Group 2B) (International Agency for Research on Cancer 2017).

A direct causal relationship between OP exposure and carcinogenesis in humans has not been established so far because the identification of molecular changes consequent to OP exposure and implicated in the etiology of cancer has proved challenging. The exact biological mechanisms underlying these associations are just starting to be revealed. This chapter summarizes epidemiological and laboratory research reports on OP exposures that have been associated with the development of different types of cancer and discuss their biological plausibility based on current research.

Cancer

Cancer or malignant neoplasm is a broad group of diseases characterized by the development of abnormal cells that show unregulated cell growth and have the ability to invade or spread to other parts of the body – metastasize, infiltrate and destroy normal body tissue. Cancer pathogenesis is a multistep process in which normal cells evolve progressively to a neoplastic state, acquiring the traits that enable them to become tumorigenic and ultimately malignant.

These traits – acquired capabilities, represented in Figure 1, are designated as the Hallmarks of Cancer (Hanahan and Weinberg 2000, 2011), and they include:

- sustaining proliferative signaling (chronic proliferation – the growth of cancer cells at a seemingly unlimited rate)
- evading growth suppressors (cancer cells do not respond to withdrawal of normal growth signals or do not respond to the anti-growth signals)
- resisting cell death (evading apoptosis – cancer cells avoid triggering internal self-destroying mechanisms, the usual process in abnormal or redundant cells)
- enabling replicative immortality (unlimited replicative potential – cancer cells do not senesce (age) and die after a limited number of cell divisions)

- inducing angiogenesis (creation of new blood vessels in order to support the growth of tumor) and
- activating invasion and metastasis (expanding into normal tissue locally or/and at distant sites in the body)

The Hallmarks of Cancer framework also includes:

- Two enabling characteristics
  - genome instability and mutation (mutations and epigenetic changes in the parent cell DNA are passed to daughter cells)
  - tumor-promoting inflammation (promotes the growth of cancer cells via the same growth signals normal cells provide to each other during the embryonic growth and wound healing, contributes to cell survival, angiogenesis, metastasis and avoiding of adaptive immunity response)
- Two emerging hallmarks
  - avoiding immune destruction (cancer cells avoid immune surveillance that would normally mark them for destruction)
  - deregulated cellular energetics – dysregulated metabolism (needed in order to support the increased anabolic and catabolic demands in proliferating cancer cells)

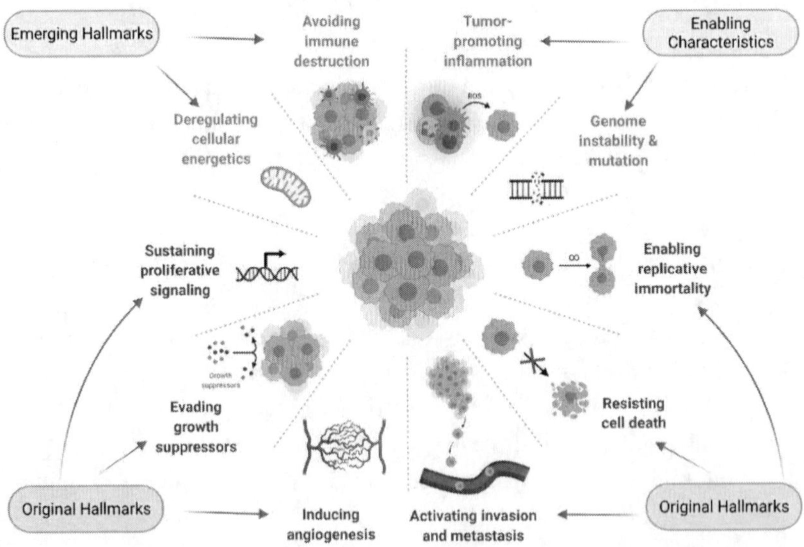

Figure 1. Hallmarks of cancer including emerging hallmarks and enabling characteristics.

The disturbed cell functions in the cancerous process are a result of the dynamic changes in the genome represented by the accumulation of many genetic and epigenetic changes within the cell, a phenomenon called genomic/genome instability (Feng and Riddle 2020; Hanahan and Weinberg 2000). The integrity and stability of the genome are maintained by complex mechanisms of DNA monitoring and repair systems in the cell, which ensures that genetic changes such as mutations are such rare events at the genome level that multiple mutations, which are characteristic for tumor cell genomes, are highly unlikely to occur within a human life span. Malfunction of the specific components of the genomic maintenance machinery, caused by mutations in genes involved, is one of the common sources of genomic instability. These mutations presumably arise via DNA damage caused by environmental agents or endogenous adducts such as oxidative damage (Wogan et al. 2004). Carcinogenic agents increase the risk of genetic damage involving changes in the expression of oncogenes (when activated, promote cell growth and survival) and the loss or inactivation of tumor suppressor genes (inhibit the progression of the cell cycle), leading to further accumulation of abnormalities and acquiring capabilities presented in Figure 1.

A number of contemporary theories on the etiology of cancer agree that cancer risk depends on random errors that occur during normal cellular replication, hereditary defects in critical genes and environmental factors including exogenous agents and lifestyle, as reviewed in (Wolf et al. 2019).

Epidemiological Evidence Linking OPs and Cancer

Most epidemiological studies to date have been cohort studies and case-control studies that have assessed the risk of cancer among agricultural workers and pest control operators in relation to general pesticide or OP exposure and, to a lesser extent, exposure to individual OPs. Many of these studies indicate that occupational exposure to OPs poses an increased risk for some types of cancer.

A number of studies have shown an association between organophosphates and non-Hodgkin's lymphoma (NHL) among men. Elevated risk for NHL was found to be associated with the use of diazinon, dichlorvos and malathion in a study involving men from Iowa and Minnesota who were newly diagnosed with NHL (Cantor et al. 1992). Another significant association between NHL incidence and exposure to organophosphorous insecticides was found in a multicenter, population-based cross-Canada study, which also reported significantly increased NHL risk after exposure to malathion as an individual compound (McDuffie et al. 2001). An increased NHL incidence in pharming men in the mid-western United States was associated with the use of several individual pesticides such as coumaphos, diazinon and fonofos (De Roos et al. 2003). In the large prospective US study – Agricultural Health Study (AHS), which included 57,310 licensed pesticide applicators from North Carolina and Iowa recruited in 1993-1997, NHL overall as well as specific NHL subtypes were associated with the use of terbufos (Alavanja et al. 2014). The association between ever use of terbufos, and increased risk of NHL was also reported by the AGRICOH Consortium – a pooled analysis of three large agricultural worker cohorts – AGRICAN (France), CNAP (Norway) and AHS (USA), after taking into account exposure to other pesticides (Leon et al. 2019). A case-control analysis of occupational exposure in Sweden reported a positive association between NHL and exposure to glyphosate (Eriksson et al. 2008). Findings from the North American Pooled Project also suggest an association between NHL and ever use of glyphosate (Pahwa et al. 2019). A significant increase in the relative risk of NHL was found to be associated with high glyphosate exposure in a meta-analysis that included the results from the AHS and another five case-control studies (Zhang et al. 2019). In an Italian multicenter case-control study, increased risk of follicular lymphoma – a type of NHL, was associated with ever use of glyphosate, while no such association was recorded for other types and subtypes of lymphoma (Meloni et al. 2021). The data from three population-based case-control studies conducted in several US states show a significant association between the use of diazinon and increased risk of small lymphocytic lymphoma in white male farmers after adjustment for other pesticide exposures (Waddell et al. 2001). Another publication reporting results from the North American Pooled Project indicate a significant association between Hodgkin lymphoma and ever use of terbufos, particularly among subjects ≤ 40 years of age, and the odds were even more elevated with a longer duration of use (Latifovic et al. 2020). In a Canadian case-control study, an increased risk of developing Hodgkin lymphoma was associated with the use of chlorpyrifos (Karunanayake et al. 2012). In the AHS, an increased risk of leukemia was associated with the highest tertile of exposure to diazinon (Beane Freeman et al. 2005). The AHS also reported a link between leukemia and long-term fonofos exposure among licensed pesticide applicators. The leukemia risk was elevated in the highest category of a lifetime and intensity-weighted exposure-days compared to the group of unexposed subjects (Mahajan et al. 2006).

Higher lung cancer risk was associated with exposure to diazinon among pest control operators from Florida. Moreover, odds ratios for lung cancer were greater for workers first licensed before age 40, and it increased for workers who were licensed for 20 or more years compared to those licensed for less than 20 years (Pesatori et al. 1994). Several publications based on the analyses from the Agricultural Health Study reported a significant association between lung cancer incidence and the exposure to chlorpyrifos and diazinon after adjusting for other lung cancer risk factors (Alavanja et al. 2004; Jones et al. 2015; Lee et al. 2004; Beane Freeman et al. 2005). Associations were also observed between occupational exposure to terbufos and lung cancer, as well as prostate cancer, leukemia and NHL. The authors also reported that the overall cancer risk was slightly increased among terbufos users (Bonner et al. 2010).

The AHS analysis on the incidence of prostate cancer among 54,412 licensed pesticide applicators shows that fonofos, malathion and terbufos are significantly associated with increased risk for aggressive prostate cancer (Koutros et al. 2012). A case-control study in Canada reported a significant association between prostate cancer risk and exposure to malathion after adjustment for potential confounding variables (Band et al. 2011). Elevated risk of prostate cancer was also found in farm workers with relatively high levels of exposure to dichlorvos, compared to workers with lower levels of exposure (Mills and Yang 2003) as well as in workers with a family history of prostate cancer who reported long term fonofos exposure (Mahajan et al. 2006). Considering pancreatic cancer, a recent US study evaluated the link between cancer risk and chronic long-term OP exposure in the non-agricultural US populations based on dialkylphosphate biomarkers in urine. The results of the study indicate a significant association between increased prostate cancer risk and higher OP exposure in male smokers aged 60–85 years (Sun, Sun, and Barr 2020).

Some other types of cancer were also found to be associated with OP exposure. In the AHS cohort, an increased risk of renal cell carcinoma was associated with exposure to chlorpyrifos (Andreotti et al. 2020), while a non-significantly elevated risk of this cancer was associated with the use of diazinon (Jones et al. 2015). An AHS analysis that has evaluated colorectal cancer incidence in pesticide applicators found a significantly increased risk for rectal cancer in the highest exposure category among chlorpyrifos users (Lee et al. 2007). A hospital-based study in Vietnam reported an increased risk of hepatocellular carcinoma among agricultural workers who were classified in the group of high use of OP pesticides (Cordier et al. 1993).

The majority of epidemiological studies investigating OP use and cancer outcomes have focused on occupational exposure among farmers or licensed pesticide applicators and have been conducted in predominantly male populations. Only a few studies so far focused on the relationship between cancer incidence and OP exposure in women and in non-occupational exposure settings.

The Agricultural Health Study is an extensive prospective health study also involving female participants – spouses of agricultural pesticide applicators. The evaluation of the cancer incidence among female subjects in the AHS found that exposure to OPs represents a risk factor for developing several hormonally related cancers such as breast, thyroid and ovary, suggesting potential for hormonally mediated effects (Lerro et al. 2015). The study included 7589 participants who self-reported personal ever use of single and overall use of OP insecticides. The Median follow-up time was 15.3 years, and during that period, 9.2% of exposed women were diagnosed with cancer. Elevated risk of breast cancer was seen in participants who reported the use of any OP, while the elevated risk of thyroid cancer and ovarian cancer were associated with malathion and diazinon, respectively (Lerro et al. 2015). According to the following report from the AHS concerning the associations between insecticide use and breast cancer risk among wives of pesticide applicators, breast cancer risk was significantly increased among women who had ever used chlorpyrifos or terbufos and among women whose husbands used fonofos (Engel et al. 2017). A non-significant risk increase was associated with the ever use of coumaphos. Risk in relation to the subjects' insecticide use was associated primarily with premenopausal breast cancer. The study found little evidence of differential risk by tumor estrogen receptor status (Engel et al. 2017). In a case-control study evaluating the risk of postmenopausal breast cancer associated with pesticide exposure in California's Central Valley, breast cancer was three times as likely to occur among women exposed to chlorpyrifos compared with those not exposed, after adjustment for exposure to other pesticides (Tayour et al. 2019). A recent study involving non-agricultural populations in the US examined the relation between OP exposure and broad health risk, including cancer risk, with a particular focus on women (Sun, Sun, and Barr 2020). Based on the data on the urinary concentration of dialkylphosphate – commonly detected OP metabolite used as an indicator of OP exposure, long-term chronic exposure to OP compounds poses a significantly higher general health risk for women than for men. Breast cancer risk and overall cancer risk in female smokers aged 60–85 years were significantly increased with higher OP exposure (Sun, Sun, and Barr 2020).

Furthermore, a growing body of evidence suggests that parental exposure to pesticides, including OPs, may contribute to childhood cancer risk. In the AHS, an increased risk of cancer was detected among children whose fathers used dichlorvos prenatally (Flower et al. 2004). In a Switzerland study, increased risk from solid tumors other than tumors in the central nervous system was found in children whose parents were occupationally exposed to pesticides (Coste et al. 2020). The results of a meta-analysis also suggest that the incidence of childhood cancers is related to parental exposure during the prenatal period (Vinson et al. 2011). Residential exposure to pesticides during childhood has also been implicated in the development of childhood cancers (Chen et al. 2015; Van Maele-Fabry, Gamet-Payrastre, and Lison 2019).

Contrary to the studies cited above, some epidemiologic studies did not find any association between occupational exposure to different OPs and increased risk for cancer. According to an earlier report from the AHS describing the analysis of the association between malathion exposure and cancer incidence, no significant association was found between lifetime days of malathion use and the incidence of all cancers combined as well as individual cancer types such as NHL and melanoma (Bonner et al. 2007). Other reports from the AHS cohort describe no significant association between glyphosate and lymphoid malignancies, including NHL and subtypes, nor any solid tumors (Andreotti et al. 2018; De Roos et al. 2005). According to a recent report from the AHS, no association was found between thyroid cancer incidence in male licensed pesticide applicators and exposure to any of the OP compounds (Lerro et al. 2021).

The presence and the magnitude of the cancer risk related to occupational exposure to different OPs have been inconsistent across studies. The reasons for the observed inconsistencies are not entirely clear, but may partially reflect the differences in study design and methodology – deficiency in adequate sample size in some of them and differences in the control for confounding factors such as multiple chemical exposures, differing exposure patterns, other environmental differences, lifestyle-related cancer risk factors as well as genetic factors.

Potential Mechanisms of OP Carcinogenicity

The principal molecular mechanism of OP toxicity is based on the inhibition of the enzyme acetylcholine esterase (AChE), which hydrolyses the neurotransmitter acetylcholine (Ach) and terminates its effect. Ach serves as a neurotransmitter in the central nervous system, and it is also being released by the terminal nerve endings of post ganglionic parasympathetic nerves and in both sympathetic and parasympathetic ganglia, as well as at myoneural junctions in skeletal muscles. Organophosphates bind irreversibly to the AChE in the plasma, red blood cells, and at the level of the synapses in both the central and peripheral nervous systems. The inhibition of AChE leads to the accumulation of Ach at synaptic junctions and overstimulation of the corresponding (muscarinic and nicotinic) receptors, and subsequent disruption of the transmission of nerve impulses in both the peripheral and central nervous system (Peter and Cherian 2000). Acute or chronic exposure to OPs is associated with multi-system manifestations involving the gastrointestinal tract, nervous system, respiratory system, cardio-vascular system, skeletal muscles and other organs (Peter and Cherian 2000).

A number of studies suggest that OP exposure can also induce oxidative stress, cytotoxicity and genotoxicity in cell lines, animal models and humans (Akbel et al. 2018; Moore, Yedjou, and Tchounwou 2010; Muniz et al. 2008; Olakkaran et al. 2020). Moreover, accumulating evidence indicates that OPs can induce or affect cellular processes crucial for cancer development and progression.

Oxidative Stress

Oxidative stress (OS) represents an imbalance between the production of reactive oxygen species (ROS) and the ability of the body to detoxify the reactive intermediates or repair the resulting damage. ROS such as superoxide anion, peroxyl radicals, hydroperoxyl radical, and hydrogen peroxide are produced as a consequence of normal cellular metabolism and are considered as part of normal oxidative metabolism when present at low to moderate concentrations. But at elevated concentrations, ROS cause tissue injuries since they are able to damage all macromolecules in the cell such as proteins, lipids, and nucleic acids, therefore influencing a variety of biochemical processes, including the function of proteins involved in DNA repair (Muniz et al. 2008; Selmi et al. 2018). Antioxidants and DNA repair processes are two major defense mechanisms to counteract the damage produced by oxidative stress. Aside from the ability to generate ROS and induce lipid peroxidation – a well-established mechanism of cellular injury, which is often used as an indicator of oxidative stress, it has been reported that OPs can also reduce the level of cellular antioxidants and the activity of antioxidant enzymes (Muniz et al. 2008).

Studies of OS in OP Exposed Humans

Oxidative DNA repair activity was found to be significantly higher in lymphocytes of pesticide applicators and farm workers than in non-exposed controls (Muniz et al. 2008; Kisby et al. 2009). The analysis of urine and serum of exposed subjects showed an increased level of oxidative stress markers. In the same study, results of *in vitro* experiments on human lymphocytes also showed increased oxidative stress and DNA damage (Muniz et al. 2008; Kisby et al. 2009). In a recent study, urine levels of OP metabolites and markers of OS – glutathione (GSH) and malondialdehyde (MDA) were compared between children in agricultural and urban communities (Sapbamrer, Hongsibsong, and Khacha-Ananda 2020). The results show that children can be exposed to OPs both outdoors and indoors and provide supporting evidence that OP exposure can cause oxidative stress in children.

Studies of OS in Human Cells *In Vitro* Following OP Exposure

Markers of oxidative stress (ROS, reduced levels of glutathione) and oxidative DNA damage were observed in human lymphocyte cell cultures treated with OPs (Muniz et al. 2008; Kisby et al. 2009). Another study showed that even at low concentrations, malathion induced ROS production and increased lipid peroxidation in human lymphocytes (Olakkaran et al. 2020). The results of this study also show that malathion affected superoxide dismutase (SOD), catalase (CAT), glutathione (GSH) and glutathione-S-transferase (GST) antioxidant activities. In human liver cancer cells HepG(2), malathion treatment induced significant production of malondialdehyde (MDA), an end product of lipid peroxidation and an indicator of OS. The authors hypothesized that increased cytotoxicity and genotoxicity observed in the treated cells was presumably mediated through OS (Moore, Yedjou, and Tchounwou 2010). A recent study also evaluated oxidative stress in HepG2 cells, but after chlorpyrifos treatment, detected alterations of reduced GSH content, glutathione peroxidase (GPx) and SOD activity, although no change in ROS production was observed (Tadee, Mahakunakorn, and Porasuphatana 2020). OS was also detected in the human prostate epithelial WPM-Y.1 cell line after glyphosate treatment (Abdel-Halim and Osman 2020). In another study, glyphosate treatment significantly increased lactate dehydrogenase (LDH) activity and MDA level compared with the untreated cells. It induced a rapid decrease in GSH content and GST activity and increased the activities of CAT, GPx, and glutathione reductase (GR) (Abdel-Halim and Osman 2020). Methyl parathion and parathion induced oxidative stress (increased MDA levels) and reduced the viability of $HepG_2$ cells in a dose-dependent manner, inducing significant genotoxicity at higher concentrations (Edwards, Yedjou, and Tchounwou 2013). Chlorpyrifos has been shown to modulate the antioxidant enzyme activity inducing redox imbalance in breast cancer cells (Ventura et al. 2015).

In Vivo and *In Vitro* Studies of OS Following OP Exposure in Non-Human Mammals

The results of a study of the effect of malathion exposure on OS in the liver and kidney tissue of prepubertal male mice show that malathion administration leads to increased lipid peroxidation and hydrogen peroxide content, decreased thiol groups level, and depletion of antioxidant enzymes (catalase – CAT, total superoxide dismutase – total SOD, Cu/Zn-SOD, Mn-SOD, Fe-SOD and glutathione peroxidase – GPx) activities (Selmi et al. 2018). Increased oxidative stress and decreased antioxidant status were also demonstrated in rats treated with malathion (Akbel et al. 2018) and diazinon (Karimani et al. 2018; Khazaie et al. 2019). Significant increases in DNA strand breaks and in levels of the reactive oxygen species superoxide anion and hydrogen peroxide were observed in rat peripheral blood lymphocytes treated with chlorpyrifos, methyl parathion, and malathion (Ojha and Srivastava 2014).

Genotoxicity

OPs have been studied for genotoxic potential in a variety of assays in exposed humans, in human cells *in vitro* and in other mammalian and non-mammalian systems *in vitro* and *in vivo*. The genotoxic effect refers to the capability of an agent to induce a cascade of events such as structural alterations in chromosomes, DNA damage and subsequent expression of mutant gene products. Chromosome aberrations, micronuclei (MN) and sister chromatid exchange (SCE) are structural chromosome changes commonly detected in the tests for chromosome damage by conventional *in vitro* cytogenetics (Phillips and Arlt 2009).

Studies of Genotoxicity in Exposed Humans

A number of studies provide evidence of the genotoxic abilities of OPs in exposed humans. DNA damage (determined by comet assay), as well as oxidative DNA repair activity, were significantly higher in lymphocytes of pesticide applicators and farm workers than in non-exposed controls (Muniz et al. 2008; Kisby et al. 2009). The results of these studies also revealed that markers of oxidative stress were elevated in the urine and serum of the exposed individuals. In the same study, results of *in vitro* experiments on human lymphocytes also showed increased oxidative stress and DNA damage (Muniz et al. 2008; Kisby et al. 2009). More serious DNA damage compared to the non-exposed groups was also found in horticultural farmers exposed to OPs (Atherton et al. 2009) and in workers who were working for a longer period in the pesticide manufacturing industry (Arshad et al. 2016). Another study shows a correlation between the extent of DNA damage and the level of OP exposure. DNA damage index in whole peripheral blood of soybean farmers from Brazil was significantly higher in the high exposure period than in the low exposure period. In both periods, DNA damage was greater than in the control group (Bernieri et al. 2020). Other studies showed that exposure to multiple pesticides including malathion, diazinon or glyphosate is associated with increased frequencies of chromosomal aberrations, SCE, MN as well as increased values of comet assay parameters in peripheral blood lymphocytes of exposed subjects (Garaj-Vrhovac and Zeljezic 2001; De Ferrari et al. 1991; Bolognesi et al. 2009). In a study conducted by Sing and associates, a significantly greater DNA damage, determined by comet assay, was observed in 115 workers exposed to OPs compared with non-exposed control subjects (Singh et al. 2011). The authors also analyzed polymorphisms in the paraoxonase (*PON1*) gene, a gene that codes for a lipoprotein-associated enzyme involved in the detoxification of OPs by hydrolyzing the bioactive oxons. The results suggest that due to polymorphisms in the *PON1* gene, there is a wide variation in the activity of the PON1 enzyme and in the extent of DNA damage induced by OP exposure (Singh et al. 2011).

Studies of Genotoxicity in Human Cells In Vitro

There is a considerable amount of literature on OPs (mostly malathion) genotoxicity in human cells *in vitro*. Some of these studies are presented below. A significant increase in micronucleated cells, concurrent with cytotoxicity and strong inhibition of proliferation, was found after the treatment of human lymphocytes with high doses (75–100 μg/mL) of malathion. Moreover, the majority of the treated cells had multiple micronuclei (Titenko-Holland et al. 1997). Interestingly, in a study that compared the effect of high purity malathion and its major metabolite malaoxon and its isomer isomalathion, which are usually present in commercial preparations of malathion, malathion did not cause significant DNA damage through the range of tested concentrations in human lymphocytes cell culture. Contrary to this, malaoxon and isomalathion induced significant DNA damage in a dose-dependent manner, with malaoxon displaying considerable cytotoxicity at the highest concentrations used (Blasiak et al. 1999). In more recent studies, cytotoxic, genotoxic and oxidative stress-inducing activities were documented in human lymphocyte cell culture after treatment with malathion (Olakkaran et al. 2020) or unspecified OPs (Kisby et al. 2009; Muniz et al. 2008). The effect of malathion was also examined in human liver cancer cells HepG2 (Moore, Yedjou, and Tchounwou 2010). MTT cell-proliferation assay was used to evaluate cell viability, malondialdehyde (MDA) production was measured as an indicator of lipid peroxidation, and comet assay was used for DNA damage assessment. The results showed that at lower levels of exposure, malathion had a mitogenic effect, while at higher concentrations, it induced cytotoxic and genotoxic effects, presumably mediated through oxidative stress (Moore, Yedjou, and Tchounwou 2010). In another study, malathion and its metabolite dimethyl thiophosphate (DMTP) increased the frequencies of MN and nucleoplasmic bridges (NPB) in human hepatic cell lines HepG2 and WRL-68 and peripheral blood mononuclear cells as determined by the Cytokinesis-Block Micronucleus assay (Hernandez-Toledano, Estrada-Muniz, and Vega 2020). Interestingly, while the effect of malathion was primarily clastogenic (inducing structural chromosome alterations), the effect of DMTP was aneuploidogenic (inducing numerical chromosome alterations) and required cytochrome P450 enzymes activity (catalyze the metabolism of malathion to its metabolites, including DMTP). This result suggests a different mechanism of MN induction by malathion metabolite DMTP compared to malathion (Hernandez-Toledano, Estrada-Muniz, and Vega 2020). As for other OP compounds, significantly increased frequency of chromosomal aberrations and micronuclei were observed in human lymphocytes after treatment with glyphosate (Santovito et al. 2018; Nagy et al. 2021). Significant oxidative stress, cytotoxicity and genotoxicity were observed in human gingival fibroblasts after treatment with methyl parathion (Argentin, Divizia, and Cicchetti 2015).

In Vivo and In Vitro Studies of OP Genotoxicity in Non-Human Mammals

Different OPs have been extensively tested in the past for genotoxicity and carcinogenicity in mice and rats, as summarized in (International Agency for Research on Cancer 2017). Some of these studies are presented below.

When the genotoxic potential of malathion was investigated in bone marrow cells and peripheral blood of rats intraperitoneally injected with malathion, the results showed an increased number of structural chromosomal aberrations and the percentages of DNA damage in treated groups (Moore, Patlolla, and Tchounwou 2011). In *in vitro* study of the effect of malathion, chlorpyrifos and methyl parathion on rat peripheral lymphocytes in culture, a significant increase in DNA strand breaks and ROS levels was observed. The greatest DNA damage and ROS production were induced by methyl parathion, followed by chlorpyrifos and malathion (Ojha and Srivastava 2014). Another study showed that both acute and chronic exposure to malathion, chlorpyrifos and methyl parathion caused significant DNA damage and histopathological changes in rat tissues such as liver, brain, kidney, and spleen (Ojha et al. 2013). The genotoxic potential of low concentrations of malathion (1.5, 0.75, 0.37, and 0.18 mg/mL) was shown and in rat hepatoma tissue culture cells. A higher frequency of chromosomal bridges indicates the clastogenic action of malathion (Bianchi, Mantovani, and Marin-Morales 2015). Malathion induced cytogenetic effect in mice fed with wheat grains with malathion residues. The cytogenetic effect (chromosomal aberrations and SCE) was evident in different mouse tissues and was dose-dependent (Amer et al. 2002). Similar results were obtained in mice when malathion was administered through the intraperitoneal route (Giri et al. 2002). In another study on mice, malathion, given either by intraperitoneal or per oral route, was found to cause a significant increase of micronuclei in mouse bone marrow cells in a dose-dependent manner (Giri, Giri, and Sharma 2011).

Effects on Cell Proliferation, Altered DNA Repair, Genome Instability, and Malignant Transformation

Malathion and parathion have been found to increase cell proliferation and induce cell transformation in non-tumorigenic breast epithelial cell line MCF-10F in a number of studies, as summarized in (Calaf, Bleak, and Roy 2021). Since the epidemiological and experimental evidence associates estrogens with the pathogenesis of breast cancer, and the human population is exposed to a mixture of estrogenic and estrogen-like agents, as well as to OP compounds, Calaf and associates investigated the effect of parathion and malathion in combination with estrogen (17β-estradiol – E2) on the malignant transformation of human breast epithelial cells MCF-10F *in vitro*. In addition, using microarray technology, they analyzed the expression of cancer-related genes in the treated cells (Calaf and Roy 2007a, b, 2008). The results of these studies showed that both alone and in combination with E2, malathion and parathion induced malignant transformation of MCF-10F cells, as confirmed by anchorage independence and invasive capabilities of treated cells. Protein expression analysis showed that parathion alone upregulated the expression of epidermal growth factor receptor (EGFR), Notch-4, proto-oncogene c-Kit, Rho-A, Rac 3, beta-catenin and mutant p53. EGFR is a receptor tyrosine kinase involved in proliferative signaling pathways. The Notch family of genes is associated with multiple signaling pathways. Rac 3 is a GTP-binding protein of the Ras superfamily, involved in the control of cell proliferation, cell polarity and cytoskeletal arrangement. Rho-A is also involved in cell polarity and motility. Beta-catenin is an important epithelial adhesion molecule in normal epithelium, and the loss of its adhesion to E-cadherin is an important step in the progression of many carcinomas. Protein p53, coded by the *TP53* gene, is an important tumor suppressor involved in the regulation of the normal cell cycle. Mutant p53 fails to mediate the effects of normal p53, which results in uncontrolled proliferation. Besides this, the analysis of gene expression at the mRNA level showed altered expression of several genes involved in cell cycle regulation (cyclins A1, A2, C, G1, G2, H; cyclin-dependent kinases (CDKs)), and in the regulation of DNA replication (minichromosome maintenance protein complex (MCM) family) (Calaf and Roy 2007a, b). In accordance with these results, in a subsequent study, human cancer microarray analysis of the expression profile in the MCF-10F cells treated with E2 in combination with either malathion or parathion showed altered expression of a number of genes associated with cell cycle progression, cell differentiation and signal transduction pathways (Calaf and Roy 2008).

All treatments were associated with upregulation of cyclin D1 (*CCND1*), cyclin-dependent kinase 4 (*CDK4*), Insulin-Like Growth Factor Binding Protein 3 and 5 (*IGFBP3* and *IGFBP5*) and keratin 18 (*KRT18*). Malathion alone, and the combination of E2 and either malathion or parathion, upregulated the *c-H-ras* oncogene, involved in key signal transduction pathways in human malignancies. Malathion alone and combination of malathion or parathion with E2 upregulated heat shock protein 27 (*HSP27*), *MCM2*, TP53 inducible protein 3. A combination of parathion and E2, as well as malathion alone, upregulated the *DVL1* gene, a gene that regulates the migration and proliferation of endothelial cells and is known to be implicated in the Wnt signaling pathway – another key signaling pathway frequently deregulated in many cancers. Tumor suppressor p53 was upregulated by malathion alone and E2 alone (Calaf and Roy 2008). p53 plays a crucial role in maintaining genomic stability by allowing the repair of damaged DNA through induction of transient cell cycle arrest in G1/S or elimination of damaged cells by triggering apoptosis. Up to 50% of human cancers have been found to have alterations in the *TP53* gene. Altered p53 can act as an oncogene, and cells with such p53 do not stop but continue to divide, although DNA damage is not repaired. This leads to further accumulation of abnormalities and increases genomic instability (Calaf and Roy 2008).

Cancer progression has been associated with an increase in genomic instability indicated by the inactivation of tumor suppressor genes and the activation of oncogenes. Loss of heterozygosity (LOH) and microsatellite instability (MSI) are common indicators of genomic instability. LOH represents a change from a heterozygous to a homozygous state due to the loss of the wild-type allele. LOH is a frequent mechanism of the inactivation of the tumor suppressor genes. MSI is defined as an alteration in the microsatellite sequence length within tumor DNA when compared with normal DNA from the individual, and it is used to detect evidence of defective repair of DNA synthesis errors. In a study by Calaf and associates, different degrees of allelic imbalance in the form of LOH or MSI were observed in the cells treated with either malathion or parathion alone or in combination with E2. The results indicated that malathion and parathion induced malignant transformation of breast cells through genomic instability altering *p53* and *H-ras* oncogene (Calaf, Echiburu-Chau, and Roy 2009).

Besides the ability to induce transformation, there is some evidence that OPs can also induce further progression of malignant transformation. In a study conducted on transformed (cancer) cells – breast cancer cell line MCF7, parathion and malathion treatment-induced progression of malignant transformation. Both agents induced mutant p53 protein expression and increased the expression of proliferating cell nuclear antigen (*PCNA*), a protein involved in DNA replication and repair (Cabello et al. 2003).

Some studies evaluated the effect of chlorpyrifos on the proliferation of estrogen-dependent and estrogen-independent breast cancer cell lines, MCF-7 and MDA-MB-231, respectively. Chlorpyrifos induced cell proliferation and modulated cell cycle progression and the expression of cyclins E and D1. Cell proliferation was induced at doses as low as 0.05 μM and mediated by estrogen receptor α (ERα) phosphorylation in MCF-7 while high dose (50 μM) induced intra-S arrest modifying checkpoints proteins, through a mechanism that may involve changes is redox balance. In MDA-MB-231, high doses induced cell cycle arrest in the G2/M phase, the authors hypothesized that this might be the consequence of chlorpyrifos capacity to alter microtubules polymerization by binding to tubulin, as some data suggest (Ventura et al. 2012). Another study showed that chlorpyrifos and its metabolite chlorpyrifos-oxon increase cell proliferation in MCF-7 and MDA-MB-231 cell lines and that this effect is mediated by mechanisms related to ERα and aromatic hydrocarbon receptor (AhR) (Moyano et al. 2020). Estrogen is known to control the growth of several types of carcinomas in experimental animals and humans. ERα is one of the two estrogen receptors, the other being ERβ, frequently differentially expressed and implicated in the pathogenesis of various human malignancies (Bozovic et al. 2021). Overexpression of AhR leads to the proliferation of breast cancer cells while its silencing reduces tumor growth and metastasis (Salisbury et al. 2014). In other studies, chlorpyrifos has been shown to induce migration and invasion in 2D and 3D cell cultures of MCF-7 and MDA-MB-231 cell lines, even at subthreshold concentrations, and to change the expression of key molecular markers involved in epithelial-mesenchymal transition such as metalloprotease MMP2, E-Cadherin, β-Catenin, and Vimentin. (Lasagna et al. 2020; Lasagna et al. 2021).

One more study investigated malathion-induced changes in gene expression. Normal human lymphocytes were treated with different concentrations of malathion, and their effect on cell viability was assessed, as well as the expression levels of cancer-associated genes (Anjitha et al. 2020). Gene expression profiles in malathion treated and control lymphocytes were analyzed using a microarray platform, and differentially expressed genes were validated by quantitative real-time PCR. A concentration-dependent reduction in cell viability occurred in malathion-treated cells compared to untreated control cells. The results suggest that at lower concentrations (50 μg/mL), malathion induces proliferation and colony formation, and it did not have a cytotoxic effect (Anjitha et al. 2020). The expression analyses showed that among 4388 genes found to have altered expression in malathion-treated cells, 15% were upregulated, and 85% were downregulated. Around 60 genes that were differentially expressed are genes linked to cancer. Some of these genes – the upregulated: CCAAT Enhancer-Binding Protein G (*CEBPG),* MHC Class I Chain-Related Protein A *(MICA)*, Paired Box 1 (*PAX1),* and Transferrin Receptor (*TFRC),* and the downregulated: Major Histocompatibility Complex, Class I A (*HLA-A)* and E *(HLA-E)* and Galectin 1 (*LGALS1)* are related to the growth – proliferation and differentiation of lymphocytes, immunoglobulin production and regulation, and hematopoiesis (Anjitha et al. 2020). Other genes with significant differential expression have been shown to act as tumor suppressors or oncogenes in a number of malignant tumors. Among the downregulated genes with potential tumor suppressor function were: Dual-Specificity Phosphatase 1 *(DUSP1)*, Glutathione S-Transferase Pi 1 *(GSTP1),* GRAM Domain Containing 4 *(GRAMD4),* HIV-1 Tat Interactive Protein 2 *(HTATIP2),* Leucine zipper tumor suppressor 2 *(LZTS2),* Ninjurin1 (*NINJ1),* Profilin-1*(PFN1),* Prohibin *(PHB),* Promyelocytic leukemia protein *(PML), FUS1* and von Hippel-Lindau (*VHL*). These genes have been shown to act as tumor suppressors by inhibiting cell proliferation (*LZTS2, FUS1, PHB1)*(Cui et al. 2013; Kondo et al. 2001; Yang, Li, and He 2018), promoting apoptosis (*GRAMD4, FUS1, PHB1*)(John et al. 2011; Kondo et al. 2001; Yang, Li, and He 2018), decreasing cell motility (migration) and invasion (*HTATIP2, NINJ1, PHB1)*(Li et al. 2009; Jang et al. 2016; Yang, Li, and He 2018), coding for detoxification enzymes (*GSTP1)*(Beer et al. 2002), and regulating angiogenesis (*VHL*) (Gossage, Eisen, and Maher 2015). They are found to be downregulated in different cancers such as lung (*LZTS2, NINJ1, HTATIP2, PML*) (Cui et al. 2013; Jang et al. 2016; Li et al. 2020; Gurrieri et al. 2004), non-Hodgkin's lymphoma (*PML*) (Gurrieri et al. 2004), prostate (*GSTP1, PML, DUSP1*) (Beer et al. 2002; Gurrieri et al. 2004; Rauhala et al. 2005), pancreatic cancer (*PFN1*) (Yao et al. 2014), breast cancer (*PML*) (Gurrieri et al. 2004), thyroid cancer (*VHL*) (Todorovic et al. 2018; Stanojevic et al. 2014) and in many other cancers as well. Among the upregulated genes in this study (Anjitha et al. 2020) were: Twist Family BHLH Transcription Factor 1 (*TWIST1*), Polycomb complex protein (*BMI1)*, neurotrophic tyrosine kinase receptor type 2 (*NTRK2*), and Wilms' tumor gene (*WT1*). *TWIST1* and *BMI1* are upregulated in various cancers and have a positive correlation with poor prognosis (Ren et al. 2016). *NTRK2* upregulation leads to enhanced proliferation and invasion and has been implicated in the pathogenesis of many cancer types (Hu et al.

2016; Khotskaya et al. 2017). *WT1* was found to be overexpressed in various types of solid tumors and in leukemia (Tatsumi et al. 2008; Tuna, Chavez-Reyes, and Tari 2005). A number of other upregulated genes were genes known to be associated with MAPK-RAS-RAF signaling pathway (Anjitha et al. 2020). Due to its crucial role – transduction of extracellular signals that regulate cell growth, differentiation, proliferation, apoptosis and migration functions, this pathway is frequently deregulated in various cancers (Stanojevic et al. 2014; Santarpia, Lippman, and El-Naggar 2012).

In an *in vivo* study, it was shown that malathion and parathion were able to induce malignant transformation of the rat mammary gland. Increased proliferation and the density of terminal end buds in the mammary gland were followed by the formation of mammary carcinoma after 28 months of exposure (Cabello et al. 2001; Calaf and Echiburu-Chau 2012). Moreover, the combination of malathion and E2 synergistically increased the number of mammary gland lobules and upregulated markers for cancer detection, such as c-myc, c-fos, CYP, and mutant p53. In another study on the experimental induced mammary gland cancer model, it was shown that malathion alone or combined with E2 significantly induced the expression of markers of epithelial-mesenchymal transition – E-cadherin, vimentin, Axl and Slug proteins, compared to the control group (Calaf et al. 2020). Epithelial-mesenchymal transition (EMT) is a phenotypic change in the epithelial cells that, under specific conditions, allows cells to acquire mesenchymal phenotype characterized by increased cell motility. EMT has a key role in tumor progression.

In a study on rats with NMU-induced mammary tumors, chlorpyrifos treatment increased tumor incidence and reduced latency period (Ventura et al. 2019). Another study in rats has shown that malathion in combination with estrogen-induced pathological lesions in lung alveolar parenchyma, bronchiolar epithelia and lymphatic tissues, in comparison to control animals or in animals treated with either agent alone (Echiburu-Chau and Calaf 2008).

Epigenetic Effects of OPs

Epigenetics may contribute to the molecular mechanisms underpinning some long-term health effects of OPs (Collotta, Bertazzi, and Bollati 2013). Epigenetics refers to modifications of DNA that do not change the DNA sequence. Several epigenetic mechanisms, including DNA methylation and microRNA expression, can be triggered by environmental factors (Zhang, Lu, and Chang 2020).

DNA methylation is a process by which methyl groups are added to the CpG dinucleotides in the DNA molecule, changing its activity without changing its sequence. DNA methylation has a crucial role in the regulation of gene expression, and it regulates many cellular processes. Aberrant DNA methylation has been found in various types of cancer. In addition to the hypermethylation (elevated methylation level) of promoters of the tumor suppressor genes leading to the inhibition of their expression, a global genomic hypomethylation (reduced methylation level) has also been related to cancer (Morgan, Davies, and Mc Auley 2018).

Evidence of the effect of OP exposure on DNA methylation, particularly in the context of cancer, is scarce. In a study involving male pesticide applicators from the AHS, unusually high pesticide exposure events were associated with hypermethylation of glutathione S-transferase pi 1 *(GSTP1)* promoter and in some groups with hypomethylation of O(6)-methylguanine-DNA methyltransferase (*MGMT)* promoter and long interspersed nuclear element-1 (*LINE-1*). *GSTP1* plays an important role in detoxification and susceptibility to cancer and other diseases. MGMT is a DNA repair protein that is involved in cellular defense against mutagenesis and toxicity from alkylating agents. Altered methylation of the *MGMT* promoter has been associated with several cancer types. Since *LINE-1* elements are present throughout the genome, their methylation status is a marker for global DNA methylation (Rusiecki et al. 2017).

More studies provided evidence that gene promoter DNA methylation levels were altered upon exposure to pesticides (Hoang et al. 2021; van der Plaat et al. 2018). In a study involving Javanese farmers exposed to pesticides, frequent methylation of p16 tumor suppressor was found (Ratna et al. 2020). A study of epigenome-wide DNA methylation in AHS pesticide applicators found that individual pesticides, malathion was one of them, were associated with specific DNA methylation patterns, suggesting that there are pesticide-specific methylation signals. In the same study, a dose-response relationship with lifetime days of use was observed (Hoang et al. 2021).

Several *in vitro* studies provided evidence of altered DNA methylation upon OP exposure. A genome-wide DNA methylation study of DNA samples from the human hematopoietic K562 cell line exposed to several OPs such as fonofos, parathion and terbufos showed that differential methylation was associated with numerous genes that are involved in carcinogenesis-related processes, suggesting that epigenetic mechanisms may contribute to pesticide-induced carcinogenesis (Zhang, Wallace, Du, Kibbe, et al. 2012). Similar findings were reported for diazinon. Significant methylation changes were found in CpG sites of 984 genes, among whom were tumor suppressors *TP53* and *PTEN* and some genes, such as *HDAC3*, associated with PI_3K/mTORC2/Akt-dependent signaling pathway, another frequently deregulated pathway in cancer (Zhang, Wallace, Du, Lin, et al. 2012).

The effect of OPs on the epigenome might also be mediated by alterations in the microRNA expression profile leading to the changes in gene expression (Collotta, Bertazzi, and Bollati 2013). MicroRNAs are small, non-coding RNAs involved in the post-transcriptional regulation of gene expression. MiRNAs can modulate target gene expression negatively by binding to the 3'-untranslated region of mRNAs, leading to their degradation or translation inhibition (Hayes, Peruzzi, and Lawler 2014). A number of studies in the past have demonstrated the importance of microRNAs in cancer biology. Aberrant expression of different microRNAs has been associated with processes such as tumor growth, invasion, angiogenesis, metastasis and immune evasion. Depending on their target genes, microRNAs can function as oncogenes or tumor suppressors (Todorovic et al. 2018; Todorovic 2020).

The effect of OP exposure on the microRNA profile and its potential implication in carcinogenesis is largely unknown. One study analyzed the epigenetic effect of dichlorvos in a porcine kidney epithelial cell line PK15. Microarray analyses showed that both microRNA and mRNA expression profiles were altered as a consequence of dichlorvos treatment (Li et al. 2011).

Further research should reveal the effect of OPs on epigenetic mechanisms such as DNA methylation and microRNA expression and whether that plays a role in the OP-induced pathogenesis of various diseases, including cancer.

Conclusion

The widespread use of OPs has led to severe environmental pollution since residual amounts of OPs have been detected in the soil, groundwater, drinking water, house dust, vegetables, grains and other food, causing environmental and public health concerns. OP metabolites have been detected in the urine of the general population. Knowledge about chronic low-dose effects of specific OPs on human health is limited and represents an area of growing scientific interest. Large-scale epidemiologic studies and basic science research of modes of action and molecular mechanisms provided some evidence for the role of OPs in the pathogenesis of cancer. As the etiology of cancer is complex and multifactorial, it is of great importance to also consider the impact of these substances when present in combination with other exogenous and endogenous agents that are implicated in the etiology of cancer, such as estrogen. Further work incorporating environmental epidemiology, population genetics and cancer biology is needed in order to establish the causality of OP agents in relation to cancer and to help implement appropriate preventative measures.

Acknowledgments

Financial support for this work, under the Research Theme "Molecular alterations as prognostic and predictive markers in human malignant tumors" – 0802103, was provided by the Ministry of Education, Science and Technological Development of the Republic of Serbia, grant number 451-03-9/2021-14/200017.

References

Abdel-Halim, K. Y., and S. R. Osman. 2020. "Cytotoxicity and Oxidative Stress Responses of Imidacloprid and Glyphosate in Human Prostate Epithelial WPM-Y.1 Cell Line." *J Toxicol* 2020:4364650. doi: 10.1155/2020/4364650.

Akbel, E., D. Arslan-Acaroz, H. H. Demirel, I. Kucukkurt, and S. Ince. 2018. "The subchronic exposure to malathion, an organophosphate pesticide, causes lipid peroxidation, oxidative stress, and tissue damage in rats: the protective role of resveratrol." *Toxicol Res (Camb)* 7 (3):503-512. doi: 10.1039/c8tx00030a.

Alavanja, M. C., M. Dosemeci, C. Samanic, J. Lubin, C. F. Lynch, C. Knott, J. Barker, J. A. Hoppin, D. P. Sandler, J. Coble, K. Thomas, and A. Blair. 2004. "Pesticides and lung cancer risk in the agricultural health study cohort." *Am J Epidemiol* 160 (9):876-85. doi: 10.1093/aje/kwh290.

Alavanja, M. C., J. N. Hofmann, C. F. Lynch, C. J. Hines, K. H. Barry, J. Barker, D. W. Buckman, K. Thomas, D. P. Sandler, J. A. Hoppin, S. Koutros, G. Andreotti, J. H. Lubin, A. Blair, and L. E. Beane Freeman. 2014. "Non-hodgkin lymphoma risk and insecticide, fungicide and fumigant use in the agricultural health study." *PLoS One* 9 (10):e109332. doi: 10.1371/journal.pone.0109332.

Ali, N., K. Shahzad, M. I. Rashid, H. Shen, I. M. I. Ismail, and S. Eqani. 2017. "Currently used organophosphate and brominated flame retardants in the environment of China and other developing countries (2000-2016)." *Environ Sci Pollut Res Int* 24 (23):18721-18741. doi: 10.1007/s11356-017-9336-3.

Amer, S. M., M. A. Fahmy, F. A. Aly, and A. A. Farghaly. 2002. "Cytogenetic studies on the effect of feeding mice with stored wheat grains treated with malathion." *Mutat Res* 513 (1-2):1-10. doi: 10.1016/s1383-5718(01)00261-3.

Andreotti, G., L. E. Beane Freeman, J. J. Shearer, C. C. Lerro, S. Koutros, C. G. Parks, A. Blair, C. F. Lynch, J. H. Lubin, D. P. Sandler, and J. N. Hofmann. 2020. "Occupational Pesticide Use and Risk of Renal Cell Carcinoma in the Agricultural Health Study." *Environ Health Perspect* 128 (6):67011. doi: 10.1289/EHP6334.

Andreotti, G., S. Koutros, J. N. Hofmann, D. P. Sandler, J. H. Lubin, C. F. Lynch, C. C. Lerro, A. J. De Roos, C. G. Parks, M. C. Alavanja, D. T. Silverman, and L.E. Beane Freeman. 2018. "Glyphosate Use and Cancer Incidence in the Agricultural Health Study." *J Natl Cancer Inst* 110 (5):509-516. doi: 10.1093/jnci/djx233.

Anjitha, R., A. Antony, O. Shilpa, K. P. Anupama, S. Mallikarjunaiah, and H. P. Gurushankara. 2020. "Malathion induced cancer-linked gene expression in human lymphocytes." *Environ Res* 182:109131. doi: 10.1016/j.envres.2020.109131.

Argentin, G., M. Divizia, and R. Cicchetti. 2015. "Oxidative Stress, Cytotoxicity, and Genotoxicity Induced by Methyl Parathion in Human Gingival Fibroblasts: Protective Role of Epigallocatechin-3-Gallate." *J Toxicol Environ Health A* 78 (19):1227-40. doi: 10.1080/15287394.2015.1079582.

Arshad, M., M. Siddiqa, S. Rashid, I. Hashmi, M. A. Awan, and M. A. Ali. 2016. "Biomonitoring of Toxic Effects of Pesticides in Occupationally Exposed Individuals." *Saf Health Work* 7 (2):156-60. doi: 10.1016/j.shaw.2015.11.001.

Atherton, K. M., F. M. Williams, F. J. Egea Gonzalez, R. Glass, S. Rushton, P. G. Blain, and E. Mutch. 2009. "DNA damage in horticultural farmers: a pilot study showing an association with organophosphate pesticide exposure." *Biomarkers* 14 (7):443-51. doi: 10.3109/13547500903137265.

Band, P. R., Z. Abanto, J. Bert, B. Lang, R. Fang, R. P. Gallagher, and N. D. Le. 2011. "Prostate cancer risk and exposure to pesticides in British Columbia farmers." *Prostate* 71 (2):168-83. doi: 10.1002/pros.21232.

Barr, D. B., R. Allen, A. O. Olsson, R. Bravo, L. M. Caltabiano, A. Montesano, J. Nguyen, S. Udunka, D. Walden, R. D. Walker, G. Weerasekera, R. D. Whitehead, Jr., S. E. Schober, and L. L. Needham. 2005. "Concentrations of selective metabolites of organophosphorus pesticides in the United States population." *Environ Res* 99 (3):314-26. doi: 10.1016/j.envres.2005.03.012.

Barr, D. B., R. Bravo, G. Weerasekera, L. M. Caltabiano, R. D. Whitehead, Jr., A.O. Olsson, S. P. Caudill, S. E. Schober, J. L. Pirkle, E. J. Sampson, R. J. Jackson, and L. L. Needham. 2004. "Concentrations of dialkyl phosphate metabolites of organophosphorus pesticides in the U.S. population." *Environ Health Perspect* 112 (2):186-200. doi: 10.1289/ehp.6503.

Beane Freeman, L. E., M. R. Bonner, A. Blair, J. A. Hoppin, D. P. Sandler, J. H. Lubin, M. Dosemeci, C.F. Lynch, C. Knott, and M. C. R. Alavanja. 2005. "Cancer Incidence among Male Pesticide Applicators in the Agricultural Health Study Cohort Exposed to Diazinon." *Am J Epidemiol* 162 (11):1070-1079. doi: 10.1093/aje/kwi321.

Beer, T. M., A. J. Evans, K. M. Hough, B. A. Lowe, J. E. McWilliams, and W. D. Henner. 2002. "Polymorphisms of GSTP1 and related genes and prostate cancer risk." *Prostate Cancer Prostatic Dis* 5 (1):22-7. doi: 10.1038/sj.pcan.4500549.

Bernieri, T., M. F. Moraes, P. G. Ardenghi, and L. Basso da Silva. 2020. "Assessment of DNA damage and cholinesterase activity in soybean farmers in southern Brazil: High versus low pesticide exposure." *J Environ Sci Health B* 55 (4):355-360. doi: 10.1080/03601234.2019.1704608.

Bhanti, M., and A. Taneja. 2007. "Contamination of vegetables of different seasons with organophosphorous pesticides and related health risk assessment in northern India." *Chemosphere* 69 (1):63-8. doi: 10.1016/j.chemosphere.2007.04.071.

Bianchi, J., M. S. Mantovani, and M. A. Marin-Morales. 2015. "Analysis of the genotoxic potential of low concentrations of Malathion on the Allium cepa cells and rat hepatoma tissue culture." *J Environ Sci (China)* 36:102-11. doi: 10.1016/j.jes.2015.03.034.

Blasiak, J., P. Jaloszynski, A. Trzeciak, and K. Szyfter. 1999. "*In vitro* studies on the genotoxicity of the organophosphorus insecticide malathion and its two analogues." *Mutat Res* 445 (2):275-83. doi: 10.1016/s1383-5718(99)00132-1.

Bolognesi, C., G. Carrasquilla, S. Volpi, K. R. Solomon, and E. J. Marshall. 2009. "Biomonitoring of genotoxic risk in agricultural workers from five colombian regions: association to occupational exposure to glyphosate." *J Toxicol Environ Health A* 72 (15-16):986-97. doi: 10.1080/15287390902929741.

Bonner, M. R., J. Coble, A. Blair, L. E. Beane Freeman, J. A. Hoppin, D .P. Sandler, and M.C.R. Alavanja. 2007. "Malathion Exposure and the Incidence of Cancer in the Agricultural Health Study." *Am J Epidemiol* 166 (9):1023-1034. doi: 10.1093/aje/kwm182.

Bonner, M. R., B. A. Williams, J. A. Rusiecki, A. Blair, L. E. Beane Freeman, J.A. Hoppin, M. Dosemeci, J. Lubin, D. P. Sandler, and M. C. R. Alavanja. 2010. "Occupational exposure to terbufos and the incidence of cancer in the Agricultural Health Study." *Cancer Causes & Control* 21 (6):871-877. doi: 10.1007/s10552-010-9514-9.

Bozovic, A., V. Mandusic, L. Todorovic, and M. Krajnovic. 2021. "Estrogen Receptor Beta: The Promising Biomarker and Potential Target in Metastases." *Int J Mol Sci* 22 (4). doi: 10.3390/ijms22041656.

Cabello, G., A. Juarranz, L. M. Botella, and G. M. Calaf. 2003. "Organophosphorous pesticides in breast cancer progression." *J Submicrosc Cytol Pathol* 35 (1):1-9.

Cabello, G., M. Valenzuela, A. Vilaxa, V. Duran, I. Rudolph, N. Hrepic, and G. Calaf. 2001. "A rat mammary tumor model induced by the organophosphorous pesticides parathion and malathion, possibly through acetylcholinesterase inhibition." *Environ Health Perspect* 109 (5):471-9. doi: 10.1289/ehp.01109471.

Calaf, G. M., T. C. Bleak, J. P. Munoz, and F. Aguayo. 2020. "Markers of epithelial-mesenchymal transition in an experimental breast cancer model induced by organophosphorous pesticides and estrogen." *Oncol Lett* 20 (4):84. doi: 10.3892/ol.2020.11945.

Calaf, G. M., T. C. Bleak, and D. Roy. 2021. "Signs of carcinogenicity induced by parathion, malathion, and estrogen in human breast epithelial cells (Review)." *Oncol Rep* 45 (4). doi: 10.3892/or.2021.7975.

Calaf, G. M., and C. Echiburu-Chau. 2012. "Synergistic effect of malathion and estrogen on mammary gland carcinogenesis." *Oncol Rep* 28 (2):640-6. doi: 10.3892/or.2012.1817.

Calaf, G. M., C. Echiburu-Chau, and D. Roy. 2009. "Organophosphorous pesticides and estrogen induce transformation of breast cells affecting p53 and c-Ha-ras genes." *Int J Oncol* 35 (5):1061-8. doi: 10.3892/ijo_00000421.

Calaf, G. M., and D. Roy. 2007a. "Gene and protein expressions induced by 17beta-estradiol and parathion in cultured breast epithelial cells." *Mol Med* 13 (5-6):255-65. doi: 10.2119/2006-00087.Calaf.

Calaf, G. M., and D. Roy. 2007b. "Gene expression signature of parathion-transformed human breast epithelial cells." *Int J Mol Med* 19 (5):741-50.

Calaf, G. M., and D. Roy. 2008. "Cancer genes induced by malathion and parathion in the presence of estrogen in breast cells." *Int J Mol Med* 21 (2):261-8. doi: 10.3892/ijmm.21.2.261.

Cantor, K. P., A. Blair, G. Everett, R. Gibson, L. F. Burmeister, L. M. Brown, L. Schuman, and F. R. Dick. 1992. "Pesticides and other agricultural risk factors for non-Hodgkin's lymphoma among men in Iowa and Minnesota." *Cancer Res* 52 (9):2447-55.

Chen, M., C. H. Chang, L. Tao, and C. Lu. 2015. "Residential Exposure to Pesticide During Childhood and Childhood Cancers: A Meta-Analysis." *Pediatrics* 136 (4):719-29. doi: 10.1542/peds.2015-0006.

Collotta, M., P. A. Bertazzi, and V. Bollati. 2013. "Epigenetics and pesticides." *Toxicology* 307:35-41. doi: 10.1016/j.tox.2013.01.017.

Cordier, S., T. B. Le, P. Verger, D. Bard, C. D. Le, B. Larouze, M. C. Dazza, T. Q. Hoang, and L. Abenhaim. 1993. "Viral infections and chemical exposures as risk factors for hepatocellular carcinoma in Vietnam." *Int J Cancer* 55 (2):196-201. doi: 10.1002/ijc.2910550205.

Coste, A., H. D. Bailey, M. Kartal-Kaess, R. Renella, A. Berthet, and B. D. Spycher. 2020. "Parental occupational exposure to pesticides and risk of childhood cancer in Switzerland: a census-based cohort study." *BMC Cancer* 20 (1):819. doi: 10.1186/s12885-020-07319-w.

Cui, Q. Z., Z. P. Tang, X. P. Zhang, H. Y. Zhao, Q. Z. Dong, K. Xu, and E. H. Wang. 2013. "Leucine zipper tumor suppressor 2 inhibits cell proliferation and regulates Lef/Tcf-dependent transcription through Akt/GSK3beta signaling pathway in lung cancer." *J Histochem Cytochem* 61 (9):659-70. doi: 10.1369/0022155413495875.

De Ferrari, M., M. Artuso, S. Bonassi, S. Bonatti, Z. Cavalieri, D. Pescatore, E. Marchini, V. Pisano, and A. Abbondandolo. 1991. "Cytogenetic biomonitoring of an Italian population exposed to pesticides: chromosome aberration and sister-chromatid exchange analysis in peripheral blood lymphocytes." *Mutat Res* 260 (1):105-13. doi: 10.1016/0165-1218(91)90086-2.

De Roos, A. J., A. Blair, J. A. Rusiecki, J. A. Hoppin, M. Svec, M. Dosemeci, D. P. Sandler, and M. C. Alavanja. 2005. "Cancer incidence among glyphosate-exposed pesticide applicators in the Agricultural Health Study." *Environ Health Perspect* 113 (1):49-54. doi: 10.1289/ehp.7340.

De Roos, A. J., S. H. Zahm, K. P. Cantor, D. D. Weisenburger, F. F. Holmes, L. F. Burmeister, and A. Blair. 2003. "Integrative assessment of multiple pesticides as risk factors for non-Hodgkin's lymphoma among men." *Occup Environ Med* 60 (9):E11. doi: 10.1136/oem.60.9.e11.

Echiburu-Chau, C., and G. M. Calaf. 2008. "Rat lung cancer induced by malathion and estrogen." *Int J Oncol* 33 (3):603-11.

Edwards, F. L., C. G. Yedjou, and P. B. Tchounwou. 2013. "Involvement of oxidative stress in methyl parathion and parathion-induced toxicity and genotoxicity to human liver carcinoma (HepG(2)) cells." *Environ Toxicol* 28 (6):342-8. doi: 10.1002/tox.20725.

Engel, L. S., E. Werder, J. Satagopan, A. Blair, J. A. Hoppin, S. Koutros, C. C. Lerro, D. P. Sandler, M. C. Alavanja, and L. E. Beane Freeman. 2017. "Insecticide Use and Breast Cancer Risk among Farmers' Wives in the Agricultural Health Study." *Environ Health Perspect* 125 (9):097002. doi: 10.1289/EHP1295.

Eriksson, M., L. Hardell, M. Carlberg, and M. Akerman. 2008. "Pesticide exposure as risk factor for non-Hodgkin lymphoma including histopathological subgroup analysis." *Int J Cancer* 123 (7):1657-63. doi: 10.1002/ijc.23589.

Feng, J. X., and N. C. Riddle. 2020. "Epigenetics and genome stability." *Mamm Genome* 31 (5-6):181-195. doi: 10.1007/s00335-020-09836-2.

Flower, K. B., J. A. Hoppin, C. F. Lynch, A. Blair, C. Knott, D. L. Shore, and D.P. Sandler. 2004. "Cancer risk and parental pesticide application in children of Agricultural Health Study participants." *Environ Health Perspect* 112 (5):631-5. doi: 10.1289/ehp.6586.

Garaj-Vrhovac, V., and D. Zeljezic. 2001. "Cytogenetic monitoring of croatian population occupationally exposed to a complex mixture of pesticides." *Toxicology* 165 (2-3):153-62. doi: 10.1016/s0300-483x(01)00419-x.

Gill, J. P. S., J. S. Bedi, R. Singh, M. N. Fairoze, R. A. Hazarika, A. Gaurav, S. K. Satpathy, A. S. Chauhan, J. Lindahl, D. Grace, A. Kumar, and M. Kakkar. 2020. "Pesticide Residues in Peri-Urban Bovine Milk from India and Risk Assessment: A Multicenter Study." *Sci Rep* 10 (1):8054. doi: 10.1038/s41598-020-65030-z.

Giri, A., S. Giri, and G. D. Sharma. 2011. "Malathion and fenvalerate induce micronuclei in mouse bone marrow cells." *Environ Mol Mutagen* 52 (8):607-13. doi: 10.1002/em.20649.

Giri, S., S. B. Prasad, A. Giri, and G. D. Sharma. 2002. "Genotoxic effects of malathion: an organophosphorus insecticide, using three mammalian bioassays *in vivo*." *Mutat Res* 514 (1-2):223-31. doi: 10.1016/s1383-5718(01)00341-2.

Gossage, L., T. Eisen, and E. R. Maher. 2015. "VHL, the story of a tumour suppressor gene." *Nat Rev Cancer* 15 (1):55-64. doi: 10.1038/nrc3844.

Gurrieri, C., P. Capodieci, R. Bernardi, P. P. Scaglioni, K. Nafa, L. J. Rush, D. A. Verbel, C. Cordon-Cardo, and P. P. Pandolfi. 2004. "Loss of the tumor suppressor PML in human cancers of multiple histologic origins." *J Natl Cancer Inst* 96 (4):269-79. doi: 10.1093/jnci/djh043.

Hanahan, D., and R. A. Weinberg. 2000. "The hallmarks of cancer." *Cell* 100 (1):57-70. doi: 10.1016/s0092-8674(00)81683-9.

Hanahan, D., and R. A. Weinberg. 2011. "Hallmarks of cancer: the next generation." *Cell* 144 (5):646-74. doi: 10.1016/j.cell.2011.02.013.

Hayes, J., P. P. Peruzzi, and S. Lawler. 2014. "MicroRNAs in cancer: biomarkers, functions and therapy." *Trends Mol Med* 20 (8):460-9. doi: 10.1016/j.molmed.2014.06.005.

Hernandez-Toledano, D. S., E. Estrada-Muniz, and L. Vega. 2020. "Genotoxicity of the organophosphate pesticide malathion and its metabolite dimethylthiophosphate in human cells *in vitro*." *Mutat Res Genet Toxicol Environ Mutagen* 856-857:503233. doi: 10.1016/j.mrgentox.2020.503233.

Hoang, T. T., C. Qi, K. C. Paul, M. Lee, J. D. White, M. Richards, S. S. Auerbach, S. Long, S. Shrestha, T. Wang, L. E. Beane Freeman, J. N. Hofmann, C. Parks, C. J. Xu, B. Ritz, G. H. Koppelman, and S. J. London. 2021. "Epigenome-Wide DNA Methylation and Pesticide Use in the Agricultural Lung Health Study." *Environ Health Perspect* 129 (9):97008. doi: 10.1289/EHP8928.

Hu, J., Y. Huang, Y. Wu, F. Liu, D. Sun, K. Wang, and H. Qu. 2016. "NTRK2 is an oncogene and associated with microRNA-22 regulation in human gastric cancer cell lines." *Tumour Biol* 37 (11):15115-15123. doi: 10.1007/s13277-016-5337-y.

International Agency for Research on Cancer, W. H. O., Centre international de recherche sur le cancer (Lyon), & Kongress. 2017. "Some organophosphate insecticides and herbicides." In. Lyon, France: *International Agency for Research on Cancer*. https://publications.iarc.fr/Book-And-Report-Series/Iarc-Monographs-On-The-Identification-Of-Carcinogenic-Hazards-To-Humans/Some-Organophosphate-Insecticides-And-Herbicides-2017 (accessed 17.06.2021).

Jaga, K., and C. Dharmani. 2005. "The epidemiology of pesticide exposure and cancer: A review." *Rev Environ Health* 20 (1):15-38. doi: 10.1515/reveh.2005.20.1.15.

Jang, Y. S., J. H. Kang, J. K. Woo, H. M. Kim, J. I. Hwang, S. J. Lee, H. Y. Lee, and S. H. Oh. 2016. "Ninjurin1 suppresses metastatic property of lung cancer cells through inhibition of interleukin 6 signaling pathway." *Int J Cancer* 139 (2):383-95. doi: 10.1002/ijc.30021.

John, K., V. Alla, C. Meier, and B. M. Putzer. 2011. "GRAMD4 mimics p53 and mediates the apoptotic function of p73 at mitochondria." *Cell Death Differ* 18 (5):874-86. doi: 10.1038/cdd.2010.153.

Jones, R. R., F. Barone-Adesi, S. Koutros, C. C. Lerro, A. Blair, J. Lubin, S. L. Heltshe, J. A. Hoppin, M. C. Alavanja, and L. E. Beane Freeman. 2015. "Incidence of solid tumours among pesticide applicators exposed to the organophosphate insecticide diazinon in the Agricultural Health Study: an updated analysis." *Occup Environ Med* 72 (7):496-503. doi: 10.1136/oemed-2014-102728.

Kapeleka, J. A., E. Sauli, O. Sadik, and P. A. Ndakidemi. 2020. "Co-exposure risks of pesticides residues and bacterial contamination in fresh fruits and vegetables under smallholder horticultural production systems in Tanzania." *PLoS One* 15 (7):e0235345. doi: 10.1371/journal.pone.0235345.

Karimani, A. D., Y. D. Mamashkhani, A. P. Moghadam Jafari, M. D. Akbarabadi, and M. P. Heidarpour. 2018. "Captopril Attenuates Diazinon-Induced Oxidative Stress: A Subchronic Study in Rats." *Iran J Med Sci* 43 (5):514-522.

Karunanayake, C. P., J. J. Spinelli, J. R. McLaughlin, J. A. Dosman, P. Pahwa, and H. H. McDuffie. 2012. "Hodgkin lymphoma and pesticides exposure in men: a Canadian case-control study." *J Agromedicine* 17 (1):30-9. doi: 10.1080/1059924X.2012.632726.

Kavvalakis, M. P., and A. M. Tsatsakis. 2012. "The atlas of dialkylphosphates; assessment of cumulative human organophosphorus pesticides' exposure." *Forensic Sci Int* 218 (1-3):111-22. doi: 10.1016/j.forsciint.2011.10.019.

Khazaie, S., M. Jafari, J. Heydari, A. Asgari, K. Tahmasebi, M. Salehi, and M. S. Abedini. 2019. "Modulatory effects of vitamin C on biochemical and oxidative changes induced by acute exposure to diazinon in rat various tissues: Prophylactic and therapeutic roles." *J Anim Physiol Anim Nutr (Berl)* 103 (5):1619-1628. doi: 10.1111/jpn.13144.

Khotskaya, Y. B., V. R. Holla, A. F. Farago, K. R. Mills Shaw, F. Meric-Bernstam, and D.S. Hong. 2017. "Targeting TRK family proteins in cancer." *Pharmacol Ther* 173:58-66. doi: 10.1016/j.pharmthera.2017.02.006.

Kisby, G. E., J. F. Muniz, J. Scherer, M. R. Lasarev, M. Koshy, Y. W. Kow, and L. McCauley. 2009. "Oxidative stress and DNA damage in agricultural workers." *J Agromedicine* 14 (2):206-14. doi: 10.1080/10599240902824042.

Kondo, M., L. Ji, C. Kamibayashi, Y. Tomizawa, D. Randle, Y. Sekido, J. Yokota, V. Kashuba, E. Zabarovsky, I. Kuzmin, M. Lerman, J. Roth, and J. D. Minna. 2001. "Overexpression of candidate tumor suppressor gene FUS1 isolated from the 3p21.3 homozygous deletion region leads to G1 arrest and growth inhibition of lung cancer cells." *Oncogene* 20 (43):6258-62. doi: 10.1038/sj.onc.1204832.

Koutros, S., L. E. Beane Freeman, J. H. Lubin, S. L. Heltshe, G. Andreotti, K. H. Barry, C. T. DellaValle, J. A. Hoppin, D. P. Sandler, C. F. Lynch, A. Blair, and M. C. R. Alavanja. 2012. "Risk of Total and Aggressive Prostate Cancer and Pesticide Use in the Agricultural Health Study." *Am J Epidemiol* 177 (1):59-74. doi: 10.1093/aje/kws225.

Lasagna, M., M. S. Hielpos, C. Ventura, M. N. Mardirosian, G. Martin, N. Miret, A. Randi, M. Nunez, and C. Cocca. 2020. "Chlorpyrifos subthreshold exposure induces epithelial-mesenchymal transition in breast cancer cells." *Ecotoxicol Environ Saf* 205:111312. doi: 10.1016/j.ecoenv.2020.111312.

Lasagna, M., C. Ventura, M. S. Hielpos, M. N. Mardirosian, G. Martin, N. Miret, A. Randi, M. Nunez, and C. Cocca. 2021. "Endocrine disruptor chlorpyrifos promotes migration, invasion, and stemness phenotype in 3D cultures of breast cancer cells and induces a wide range of pathways involved in cancer progression." *Environ Res* 204 (Pt A):111989. doi: 10.1016/j.envres.2021.111989.

Latifovic, L., L. E. B. Freeman, J. J. Spinelli, M. Pahwa, L. Kachuri, A. Blair, K. P. Cantor, S. H. Zahm, D. D. Weisenburger, J. R. McLaughlin, J. A. Dosman, P. Pahwa, S. Koutros, P.A. Demers, and S. A. Harris. 2020. "Pesticide use and risk of Hodgkin lymphoma: results from the North American Pooled Project (NAPP)." *Cancer Causes Control* 31 (6):583-599. doi: 10.1007/s10552-020-01301-4.

Lee, W. J., A. Blair, J. A. Hoppin, J. H. Lubin, J. A. Rusiecki, D. P. Sandler, M. Dosemeci, and M. C. Alavanja. 2004. "Cancer incidence among pesticide applicators exposed to chlorpyrifos in the Agricultural Health Study." *J Natl Cancer Inst* 96 (23):1781-9. doi: 10.1093/jnci/djh324.

Lee, W. J., D. P. Sandler, A. Blair, C. Samanic, A. J. Cross, and M. C. Alavanja. 2007. "Pesticide use and colorectal cancer risk in the Agricultural Health Study." *Int J Cancer* 121 (2):339-46. doi: 10.1002/ijc.22635.

Leon, M. E., L. H. Schinasi, P. Lebailly, L. E. Beane Freeman, K. C. Nordby, G. Ferro, A. Monnereau, M. Brouwer, S. Tual, I. Baldi, K. Kjaerheim, J. N. Hofmann, P. Kristensen, S. Koutros, K. Straif, H. Kromhout, and J. Schuz. 2019. "Pesticide use and risk of non-Hodgkin lymphoid malignancies in agricultural cohorts from France,

Norway and the USA: a pooled analysis from the AGRICOH consortium." *Int J Epidemiol* 48 (5):1519-1535. doi: 10.1093/ije/dyz017.

Lerro, C. C., L. E. Beane Freeman, C. T. DellaValle, G. Andreotti, J. N. Hofmann, S. Koutros, C. G. Parks, S. Shrestha, M. C. R. Alavanja, A. Blair, J. H. Lubin, D. P. Sandler, and M. H. Ward. 2021. "Pesticide exposure and incident thyroid cancer among male pesticide applicators in agricultural health study." *Environ Int* 146:106187. doi: 10.1016/j.envint.2020.106187.

Lerro, C. C., S. Koutros, G. Andreotti, M. C. Friesen, M. C. Alavanja, A. Blair, J. A. Hoppin, D. P. Sandler, J. H. Lubin, X. Ma, Y. Zhang, and L. E. Beane Freeman. 2015. "Organophosphate insecticide use and cancer incidence among spouses of pesticide applicators in the Agricultural Health Study." *Occup Environ Med* 72 (10):736-44. doi: 10.1136/oemed-2014-102798.

Lewandowska, A. M., M. Rudzki, S. Rudzki, T. Lewandowski, and B. Laskowska. 2019. "Environmental risk factors for cancer - review paper." *Ann Agric Environ Med* 26 (1):1-7. doi: 10.26444/aaem/94299.

Li, J., J. He, Y. Li, Y. Liu, W. Li, N. Wu, L. Zhang, Y. Zhang, and Z. Niu. 2019. "Assessing the threats of organophosphate esters (flame retardants and plasticizers) to drinking water safety based on USEPA oral reference dose (RfD) and oral cancer slope factor (SFO)." *Water Res* 154:84-93. doi: 10.1016/j.watres.2019.01.035.

Li, M., J. Li, X. Guo, H. Pan, and Q. Zhou. 2020. "Absence of HTATIP2 Expression in A549 Lung Adenocarcinoma Cells Promotes Tumor Plasticity in Response to Hypoxic Stress." *Cancers (Basel)* 12 (6). doi: 10.3390/cancers12061538.

Li, S., X. Q. Ran, L. Xu, and J. F. Wang. 2011. "microRNA and mRNA expression profiling analysis of dichlorvos cytotoxicity in porcine kidney epithelial PK15 cells." *DNA Cell Biol* 30 (12):1073-83. doi: 10.1089/dna.2011.1267.

Li, X., Y. Zhang, S. Cao, X. Chen, Y. Lu, H. Jin, S. Sun, B. Chen, J. Liu, J. Ding, K. Wu, and D. Fan. 2009. "Reduction of TIP30 correlates with poor prognosis of gastric cancer patients and its restoration drastically inhibits tumor growth and metastasis." *Int J Cancer* 124 (3):713-21. doi: 10.1002/ijc.23967.

Mahajan, R., A. Blair, C. F. Lynch, P. Schroeder, J. A. Hoppin, D. P. Sandler, and M. C. Alavanja. 2006. "Fonofos exposure and cancer incidence in the agricultural health study." *Environ Health Perspect* 114 (12):1838-42. doi: 10.1289/ehp.9301.

McDuffie, H. H., P. Pahwa, J. R. McLaughlin, J. J. Spinelli, S. Fincham, J. A. Dosman, D. Robson, L. F. Skinnider, and N. W. Choi. 2001. "Non-Hodgkin's lymphoma and specific pesticide exposures in men: cross-Canada study of pesticides and health." *Cancer Epidemiol Biomarkers Prev* 10 (11):1155-63.

Meloni, F., G. Satta, M. Padoan, A. Montagna, I. Pilia, A. Argiolas, S. Piro, C. Magnani, A. Gambelunghe, G. Muzi, G. M. Ferri, L. Vimercati, R. Zanotti, A. Scarpa, M. Zucca, S. De Matteis, M. Campagna, L. Miligi, and P. Cocco. 2021. "Occupational exposure to glyphosate and risk of lymphoma:results of an Italian multicenter case-control study." *Environ Health* 20 (1):49. doi: 10.1186/s12940-021-00729-8.

Mills, P. K., and R. Yang. 2003. "Prostate cancer risk in California farm workers." *J Occup Environ Med* 45 (3):249-58. doi: 10.1097/01.jom.0000058339.05741.0c.

Moore, P. D., A. K. Patlolla, and P.B . Tchounwou. 2011. "Cytogenetic evaluation of malathion-induced toxicity in Sprague-Dawley rats." *Mutat Res* 725 (1-2):78-82. doi: 10.1016/j.mrgentox.2011.07.007.

Moore, P. D., C. G. Yedjou, and P. B. Tchounwou. 2010. "Malathion-induced oxidative stress, cytotoxicity, and genotoxicity in human liver carcinoma (HepG2) cells." *Environ Toxicol* 25 (3):221-6. doi: 10.1002/tox.20492.

Morgan, A. E., T. J. Davies, and M. T. Mc Auley. 2018. "The role of DNA methylation in ageing and cancer." *Proc Nutr Soc* 77 (4):412-422. doi: 10.1017/S0029665118000150.

Moyano, P., J. Garcia, J. M. Garcia, A. Pelayo, P. Munoz-Calero, M. T. Frejo, M. J. Anadon, M. Lobo, and J. Del Pino. 2020. "Chlorpyrifos-induced cell proliferation in human breast cancer cell lines differentially mediated by estrogen and aryl hydrocarbon receptors and KIAA1363 enzyme after 24 h and 14 days exposure." *Chemosphere* 251:126426. doi: 10.1016/j.chemosphere.2020.126426.

Muniz, J. F., L. McCauley, J. Scherer, M. Lasarev, M. Koshy, Y. W. Kow, V. Nazar-Stewart, and G. E. Kisby. 2008. "Biomarkers of oxidative stress and DNA damage in agricultural workers: a pilot study." *Toxicol Appl Pharmacol* 227 (1):97-107. doi: 10.1016/j.taap.2007.10.027.

Nagy, K., R. Argaw Tessema, I. Szasz, T. Smeirat, A. Al Rajo, and B. Adam. 2021. "Micronucleus Formation Induced by Glyphosate and Glyphosate-Based Herbicides in Human Peripheral White Blood Cells." *Front Public Health* 9:639143. doi: 10.3389/fpubh.2021.639143.

Ojha, A., and N. Srivastava. 2014. "*In vitro* studies on organophosphate pesticides induced oxidative DNA damage in rat lymphocytes." *Mutat Res Genet Toxicol Environ Mutagen* 761:10-7. doi: 10.1016/j.mrgentox.2014.01.007.

Ojha, A., S. K. Yaduvanshi, S. C. Pant, V. Lomash, and N. Srivastava. 2013. "Evaluation of DNA damage and cytotoxicity induced by three commonly used organophosphate pesticides individually and in mixture, in rat tissues." *Environ Toxicol* 28 (10):543-52. doi: 10.1002/tox.20748.

Olakkaran, S., A. Kizhakke Purayil, A. Antony, S. Mallikarjunaiah, and G. Hunasanahally Puttaswamygowda. 2020. "Oxidative stress-mediated genotoxicity of malathion in human lymphocytes." *Mutat Res Genet Toxicol Environ Mutagen* 849:503138. doi: 10.1016/j.mrgentox.2020.503138.

Pahwa, M., L. E. Beane Freeman, J. J. Spinelli, A. Blair, J. R. McLaughlin, S. H. Zahm, K.P. Cantor, D. D. Weisenburger, P. P. Punam Pahwa, J. A. Dosman, P. A. Demers, and S. A. Harris. 2019. "Glyphosate use and associations with non-Hodgkin lymphoma major histological sub-types: findings from the North American Pooled Project." *Scand J Work Environ Health* 45 (6):600-609. doi: 10.5271/sjweh.3830.

Pesatori, A. C., J. M. Sontag, J. H. Lubin, D. Consonni, and A. Blair. 1994. "Cohort mortality and nested case-control study of lung cancer among structural pest control workers in Florida (United States)." *Cancer Causes Control* 5 (4):310-8. doi: 10.1007/BF01804981.

Peter, J. V., and A. M. Cherian. 2000. "Organic insecticides." *Anaesth Intensive Care* 28 (1):11-21. doi: 10.1177/0310057X0002800102.

Phillips, D. H., and V. M. Arlt. 2009. "Genotoxicity: damage to DNA and its consequences." *EXS* 99:87-110. doi: 10.1007/978-3-7643-8336-7_4.

Pierre, A. A. E. M. L. 2021. "*Organophosphates.*" StatPearls Publishing, Last Modified 2021 Jul 29, accessed October 14. https://www.ncbi.nlm.nih.gov/books/NBK499860/.

Poet, T. S., A. A. Kousba, S. L. Dennison, and C. Timchalk. 2004. "Physiologically based pharmacokinetic/pharmacodynamic model for the organophosphorus pesticide diazinon." *Neurotoxicology* 25 (6):1013-30. doi: 10.1016/j.neuro.2004.03.002.

Ratna, M. G., D. A. A. Nugrahaningsih, E. N. Sholikhah, E. K. Dwianingsih, and R.G. Malueka. 2020. "The association between PON1 and GSTM1 genetic variation with methylation of p16 gene promoter among Javanese farmers exposed to pesticides at Magelang Regency, Central Java, Indonesia." *Heliyon* 6 (5):e03993. doi: 10.1016/ j.heliyon.2020.e03993.

Rauhala, H. E., K. P. Porkka, T. T. Tolonen, P. M. Martikainen, T. L. Tammela, and T. Visakorpi. 2005. "Dual-specificity phosphatase 1 and serum/glucocorticoid-regulated kinase are downregulated in prostate cancer." *Int J Cancer* 117 (5):738-45. doi: 10.1002/ijc.21270.

Ren, H., P. Du, Z. Ge, Y. Jin, D. Ding, X. Liu, and Q. Zou. 2016. "TWIST1 and BMI1 in Cancer Metastasis and Chemoresistance." *J Cancer* 7 (9):1074-80. doi: 10.7150/jca. 14031.

Rusiecki, J. A., L. E. Beane Freeman, M. R. Bonner, M. Alexander, L. Chen, G. Andreotti, K. H. Barry, L. E. Moore, H. M. Byun, F. Kamel, M. Alavanja, J. A. Hoppin, and A. Baccarelli. 2017. "High pesticide exposure events and DNA methylation among pesticide applicators in the agricultural health study." *Environ Mol Mutagen* 58 (1):19-29. doi: 10.1002/em.22067.

Salisbury, T. B., J. K. Tomblin, D. A. Primerano, G. Boskovic, J. Fan, I. Mehmi, J. Fletcher, N. Santanam, E. Hurn, G. Z. Morris, and J. Denvir. 2014. "Endogenous aryl hydrocarbon receptor promotes basal and inducible expression of tumor necrosis factor target genes in MCF-7 cancer cells." *Biochem Pharmacol* 91 (3):390-9. doi: 10.1016/j.bcp.2014.06.015.

Santarpia, L., S. M. Lippman, and A. K. El-Naggar. 2012. "Targeting the MAPK-RAS-RAF signaling pathway in cancer therapy." *Expert Opin Ther Targets* 16 (1):103-19. doi: 10.1517/14728222.2011.645805.

Santovito, A., S. Ruberto, C. Gendusa, and P. Cervella. 2018. "*In vitro* evaluation of genomic damage induced by glyphosate on human lymphocytes." *Environ Sci Pollut Res Int* 25 (34):34693-34700. doi: 10.1007/s11356-018-3417-9.

Sapbamrer, R., S. Hongsibsong, and S. Khacha-Ananda. 2020. "Urinary organophosphate metabolites and oxidative stress in children living in agricultural and urban communities." *Environ Sci Pollut Res Int* 27 (20):25715-25726. doi: 10.1007/s11356-020-09037-z.

Selmi, S., K. Rtibi, D. Grami, H. Sebai, and L. Marzouki. 2018. "Malathion, an organophosphate insecticide, provokes metabolic, histopathologic and molecular disorders in liver and kidney in prepubertal male mice." *Toxicol Rep* 5:189-195. doi: 10.1016/j.toxrep.2017.12.021.

Singh, S., V. Kumar, S. Thakur, B. D. Banerjee, R. S. Rautela, S. S. Grover, D. S. Rawat, S. T. Pasha, S. K. Jain, R. L. Ichhpujani, and A. Rai. 2011. "Paraoxonase-1 genetic polymorphisms and susceptibility to DNA damage in workers occupationally exposed to organophosphate pesticides." *Toxicol Appl Pharmacol* 252 (2):130-7. doi: 10.1016/j.taap.2011.01.014.

Stanojevic, B., V. Saenko, L. Todorovic, N. Petrovic, D. Nikolic, V. Zivaljevic, I. Paunovic, M. Nakashima, S. Yamashita, and R. Dzodic. 2014. "Low VHL mRNA expression is associated with more aggressive tumor features of papillary thyroid carcinoma." *PLoS One* 9 (12):e114511. doi: 10.1371/journal.pone.0114511.

Sun, H., M. L. Sun, and D. B. Barr. 2020. "Exposure to organophosphorus insecticides and increased risks of health and cancer in US women." *Environ Toxicol Pharmacol* 80:103474. doi: 10.1016/j.etap.2020.103474.

Sung, H., J. Ferlay, R. L. Siegel, M. Laversanne, I. Soerjomataram, A. Jemal, and F. Bray. 2021. "Global Cancer Statistics 2020: GLOBOCAN Estimates of Incidence and Mortality Worldwide for 36 Cancers in 185 Countries." *CA Cancer J Clin* 71 (3):209-249. doi: 10.3322/caac.21660.

Tadee, A., P. Mahakunakorn, and S. Porasuphatana. 2020. "Oxidative stress and genotoxicity of co-exposure to chlorpyrifos and aflatoxin B1 in HepG2 cells." *Toxicol Ind Health* 36 (5):336-345. doi: 10.1177/0748233720928169.

Tatsumi, N., Y. Oji, N. Tsuji, A. Tsuda, M. Higashio, S. Aoyagi, I. Fukuda, K. Ito, J. Nakamura, S. Takashima, Y. Kitamura, S. Miyai, T. Jomgeow, Z. Li, T. Shirakata, S. Nishida, A. Tsuboi, Y. Oka, and H. Sugiyama. 2008. "Wilms' tumor gene WT1-shRNA as a potent apoptosis-inducing agent for solid tumors." *Int J Oncol* 32 (3):701-11.

Tayour, C., B. Ritz, B. Langholz, P. K. Mills, A. Wu, J. P. Wilson, K. Shahabi, and M. Cockburn. 2019. "A case-control study of breast cancer risk and ambient exposure to pesticides." *Environ Epidemiol* 3 (5):e070. doi: 10.1097/EE9.0000000000000070.

Titenko-Holland, N., G. Windham, P. Kolachana, F. Reinisch, S. Parvatham, A. M. Osorio, and M. T. Smith. 1997. "Genotoxicity of malathion in human lymphocytes assessed using the micronucleus assay *in vitro* and *in vivo*: a study of malathion-exposed workers." *Mutat Res* 388 (1):85-95. doi: 10.1016/s1383-5718(96)00140-4.

Todorovic, L., Mandusic, V.,Vucetic-Tadic, V.,Zivaljevic, V., Paunovic, I., Stanojevic, B. 2020. "Altered expression of microRNA-30a-3p in papillary thyroid cancer and its association with clinicopathological characteristics." *Archives of Biological Sciences* 72 (1):31-36. doi: 10.2298/ABS191004063T.

Todorovic, L., B. Stanojevic, V. Mandusic, N. Petrovic, V. Zivaljevic, I. Paunovic, A. Diklic, V. Saenko, and S. Yamashita. 2018. "Expression of VHL tumor suppressor mRNA and miR-92a in papillary thyroid carcinoma and their correlation with clinical and pathological parameters." *Med Oncol* 35 (2):17. doi: 10.1007/s12032-017-1066-3.

Tuna, M., A. Chavez-Reyes, and A. M. Tari. 2005. "HER2/neu increases the expression of Wilms' Tumor 1 (WT1) protein to stimulate S-phase proliferation and inhibit apoptosis in breast cancer cells." *Oncogene* 24 (9):1648-52. doi: 10.1038/sj.onc.1208345.

van der Plaat, D. A., K. de Jong, M. de Vries, C. C. van Diemen, I. Nedeljkovic, N. Amin, H. Kromhout, R. Vermeulen, D. S. Postma, C. M. van Duijn, H. M. Boezen, and J. M. Vonk. 2018. "Occupational exposure to pesticides is associated with differential DNA methylation." *Occup Environ Med* 75 (6):427-435. doi: 10.1136/oemed-2017-104787.

Van Maele-Fabry, G., L. Gamet-Payrastre, and D. Lison. 2019. "Household exposure to pesticides and risk of leukemia in children and adolescents: Updated systematic review and meta-analysis." *Int J Hyg Environ Health* 222 (1):49-67. doi: 10.1016/j.ijheh.2018.08.004.

Ventura, C., M. Nunez, N. Miret, D. Martinel Lamas, A. Randi, A. Venturino, E. Rivera, and C. Cocca. 2012. "Differential mechanisms of action are involved in chlorpyrifos effects in estrogen-dependent or -independent breast cancer cells exposed to low or high concentrations of the pesticide." *Toxicol Lett* 213 (2):184-93. doi: 10.1016/j.toxlet.2012.06.017.

Ventura, C., A. Venturino, N. Miret, A. Randi, E. Rivera, M. Nunez, and C. Cocca. 2015. "Chlorpyrifos inhibits cell proliferation through ERK1/2 phosphorylation in breast cancer cell lines." *Chemosphere* 120:343-50. doi: 10.1016/j.chemosphere.2014.07.088.

Ventura, C., C. D. Zappia, M. Lasagna, W. Pavicic, S. Richard, A. D. Bolzan, F. Monczor, M. Nunez, and C. Cocca. 2019. "Effects of the pesticide chlorpyrifos on breast cancer disease. Implication of epigenetic mechanisms." *J Steroid Biochem Mol Biol* 186:96-104. doi: 10.1016/j.jsbmb.2018.09.021.

Vinson, F., M. Merhi, I. Baldi, H. Raynal, and L. Gamet-Payrastre. 2011. "Exposure to pesticides and risk of childhood cancer: a meta-analysis of recent epidemiological studies." *Occup Environ Med* 68 (9):694-702. doi: 10.1136/oemed-2011-100082.

Waddell, B. L., S. H. Zahm, D. Baris, D. D. Weisenburger, F. Holmes, L. F. Burmeister, K. P. Cantor, and A. Blair. 2001. "Agricultural use of organophosphate pesticides and the risk of non-Hodgkin's lymphoma among male farmers (United States)." *Cancer Causes Control* 12 (6):509-17. doi: 10.1023/a:1011293208949.

WHO. 2020. *World Cancer Report: Cancer Research for Cancer Prevention.* Lyon, France: International Agency for Research on Cancer.

Wogan, G. N., S. S. Hecht, J. S. Felton, A. H. Conney, and L. A. Loeb. 2004. "Environmental and chemical carcinogenesis." *Semin Cancer Biol* 14 (6):473-86. doi: 10.1016/j.semcancer.2004.06.010.

Wolf, D. C., S. M. Cohen, A. R. Boobis, V. L. Dellarco, P. A. Fenner-Crisp, A. Moretto, T. P. Pastoor, R. S. Schoeny, J. G. Seed, and J. E. Doe. 2019. "Chemical carcinogenicity revisited 1: A unified theory of carcinogenicity based on contemporary knowledge." *Regul Toxicol Pharmacol* 103:86-92. doi: 10.1016/j.yrtph.2019.01.021.

Yang, J., B. Li, and Q. Y. He. 2018. "Significance of prohibitin domain family in tumorigenesis and its implication in cancer diagnosis and treatment." *Cell Death Dis* 9 (6):580. doi: 10.1038/s41419-018-0661-3.

Yao, W., S. Ji, Y. Qin, J. Yang, J. Xu, B. Zhang, W. Xu, J. Liu, S. Shi, L. Liu, C. Liu, J. Long, Q. Ni, M. Li, and X. Yu. 2014. "Profilin-1 suppresses tumorigenicity in pancreatic cancer through regulation of the SIRT3-HIF1alpha axis." *Mol Cancer* 13:187. doi: 10.1186/1476-4598-13-187.

Yu, Y., S. Hu, Y. Yang, X. Zhao, J. Xue, J. Zhang, S. Gao, and A. Yang. 2017. "Successive monitoring surveys of selected banned and restricted pesticide residues in vegetables from the northwest region of China from 2011 to 2013." *BMC Public Health* 18 (1):91. doi: 10.1186/s12889-017-4632-x.

Zhang, L., Q. Lu, and C. Chang. 2020. "Epigenetics in Health and Disease." *Adv Exp Med Biol* 1253:3-55. doi: 10.1007/978-981-15-3449-2_1.

Zhang, L., I. Rana, R. M. Shaffer, E. Taioli, and L. Sheppard. 2019. "Exposure to glyphosate-based herbicides and risk for non-Hodgkin lymphoma: A meta-analysis and supporting evidence." *Mutat Res Rev Mutat Res* 781:186-206. doi: 10.1016/j.mrrev.2019.02.001.

Zhang, X., A. D. Wallace, P. Du, W. A. Kibbe, N. Jafari, H. Xie, S. Lin, A. Baccarelli, M. B. Soares, and L. Hou. 2012. "DNA methylation alterations in response to pesticide exposure *in vitro*." *Environ Mol Mutagen* 53 (7):542-9. doi: 10.1002/em.21718.

Zhang, X., A. D. Wallace, P. Du, S. Lin, A. A. Baccarelli, H. Jiang, N. Jafari, Y. Zheng, H. Xie, M. B. Soares, W. A. Kibbe, and L. Hou. 2012. "Genome-wide study of DNA methylation alterations in response to diazinon exposure *in vitro*." *Environ Toxicol Pharmacol* 34 (3):959-68. doi: 10.1016/j.etap.2012.07.012.

Chapter 6

Organophosphates: Fate in the Environment, Biomonitoring, Analysis, and Legislation

Lejla Klepo*

Department of Chemistry, Faculty of Science, University of Sarajevo, Sarajevo, Bosnia and Herzegovina

Abstract

The presence of organophosphorus (OP) compounds in soil, atmosphere, groundwater, surface water, and river sediments originates from using those compounds in pest management on vegetation and animals, as well as in disposal of cattle and sheep dips on farmland. The environmental persistence of OP pesticides is thought to be stricken by transporting methods combined with chemical and organic degradation methods. The introduction of OP residues in remote environments demonstrates that those pesticides may be transported long distances via air and water. Today, it is known that the fate of the OP in the environment is affected by transport (sorption, volatilization, leaching, runoff), chemical (solubility, oxidation/reduction, hydrolysis, photolysis), and biological processes.

Biological tracking is an important factor in any complete evaluation of exposure to OP. Historically, biological impact tracking (BIM) through the extent of the reduction of blood cholinesterase activity became the method of choice. It involved the measurement of plasma cholinesterase (ChE) and erythrocyte acetylcholinesterase (AChE), reflecting the influence of OPs' intake on inhibition of mentioned blood enzymes as surrogates of an AChE in neural tissue and neuromuscular junctions. An alternative or complementary approach to the biological control of OPs is based on the analysis of metabolites in the urine. These

* Corresponding Author's Emails: lklepo@pmf.unsa.ba; klepolejla@gmail.com.

In: Organophosphates
Editor: Tamara Lazarević-Pašti
ISBN: 978-1-68507-652-8

methods detect specific metabolites for the particular OP or non-specific metabolites common for various OPs – the dialkyl phosphates (DAPs). Once ingested, OPs no longer accumulate substantially in the human body, and over 75% is rapidly metabolized to DAPs. After, they are excreted through the urine inside 6 to 24 h after exposure.

Depending on the type of sample, different methods of analysis are used. Most analytical methods for the analysis of OP pesticides are useful for determination in the plant, water, soil, or food. However, they are not directly applicable to biological samples taken in an emergency or during an autopsy. A solid-phase microextraction (SPME) method has been used as a variable solvent-free alternative to conventional liquid-liquid extraction. The Quick, Easy, Cheap, Effective, Rugged, and Safe method coined the "QuEChERS" today is also the best known and most widespread method for the analysis of pesticide residues.

Legislation related to pesticides can be divided into two parts:

1. part relating to registration (permission to place into circulation and use);
2. the part related to the regulation of the maximum allowed amount of pesticide residues in food.

No plant protection product can be registered in the EU until it is scientifically determined that:

- has no adverse effects on consumers, farmers, or passive observers;
- does not cause adverse effects on the environment;
- has sufficient effectiveness against targeted pests.

Keywords: organophosphates, legislation, QuEChERS

Introduction

The esterification of alcohols with phosphoric acid has started organophosphorus (OP) chemistry around 1820. Although several OP compounds were synthesized in the early 1900s, the potential for their toxicity was not recognized until the 1930s. B. C. Saunders in England and Gerhard Schrader in Germany had developed many highly poisonous chemicals for use as chemical warfare agents by 1940. Sarin and soman, both phosphorofluoridates, and tabun, a phosphorocyanidate, were the most prominent. Tepp, dimefox, schradan (OMPA), and parathion, among the first commercial OP insecticides, were developed by Schrader's company in 1937,

1940, and 1944. Following WWII, when Schrader's research documents were taken, interest in OP pesticides increased. Even though all of the early compounds were excellent insecticides, they were all extremely harmful to animals. Malathion, one of the safest OPs ever manufactured, was introduced by American Cyanamid in 1950. More than 50,000 OP compounds were predicted to have been created by 1959. Over 200 OP insecticides were on the market in 1970. Though pest resistance and the introduction of new, safer insecticides have reduced the use of OPs and hampered the development of new products, they remain an essential class of pest control agents that will likely continue to do so for another decade or more (Chambers Howard, Meek, and Chambers Janice, 2010).

Today, OP compounds are one of the most widely used chemical classes in crop and livestock protection, accounting for roughly 34% of global insecticide sales. Approximately 150 distinct OP compounds have been employed to protect crops, livestock, and human health during the last 60 years. In the United States, organophosphate pesticides are used on a wide range of crops, with corn and cotton accounting for the majority of total organophosphate insecticide usage. According to the Environmental Protection Agency (EPA), these two crops received nearly half of organophosphate insecticides. To protect high-value crops from pest damage, pesticide use on crops such as fruits, vegetables, and specialty crops may be quite extensive. Consumers who eat fresh fruits and vegetables may be exposed to more pesticides due to these crops. In recent years, OP compounds have become Australia's most extensively utilized pesticide class. The most often used OP pesticides include parathion methyl, chlorpyrifos, dimethoate, and profenophos, in addition to fenamiphos (nematicide), isofenphos, and coumaphos (insecticides)(Casida and Quistad 2004; Singh and Walter 2006, Cáceres et al. 2009).

Organophosphate pesticides are used to protect crops from pests; however, the majority of their unused portion, as well as their by-products, is wasted, and remains are contaminating the soil, resulting in loss of fertility, soil acidification, nitrate leaching, weed species resistance, and biodiversity loss (Mulla et al. 2020). Because several members of the OP pesticide group are highly harmful to aquatic invertebrates, the environmental fate of these chemicals is a major concern. Many studies have been published on the impacts of OP pesticides like fenamiphos and isofenphos on water and food pollution, as well as the effects on animal and human health. Furthermore, the use of these substances causes the poisoning of around 200,000 people each year worldwide. The residential usage of organophosphates such as chlorpyrifos and diazinon has been reduced in the United States to avoid exposure situations from highly toxic compounds (Kalvaci et al., 2009). Over the last ten years, Guiatart et al. (2009) investigated the poisoning of European wildlife by several xenobiotics, including organophosphorus pesticides. In France, cholinesterase inhibitors were to blame for 42.5 percent of all wild bird deaths. Organophosphates such as methamidophos, parathion, methyl phorate, and fenthion have been linked to 50 percent of all wild bird deaths in Greece. Other organophosphates were found in the gastrointestinal contents of three raptors including malathion and fenthion, which are currently employed on olive trees and vineyards. Furthermore, organophosphates have been linked to the poisoning of wild mammals in northern Greece (Cáceres et al., 2009).

Fate in the Environment

The agricultural business has become more reliant on organophosphate and carbamate pesticides, which are considered harmless due to their quick degradation in the environment. Both of these pesticide classes are anticholinesterases, and because of the scale of their use, they are the compounds most frequently linked to toxicity in domestic and wildlife animals. Pesticide residues in food cost about 2 million pounds per year in the United Kingdom alone, and the water industry spends over 880 million pounds per year treating pesticide-polluted water (Ragnarsdottir, 2000). Organophosphates are distributed in a variety of ways throughout the environment (Edwards and Tchounwou, 2005). The use of OP compounds as pest control agents on crops and animals, as well as the disposal of cow and sheep dips on farms, has resulted in the accumulation of OP compounds in soil, groundwater, surface water (river, lakes, oceans), and the atmosphere (Cáceres et al. 2009). Organophosphates can be detected in air, snow, fog, rainwater and were also found in pine needles in the Sierra Nevada mountains (Aston and Seiber, 1996), which are miles away from the agricultural spraying area of California. These findings show that OPs are carried over long distances by water and the atmosphere. Therefore, estimating the fate and transport processes of OP is essential for understanding their environmental behavior (Ragnarsdottir, 2000). Although transport processes, as well as chemical and biological degradation processes, are known to affect the environmental persistence of OP insecticides, the presence of OP residues in remote environments shows that these pesticides can be transported via air and water on long distances. Such discoveries indicate the role of learning more about their fate and transportation in the environment (Cáceres et al., 2009).

Transport Processes

Sorption

The transfer of a solute between a fluid and a solid phase is referred to as sorption. The behavior of both the solute and the solvent must be considered in this procedure. Sorption is determined by three groups of factors that frequently interact in a complicated manner: the solute's molecular properties, soil composition, and the experimental conditions under which the sorption is investigated.

Pesticides' behavior and their fate in the environment are influenced by their sorption behavior (Caceres et al., 2002; Wu et al., 2006). When a pesticide molecule comes in contact with soil constituents, a pseudo-equilibrium is generally achieved within hours. It is known that OPinsecticides bind to mineral surfaces or organic materials in soils. Mineral surfaces with a pH-dependent surface charge are responsible for the adsorption of solutes, such as OP compounds, to soil particles. In aqueous solutions, a mineral surface with a surface charge, which is pH-dependent, causes solute adsorption on soil particles. Mineral surfaces are generally positively charged in acidic waters and negatively charged in alkaline waters. As sorption occurs in an aqueous media, the interactions between water and solutes are crucial in this process. OPs that are less water-soluble and thus more hydrophobic are usually stronger sorbed to soil (Cáceres et al., 2009; Kaushal, Khatri, and Arya, 2021). Other processes that are significant for dissolved ions include adsorption (diffusion into mineral surfaces), precipitation or co-precipitation of mineral phases, as well as ion exchange. Polar organic molecules are exchanged for interlayer cations in the latter method, the principal sorption process for clay minerals (Collins, Ragnarsdóttir and Sherman, 1999; Ragnarsdottir, 2000).

OPs are thus adsorbed to mineral and organic matter surfaces. However, it is unclear whether this adsorption makes the OPs more susceptible to biodegradation. While soil surfaces frequently determine the time required for bioavailability, it is uncertain. Adsorption, on the other hand, lowers OPs' insecticidal activity (Ragnarsdottir, 2000).

The most important soil mineral components that affect adsorption are clays and iron oxides. It is due to their abundance as well as their surface properties. Water content, acidity/basicity, and the ionic composition of the surfaces all influence the adsorption on clays. Compensating cations can act in three ways:

- They can compete with positively charged organic molecules for adsorption sites (for example, the herbicides paraquat and atrazine),
- they can act directly as adsorption sites by forming coordination bonds, or
- they can form hydroxides on the clay surface (for example, Al^{3+} and Fe^{3+}). As a result, mineral adsorption capability is greatly increased (Caceres et al., 2009).

Aside from the type of the adsorbent and pesticide, adsorption rates are influenced by several factors. Temperature, ionic strength of the solution, and soil–water ratios are among these variables. Direct and indirect measurements are used to determine the rate of sorption; both require equilibrating a soil sample with an aqueous pesticide solution of a known initial concentration. The direct approach involves measuring both equilibria and sorbed concentrations. Once equilibration is achieved, the sorbed pesticide is separated from the soil using an appropriate solvent and directly quantified. The indirect technique measures the amount of sorbed pesticide by measuring the change in solution concentration caused by pesticide sorption in soil. In this case, the amount of sorbed solute is presumed to be equal to the amount lost from the solution (Caceres et al. 2009).

The sorption isotherms can be used to characterize sorption. For the design of any adsorption system, adsorption isotherms are the basic requirement. Isotherms provide the relation between the amount of adsorbate removed from the liquid phase at constant temperature *per* unit mass of adsorbent (Ayuba et al. 2020). Two most used isotherms are Langmuir and Feundlich. The Freundlich equation establishes a nonlinear connection between the sorbed amount of sorbent and its equilibrium solution concentration and represents the most frequently used expression of a sorption isotherm (Wauchope et al., 2002; Van Scoy, Pennell, and Zhang, 2016).

Volatilization

The volatilization process aids the dispersion of certain OP compounds from the surface of damp soil, leaves, or water. The intensity of this process is determined by several factors, including the compound's vapor pressure, solubility, adsorption affinity, persistence, as well as ambient conditions, like temperature, moisture level, and air movements (Cáceres et al., 2009). The adsorption of volatile organic pesticides is a predominant transport mechanism in the vadose zone, influenced by soil moisture content (Ragnarsdottir, 2000). With a lower volatilization rate and larger soil surface area, organic particles become more accessible for adsorption. Organic chemical volatilization from water to the atmosphere is a first-order rate process that occurs near or above the surface (Ragnarsdottir, 2000). It is particularly relevant for organics with high vapor pressure but low water solubilities. Since they tend to sorb to soil particulate matter, they are not readily volatilized into the atmosphere (Weber 1972; Ragnarsdottir, 2000).

Leaching

The migration of pesticide residues into the soil profile and potentially into groundwater via percolating water is called leaching. Depending on the toxicity and quantity of the chemicals, biota exposure to pesticides in groundwater may be an issue. Chemical characteristics, sorption behavior, persistence, and environmental variables like rainfall and soil porosity influence the leaching potential. According to research conducted by the US Department of Agriculture's Soil Conservation Service on pesticide mobility, most pesticides examined have a minor to medium leaching potential (Cáceres et al., 2009).

Runoff

Pesticide dissipation occurs primarily through the movement of pesticides over the surface of treated fields by moving water or sediment. Pesticides that migrate to waterways from application locations may impact aquatic non-target creatures. Therefore runoff is a concern (Kaushal, Khatri, and Arya, 2021). Pesticides with a solubility of more than 10 mg/L will flow mostly through the solution phase, while pesticides with a lower solubility will move primarily by sorption to eroding soil particles. The degree of solubility, as well as environmental conditions like rainfall and soil hydraulic conductivity, drive the movement of somewhat soluble chemicals. The movement of more intensely sorbed compounds is mostly a result of erosion and is highly influenced by erosion management measures. In research conducted in Iowa, United States, the amounts of chlorpyrifos carried by runoff to a pond after three continuous treatments were barely 0.03%.

Similarly, runoff from corn fields included about 0.1% of a granular diazinon application. Erosion did not occur in turf (an irrigated environment) because there were no soil particles that heavily sorbedOPsaschlorpyrifos could attach for transportation. Other extremely water-soluble OP compounds, such as fenamiphos, can move to surface water bodies because of limited soil matrix binding. (Cáceres et al. 2002).

Chemical Processes

Various chemical processes impact the concentration of dissolved ingredients and influence the persistence of OP pesticides in the environment, including hydrolysis, solubility, oxidation/reduction, and photolysis (Cáceres et al., 2009; Dyguda-Kazimierowicz, Roszak, and Sokalski, 2014).

Hydrolysis

Hydrolysis, which may result in detoxification, is a key process in reducing organophosphate pesticide persistence in the environment for OPs dissolved in water. The most frequent reaction is cleavage at the phosphate ester bond due to base catalysis (Racke, 1992). OP hydrolysis is almost always a first-order reaction. As the pH-dependent breakdown process is also seen in OPs, the hydrolysis reaction for OPs is reliant on the concentration of OH^- in second-order rate mechanism (Cáceres et al., 2009; Ragnarsdottir 2000).

The hydrolysis of OP compounds is influenced by both pH and temperature. Hydrolysis of OP structures occurs most commonly above pH 7 and 25°C, but it can also occur at pH 5–9 in groundwaters. In general, for many OP pesticides, the higher the pH, the faster they hydrolyze. For each pH unit, the OP hydrolysis rate increases by around a factor of ten (Schwarzenbach, Gschwend, and Imboden, 1993). If OP pesticides are introduced into groundwater systems, the change in hydrolysis rates must be assessed and considered when calculating the pesticide's half-life (Cáceres et al., 2009; Kumar et al., 2018).

Non-Hydrolytic Reactions

Non-hydrolytic reactions such as biotransformation or chemical reactions can alter an OP compound's anticholinesterase activity, allowing it to transform non-toxic OP compounds into toxic ones. Alternatively, one toxic compound can be converted to another (activation reactions). However, toxic OP compounds are also converted into compounds that are no longer acetylcholinesterase inhibitors (detoxification reactions) (Bosak, 2006).

Solubility

Although OPs are most commonly found in surface soils, they have also been found in groundwater and flooding. Low water solubility, a high water-oil partition coefficient, and low vapor pressure are common characteristics of OPPs (Riches, 2015). Their solubility is tenths of thousands of parts *per* million (ppm), which is eight orders of magnitude higher than the drinking water limit. OPs can also be detected in rivers as a result of agriculture runoff or unintentional releases. Because the solubility of many OPs has not been properly studied, solubility values reported in the literature vary by a factor of up to 30 for some OPs. Rather, the solubility is calculated using either the chemical structure or the octanol-water partition coefficient (Ragnarsdottir, 2000).

Oxidation/Reduction

Singlet oxygen, alkyl peroxyl radicals (monovalent aliphatic hydrocarbon radicals with O_2 groups), hydrogen peroxide (H_2O_2), metal salts, and azo compounds can all cause organic molecules to oxidize. OP pesticides can be oxidized or reduced, generating metabolites. Oxidation and reduction reactions can occur as a direct result of microbial activity, sunlight, contact with soil components, or as an indirect result of microbial activity. In natural and atmospheric waters, free radical oxidation is part of the photolytic cycle. Some OP pesticides are used on crops because they are systemic, yet their oxidized products are powerful acetyl-cholinesterase inhibitors. Both inorganic and biological processes contribute to the reduction in anaerobic conditions. In the presence of bacteria, this is just a significant OP fate process. Parathion, for example, can be reduced (to amino parathion) or oxidized (to paraoxon). The environmental impact of OPpesticide oxidation is that the metabolites formed, such as oxons from phosphorothioates, are more harmful than the parent compounds but are less stable (Chambers and Levi 1992; Ragnarsdottir, 2000). Organophosphates are transformed into oxons in the body by replacing the P=S bond with a P=O bond. Upon this conversion, the molecule binds permanently to the active site of the enzyme acetylcholinesterase and inhibits it (Miodovnik, 2019).

Photolysis

The photochemical reaction can take place either directly or indirectly. When a chemical absorbs light and subsequently conducts a transformation reaction from an excited state, it is called direct photolysis. Rearrangement, dissociation, and oxidation are examples of such changes. Indirect photolysis is a second-order rate process during which naturally occurring chemicals in the aquatic environment absorb sunlight to create excited chemical species or radicals, which subsequently react with the pollutant chemical (Cáceres et al., 2009; Changlong and Linden, 2010; Shrikant and Khambete, 2015). The photolysis of OP compounds in surface waters and the atmosphere is a crucial mechanism for their dissipation. Photolysis occurs when chemicals are bombarded by solar light having wavelengths >290 nm(Cáceres et al., 2009; Ragnarsdottir 2000).

Biological Processes

Enzyme-catalyzed OP transformations and chemical build-up in the food chain are examples of biological processes. Pesticide breakdown in the subsurface vadose zone is typically believed to be largely microbiological. It stands for OP insecticides as well. It's worth mentioning that the majority of insecticidal residue microbial metabolism is considered accidental. It means that these compounds are metabolized by non-specific enzyme systems found in soils unless they are extremely harmful to microorganisms (Ragnarsdottir, 2000).

Biodegradation

Microorganisms are preferred for degrading or removing organophosphate pollutants in laboratory conditions because their degradation rate is faster than chemical hydrolysis, which is about ten times faster than photolysis (Dar, Kaushik, and Villareal, 2020). Under certain environmental circumstances, the rate of biodegradation is dependent on the microbial biomass amount and the concentration of the OP. Chemical substrates can be used as an energy source by microorganisms. Organophosphates are primarily degraded by microorganisms that use them as carbon, nitrogen, or phosphorus sources. Biodegradation rates are a function of cell growth rate in such circumstances. Cometabolism and catabolism are the two types of microbial pesticide breakdown. Incomplete microbial breakdown of the pesticide and consequent temporary accumulation of some of its metabolites are common outcomes of cometabolism. Because the microorganisms that begin the cometabolic degradation process get no benefit from their actions, it is sometimes referred to as accidental degradation. A parent pesticide or main metabolite is degraded in catabolic processes, benefiting the microbe, using the chemical as a carbon/energy source or food. Microorganisms' capacity to digest pesticides may potentially result in an adaptive response. Repeated exposure of soil or aquatic microorganisms to certain OP pesticides has been shown to result in the growth of pesticide-degrading microorganisms, primarily bacteria, with increased biodegradation capability. Some OP compounds are sensitive to these phenomena, including diazinon, coumaphos, fensulfothion, and chlorfenvinphos.

Bioaccumulation

Because OPs are often hydrophobic compounds that partition into fat and lipid tissues, bioaccumulation of OPs in living organisms is critical. Inorganic chemicals are also partitioned into bones (which are made up of the mineral hydroxyapatite) and bone marrow in the process of bioaccumulation. The bioconcentration factor (BCF), which is equivalent to the concentration of a chemical in tissue, is commonly used to assess bioaccumulation.

The fact that a chemical's concentration is generally higher in fatty tissues than in leaner tissue complicates bioconcentration statistics. The rate of uptake and the time it takes to achieve equilibration in various organs (and species) is also affected by the uptake pathway (e.g., dietary, skin absorption, etc.). BCF is related to the hydrophobicity of the organic chemical; the higher the hydrophobicity, the higher the BCF (Ragnarsdottir, 2000).

When used as an insecticide, methyl parathion degrades quickly in sediment and water, mostly by photolysis and biodegradation by microorganisms. Rain, fog, and wind can all carry methyl parathion to nearby regions. Because of its rapid hydrolysis and its physical and chemical characteristics, methyl parathion does not pass through the food chain. Methyl parathion has an environmental half-life of days or weeks when appropriately administered in agricultural settings exposed to water and UV radiation; however, when applied inside, the half-life rises significantly (Edwards and Tchounwou, 2005). According to studies, environmental variables, rather than biological factors, influenced the breakdown of methyl parathion (Athanasopoulos, Kyriakidis, and Stavropoulos 2004).

Chlorpyrifos' environmental destiny has been extensively researched. Chemical hydrolysis and microbiological activity are both involved in soil degradation. Chlorpyrifos has a half-life of 10 to 120 days (Table 1) in soil, with 3,5,6-trichloro-2-pyridinol (TCP) being the main breakdown product. Different environmental variables, the most important of which are soil pH, temperature, moisture content, organic carbon content, and pesticide formulation, have been blamed for the wide variance in half-life. Chemical hydrolysis was at first considered responsible for the rapid rate of chlorpyrifos breakdown in alkaline pH soils. In recent studies, it was discovered that chlorpyrifos breakdown was relatively slow in acidic soils, but the rate of degradation accelerated dramatically as the soil pH increased.

Table 1. History, toxicity, and half-life of some organophosphorus pesticides

Name	Type	Year of introduction	Mammalian LD50 (mg kg^{-1})	Half-life Soil (days)
Chlorpyrifos	Insecticide	1965	135–163	10–120
Coumaphos	Acaricide	1952	16–41	24–1400
Diazinon	Insecticide	1953	80–300	11–21
Dicrotophos	Insecticide	1965	15–22	45–60
Dimethoate	Insecticide	1955	160–387	2–41
Ethoprophos	Nematicide	1966	146–170	3–30
Fenamiphos	Nematicide	1967	6–10	28–90
Fenitrothion	Insecticide	1959	1700	12–28
Glyphosate	Herbicide	1971	3530–5600	30–174
Methyl parathion	Insecticide	1949	3–30	25–130
Monocrotophos	Insecticide	1965	18–20	40–60
Parathion	Insecticide	1947	2–10	30–180

In biologically active soil, parathion degrades quickly. An increase in the concentration of parathion added to soils resulted in a corresponding rise in the bacterial population. In the rhizosphere of rice seedlings, flood soil conditions favored parathion hydrolysis and the release of $^{14}CO_2$ from ring-labeled parathion.

Several kinds of bacteria that can hydrolyze parathion have been identified from parathion enrichment or other organophosphate enriched environments. Also, the most significant transformation mechanisms regulating glyphosate persistence in the soil is microbial degradation. The activity and biomass of soil microorganisms were shown to be linked to glyphosate mineralization. The main metabolite aminomethylphosphonic acid is produced by the microbial breakdown of glyphosate, leading to CO_2, phosphate, and water formation. Although the complete breakdown mechanisms for parathion and glyphosate are known, the pathways for numerous additional organophosphorus chemicals are still unknown. Degradation products of numerous chemicals are pollutants that can harm the environment and non-target species. Therefore this field of research requires coordinated efforts.

Several bacterium species (Table 2) that can digest these chemicals in liquid medium and soils have been identified and described in the literature (Singh and Walker, 2006).

Table 2. Microorganisms isolated for the degradation of organophosphorus compounds (Karpouzas, and Singh, 2006; Singh and Walker, 2006)

Compound	**Microorganisms**		**Mode of degradation**
Chlorpyrifos	Bacteria	*Enterobacter* sp.	Catabolic (C, P)
		Flavobacterium sp. ATCC27551	Co-metabolic
		Pseudomonas diminuta	Co-metabolic
		Micrococcus sp.	Co-metabolic
	Fungi	*Phanerochaete chrysosporium*	Catabolic (C)
		Hypholama fascicularae	ND.
		Coriolus versicolor	ND.
		Aspergillus sp.	Catabolic (P)
		Trichoderma harzianum	Catabolic (P)
		Penicillium brevicompactum	Catabolic (P)
Coumaphos		*Nocardiodes simplex* NRRL B24074	Co-metabolic
		Agrobacterium radiobacter P230	Co-metabolic
		Pseudomonas monteilli	Co-metabolic
		Flavobacterium sp.	Co-metabolic
		Pseudomonas diminuta	Co-metabolic)
		Nocardia strain B-1	Catabolic (C)
Diazon		*Flavobacterium* sp.	Catabolic (P)
		Pseudomonas spp.	Co-metabolic
		Arthrobacter spp.	Co-metabolic
Fenitrothion		*Flavobacterium* sp.	Co-metabolic
		Arthrobacter aurescenes TW17	Catabolic (C)
		Burkholderia sp. NF100	Catabolic (C)
Glyphosate	Bacteria	*Pseudomonas* ssp.	Catabolic (P)
		Alcaligene sp.	Catabolic (P)
		Bacillus megaterium 2BLW	Catabolic (P)
		Rhizobium sp.	Catabolic (P)
		Agrobacterium sp.	Catabolic (P)
		Arthrobacter sp. GLP	Catabolic (P)
		Arthrobacter atrocyaneus	Catabolic (P)
		Geobacillus caldoxylosilyticus T20	Catabolic (P)
		Flavobacterium sp.	Catabolic (P)
	Fungi	*Penicillium citrium*	Co-metabolic
		Penicillium natatum	Catabolic (P)
		Penicillium chrysogenum	Catabolic (N)
		Trichoderma viridae	Catabolic (P)
		Scopulariopsis spand	Catabolic (P)
		Aspergillus niger	Catabolic (P)
		Alternaria alternata	Catabolic (N)

Table 2. (Continued)

Compound	Microorganisms		Mode of degradation
Methyl parathion		Pseudomonas sp.	Co-metabolic
		Bacillus sp.	Co-metabolic
		Plesimonas spM6	Co-metabolic
		Pseudomonas putida	Catabolic (C)
		Pseudomonas sp. A3	Catabolic (C, N)
		Pseudomonas sp. WBC	Catabolic (C, N)
		Flavobacterium balustinum	Catabolic (C)
Monocrotophos		*Pseudomonas* spp.	Catabolic (C)
		Bacillus spp.	Catabolic (C)
		Arthrobacter spp.	Catabolic (C)
		Pseudomonas mendocina	Catabolic (C)
		Bacillus megaterium	Catabolic (C)
		Arthrobacter atrocyaneus	Catabolic (C)
		Pseudomonas aeruginosa F10B	Catabolic (P)
		Clavibacter michiganense SBL11	Catabolic (P)
Parathion	Bacteria	*Flavobacterium* sp. ATCC27551	Co-metabolic
		Pseudomonas diminuta	Co-metabolic
		Pseudomonas stutzeri	Co-metabolic
		Arthrobacter spp.	Co-metabolic
		Agrobacterium radiobacter	Co-metabolic
		Bacillus spp.	Co-metabolic
		Pseudomonas sp.	Catabolic (C, N)
		Pseudomonas spp.	Catabolic (P)
		Arthrobacter sp.	Catabolic (C)
		Xanthomonas sp.	Catabolic (C)

Biomonitoring of Organophosphates

Biomonitoring is a method for estimating human pesticide exposure by detecting biomarkers (parent chemicals and their metabolites) in biological tissues and fluids such as blood, urine, amniotic fluid, and hair. It can be applied for all types of exposure (inhalation, ingestion, dermal absorption). It is used in epidemiological research in health sciences, toxicology, and analytical chemistry to prove that exposure has happened. The potential presence of several biomarkers in biological samples might offer information about the exposure period and intensity (Katsikantamia et al., 2017).

Pesticides can enter the body through the skin, inhalation, or oral ingestion. The majority of pesticide toxicity research is motivated by the fear of oral exposure through food. On the other hand, the oral route does not usually apply to agricultural laborers exposed largely through their skin, whether by touching pesticides or reentering treated fields. When spraying pesticides on fields, the skin is the most exposed organ. Farmers are also exposed to pesticides while mixing pesticides, loading pesticides, cleaning equipment, and discarding empty containers (Kapka-Skrzypczak et al., 2011).

Biomonitoring data is used in epidemiological research to estimate the risk of exposure. The findings are used to identify exposure patterns across different demographic groups, as well as temporal trends in exposure and the effectiveness of intervention methods. Biomonitoring is connected to health impacts, clinical symptoms, and illnesses employing statistical analysis and modeling, and it provides the assessment of possible environmental and occupational health risks (Kapka-Skrzypczak et al. 201; Katsikantamia et al. 2017).

Human Monitoring of OPs Exposure

When OPs reach the human body, they are oxidized, degraded, and eliminated through bodily fluids. The more reactive oxon may attach to the cholinesterase enzyme or be hydrolyzed to a dialkyl phosphate (DAP) and a hydroxylated organic component unique for each pesticide in normal metabolic processes. These metabolic products are extremely polar and are eliminated through urine in either free form or bound to sugars or sulfates. Besides specific, there are non-specific metabolites of OPs - DAPs. OPs' non-specific metabolites include dimethyl phosphates (DMPs) like dimethyl phosphate (DMP), dimethyl thiophosphate (DMTP), and dimethyl dithiophosphate (DMDTP), and diethyl phosphates (DEPs) such as diethyl phosphate (DEP), diethyl thiophosphate (DETP), and diethyl dithiophosphate (DEDTP). As non-specific OP metabolites, DAPs are the most widely utilized markers for biomonitoring and estimating cumulative OP exposure in people. Blood and urine tests are commonly used to determine acute pesticide exposure (Valcke et al. 2006; Koureas et al. 2012).

Urinary Markers of Pesticide Exposure

Urine is the most often utilized fluid for biological monitoring of OPpesticide exposure, owing to its simplicity of collection and widespread availability (Valcke et al. 2006). On the other hand, urinary measurements have numerous limitations, such as the variable concentrations of endogenous and exogenous substances and the temporal fluctuation of the amount of collected urine. Some studies claim that early morning void collections yields are more consistent or representative data than spot samples taken at other times of the day. When utilizing spot urine samples, creatinine or a specific mass marker should be determined to normalize concentration values and rule out overdiluted or overconcentrated samples.

As indicated before, many organophosphorus pesticides have similar urine metabolites that make it difficult to determine which pesticide(s) an individual was exposed to (DAPs). Alkyphosphates are eliminated as sodium or potassium salts in the urine. A pesticide can be removed quickly or slowly in urine. The sample run period is critical in the design of biomonitoring research, and the amount of time required will vary based on the pesticide.

Pesticides and Their Metabolites in Blood

The most specific sign of exposure to a certain organophosphate pesticide is the measurement of the intact pesticide in the blood (Kapeleka et al., 2019). However, this type of test is complicated and may be limited by the pesticide's instability in blood. Because some pesticides in the body have relatively short half-lives, levels measured in blood taken at one point during the day may not represent internal levels at another point throughout the day, much less over longer periods. As a result, correlations between measured levels of a particular chemical and a biochemical marker or other health endpoints may be very unstable, with no consistent link between the compound's internal level and the health endpoint over time. The primary drawback of blood measures is the requirement for and risk associated with blood collection by intrusive procedures such as venipuncture. Due to the invasive nature of venipuncture, researchers' ability to acquire samples from youngsters and pregnant women, who may be particularly sensitive to pesticides, is limited. Blood samples are often difficult to obtain in big environmental and occupational investigations. Furthermore, the volume of blood accessible for examination is frequently restricted. The intrinsically low quantities of OP pesticides present in the blood relative to urinary metabolite concentrations further complicates analysis. As a result, ultrasensitive analytical methods are necessary for this purpose. The benefit of utilizing blood as a biomonitoring matrix is that the blood readings are pesticide-specific since the parent component is often quantified. Because the blood volume in a human's body is constant, calculating the body burden (the quantity of chemicals concerning the amount of blood in the body) is easier than measuring the chemical or its metabolite in urine. Adducts to DNA, hemoglobin, and albumin may all be measured using blood as a matrix. Adducts are essential early indicators of effect, giving more specific information about a particular health endpoint, such as cancer. Furthermore, because an adduct's lifetime in the body is primarily determined by the biomolecule's lifetime, adducts confirm exposure over a longer period.

Saliva as a Matrix

Although biological matrices such as blood and urine have traditionally been used in biomonitoring, other matrices, such as saliva, are simple and widely available fluids. Saliva has been used to test a wide range of biomarkers, medicines, and environmental pollutants, including narcotics, hormones, chemotherapeutics, heavy metals, and pesticides. However, estimating an internal dosage from a 'spot' saliva sample needs a thorough understanding of the chemical's pharmacokinetics and the connection between the chemical's concentration in the saliva and the chemical's concentration in the blood. Several field studies have been carried out to assess saliva biomonitoring of OPs. According to the findings, pesticide levels in saliva can be significantly lower than in blood, depending on the degree of protein binding. Nonetheless, due to the ease of collecting and analysis and the potential accuracy of salivary concentrations as a tissue availability indication, measuring pesticides in the saliva is promising (McKelvey et al., 2014; Ross and Ginevan, 2014).

Sweat in Biomonitoring Studies

According to the authors, there has only been one research conducted in 1985 to see if sweat might be used to monitor pesticide levels in farmworkers. The test didn't find its application because of a relatively small number of samples, large standard errors in regression line slope, etc. (Rosenberg, Queen, and Stamper, 1985).

Pesticides in Hair

Hair seems like a more promising matrix than urine or blood because of the larger exposure window and better detection rates of the chemicals of interest. Margariti and Tsatsakis published the first study on measuring DAPs in human hair in 2009. Children who live in an area where intensive agriculture is practiced are likely to be exposed to OP pesticides through various routes (Margariti and Tsatsakis 2009).

Hair dialkyl phosphate metabolites levels are influenced by a variety of factors, including the consumption of foods containing OP-generating dialkyl phosphate metabolites, pesticide usage in agriculture or at home, and socioeconomic position (e.g., family income and parental education level). Hair testing of dialkyl phosphate metabolites (and even parent OPpesticides) may give more information for human biomonitoring than urine or blood testing since hair is simpler to collect, handle, store, and transport than urine or blood (Hernández et al. 2019). As a result, chemicals in hair indicate long-term exposure, but chemicals in urine and blood show short-term contact (Katsikantamia et al., 2017).

Methods for Analysis of Organophosphates

The analysis of OP pesticides includes several phases, including extraction, cleanup or interference removal, pesticide residue detection, and confirmation of their identity using various methods. In the area of pesticide residue analysis, sample pretreatment is one of the most essential and crucial operations. Traditionally, liquid-liquid extraction (LLE) and solid–phase extraction (SPE) have been used to determine trace amounts of pesticide residues. Liquid-liquid extraction (LLE) is still the most common sample preparation approach (Ma et al., 2009). The primary disadvantages of LLE are its expensive cost, extended run time, and vast amounts of hazardous organic solvents it needs. Solid-phase extraction takes less time than LLE, although it still needs column conditioning and organic solvent elution. In the last decade, pesticides in soil samples have been determined using more environmentally friendly procedures such as supercritical fluid extraction, microwave-assisted extraction, and rapid solvent extraction. These methods benefit from requiring fewer organic solvents and having a shorter extraction time, but the instruments required are costly. As a result, there is a growing need to create a fast, simple, and sensitive sample pretreatment approach for detecting such OPs in soil (Wang et al., 2007; Yang et al., 2011).

Recently the trends for sample preparation methods have been towards:

- enhanced specificity or selectivity in extraction;
- a more ecologically friendly approach with less solvent consumption and organic waste;
- increased possibility in automation or online solutions to reduce manual activities.

Miniaturized sample preparation techniques have recently been developed as alternatives to traditional sample preparation processes, such as solid-phase microextraction (SPME) and liquid-phase microextraction (LPME) (Bouaid et al. 2001; Doong, Ruey, and Liao, 2001). For the determination of pesticides in soil, SPME, a more modern approach, a simple and organic solvent-free technology, has been employed (Zhang and Kee Lee 2008). On the other hand, SPME fibers are costly, brittle, and have a short lifespan; in addition, sample carry-over is an issue. Liquid-phase microextraction is a relatively recent sample preparation technique. Compounds from diluted samples have been pre-concentrated using liquid-phase microextraction.

All extraction methods use the same fundamental ideas to concentrate analyte selectivity in a single phase. Still, any analyte will be distributed across two phases according to the distribution coefficient, temperature, and relative volumes of the phases. On the other hand, the extraction rates are determined by temperature and diffusion rates in the two phases since they are dependent on migration kinetics. It's common to strike a compromise between extracting all of the soluble organic components and extracting only the compound of interest. Grinding, mixing, agitating, stirring, chopping, crushing, pressing, and pulverizing solid samples can be done directly or after drying as improved solvent extraction techniques, followed by solvent or liquid extraction.

QuEChERS (quick, easy, cheap, effective, rugged, and safe) methods without buffer or with acetate or citrate buffer or other modified QuEChERS methods remain among the most popular approaches to sample extraction and cleanup of food commodities. It is a quick and easy alternative to LLE that produces high-quality findings with less labor-intensive sample preparation stages and reduced solvent and glassware use. This approach has a lot of advantages when it comes to analyzing polar pesticides. The principal step in extracting a homogenized sample is to mix it with an equal amount of acetonitrile by hand-shaking or vortex rotation. It should result in a final extract that is suitably concentrated owing to the lack of solvent evaporation. To promote partitioning of the analytes between the aqueous residue and the solvent, a combination of 4 g anhydrous magnesium sulfate ($MgSO_4$) and 2 g sodium chloride (NaCl) is mixed into the sample. It offers well-defined phase separation without dilution and hazardous non-polar organic solvent. After shaking and centrifugation, a rapid technique called dispersive solid-phase extraction (DSPE) is used to clean up and remove residual water. A primary-secondary amine (PSA) adsorbent (a weak anion-exchanger that removes fatty acids, sugars, and other matrix co-extractives that form hydrogen bonds) and extra anhydrous MgSO4 are blended with the sample. The adsorbent is immediately introduced to the extract in the DSPE technique, and the cleanup is simply done by shaking and centrifugation. This technique is faster than standard SPE and allows the removal of residual water as well as a variety of polar matrix components such as organic acids, polar pigments, and sugar (Sobhanzadeh et al., 2009). Today, QuEChERS procedures are still modified to enable better analysis of pesticides in food samples.

Recent work of Yang et al. 2011 aimed to determine organophosphates in soil samples. It resulted in the development of a simple, fast, and sensitive dispersive liquid-liquid microextraction technique combined with gas chromatography–flame photometric detection. In comparison to traditional techniques, dispersive liquid-liquid microextraction offers high reproducibility and recovery, as well as the advantages of simplicity, speed, and decreased organic solvent use.

Ultrasound-assisted solvent extraction is a good alternative for organic compound extraction from various matrices because it increases both pressure (which favors penetration and transport) and temperature (which improves solubility and diffusivity), resulting in a more efficient contact between the solid and the solvent. Multiple extractions may be carried out at the same time, and the process is very affordable when compared to other current extraction procedures because it does not require specialized laboratory equipment (Bidari et al. 2011)

Pesticide detection technologies that are quick and easy to use are becoming increasingly popular. Because biosensors are rapid and simple to use, they are used frequently. Simultaneously, they provide rapid reaction and may be evaluated with suitable parameters under the law, equivalent to chromatographic methods. The primary drawback of OP detecting biosensors available today on the market is that they only measure the whole group of pesticides present in the sample rather than for each particular target analyte (Boneva, Yaneva, and Danalev 2021).

Gas chromatography with mass spectrometry (GC-MS or GC-MS/MS) techniques are commonly used to evaluate organochlorines (OCs), organophosphorus pesticides (OPs), and pyrethroids. The GCcoupled with mass spectrometric detectors, including single quadrupole, ion trap, and triple quadrupole mass spectrometers, has been adapted for analysis of pesticide residues in foods of animal origin (Boneva, Yaneva, and Danalev 2021). On the other hand, liquid chromatography with MS/MS is more commonly used to test azole fungicides, carbamates, neonicotinoids, phenylureas, and strobilurin fungicides. Over the last ten years, the use of liquid chromatography-electrospray ionization-mass spectrometry (LC-ESI+-MS/MS) for OP analysis has also grown (Raina-Fulton, Dunn and Xie, 2017).

Legislation for Organophosphates

Organophosphate insecticides were chosen as the first class of pesticides for reassessing existing levels for allowable *pesticide* residues on *food* by the US Environmental Protection Agency (EPA). By August 2006, chemical-specific evaluations of virtually all OP pesticides had been completed, resulting in the termination of all but a few OP pesticides' residential usages (Clune et al., 2021). In the early 2000s, most organophosphates were banned in residential use due to their relation to brain damage in children. One of the most well-known is chlorpyrifos, a popular insecticide prohibited by the EPA in 2016 but was later reversed, and it is still available today. The European Union (EU) stated on December 6, 2019, that sales of chlorpyrifos would be prohibited after January 31, 2020. The Standing Committee on Plants, Animals, Food, and Feed (PAFF Committee) approved two draft Implementing Regulations that would prevent chlorpyrifos and chlorpyrifos-methyl permits from being renewed (Zikankuba et al. 2019; Backstrom and Garson 2020). Despite all mentioned, it is still in use.

In total, 23 nations have prohibited or limited the use of parathion, while 50 countries have made it illegal to import it. In the United States, it was outlawed in 2000, and it hasn't been used since 2003.

The organophosphate pesticide omethoate was prohibited in Europe in 2002 because of its severe toxicity to the environment (for example, bees and aquatic life) and people (suspected mutagen, carcinogen, and neurotoxic). However, omethoate is the primary metabolite of dimethoate, an organophosphate pesticide that was recently banned (2019). As a result, omethoate residues remained in food, and even though MRLs were reduced, omethoate residues are still allowed in select foods (cherries, olives, and spring onions) today. The majority of the omethoate was found in food manufactured mostly in Europe but also in third countries, indicating either the usage of dimethoate or illegal omethoate use. According to the European Food Safety Authority (EFSA), omethoate residues in food are found to be above regulatory limits on many occasions (EFSA 2020).

Malathion is an organophosphate insecticide designated as possibly carcinogenic by the International Agency for Research on Cancer (IARC) in 2015. It is suspected of being harmful to reproduction, and scholarly research suggests it is also neurotoxic. Furthermore, it is extremely harmful to the environment. Although malathion's toxicity is recognized to some extent in Europe, and its usage is limited in many countries to greenhouses only, residues are allowed in a variety of foods. Malathion was found in 201 samples, 86 percent of which were imported, indicating that the EU imports food produced with this hazardous chemical. The irony is that many of the pesticides that have been prohibited might have been manufactured in the EU. When a pesticide is prohibited in the EU, European firms continue to manufacture and sell their goods in nations with weaker human health and environmental legislation. According to new research by Public Eye and Greenpeace Unearthed, 41 prohibited pesticides were declared for export from the EUin 2018, with the majority of them coming from only seven nations.

Between 2012 and 2016, restrictions regulating agricultural use of organophosphorus pesticides were shown to lower urine DAP metabolites in urban pregnant women and their newborns in Israel (Waras et al. 2020).

Conclusion

To summarize the OP's destiny and transportation, the foregoing review of OP fate and transport mechanisms shows that the most significant processes are sorption, volatilization, water-solubility, oxidation/reduction, hydrolysis, photolysis, and biodegradation. The amount of OPs adsorbed is determined by the mineralogical composition and organic matter concentration. The process of OP volatilization is particularly significant in the vadose zone and is influenced by soil moisture. Because OP solubility is an important physicochemical characteristic for understanding OP transport in groundwaters and surface waters, additional research is needed to obtain precise solubility values. Oxidation is crucial for OP breakdown in animals and insects, but a reduction is vital in soils with bacteria.

In groundwater, hydrolysis is the major OP breakdown route, whereas photo-lysis occurs only in surface water and the atmosphere. In soils, biodegradation is the major fate process. However, degradation rates vary depending on the soil. Biodegradation is one order of magnitude faster than hydrolysis, which is one order of magnitude faster than photolysis. Our ability to properly anticipate the behavior of OPs in the natural environment is hampered by a lack of accurately determining OPwater solubility and relative degradation rates for the mechanisms mentioned above (Ragnarsdottir, 2000).

References

Ayub Asif, Raza A. Zulfiqar, Majeed I. Muhammad, Tariq R. Muhammad., and AyeshaIrfan, 2020. Development of sustainable magnetic chitosan biosorbent beads for kinetic remediation of arsenic contaminated water. *International Journal of Biological Macromolecules*. doi:10.1016/j.ijbiomac.2020.06.287.

Anastassiades Michelangelo, Lehotay J. Steven, Stajnbaher Darinka, and Schenck J. Frank. 2003. "Fast and easy multiresidue method employing acetonitrile extraction/ partitioning and "dispersive solid-phase extraction" for the determination of pesticide residues in produce." *Journal of AOAC International*, 86(2): 412-31. doi: 10.1093/ jaoac/86.2.412.

Aston S. Linda and Seiber N. James. 1996. "Exchange of airborne organophosphorus pesticides with pine needles." *Journal of Environmental Science and Health, Part B*, 31(4):671-698. doi: 10.1080/ 03601239609373034.

Athanasopoulos E. Panagiotis, Kyriakidis V. Nikolaos, and Stavropoulos Panagiotis. 2004. "A study of the environmental degradation of pesticides azinphos methyl and parathion methyl." *Journal of Environmental Science Health, Part B*, 39(2): 297–309. doi: 10.1081/PFC-120030243.

Backstrom D. Timoty, and Garson N. Kelly. 2020. "European Union to ban chlorpyrifos an after January 31, 2020." *National Law Review*, XI(213).

Bidari Azraz, Ganjali M. Reza, Norouzi Parviz, Hosseini M. Reza Milani and Assadi Yaghoub. 2011. "Sample preparation method for the analysis of some organophosphorus pesticides residues in tomato by ultrasound-assisted solvent extraction followed by dispersive liquid–liquid microextraction." *Food Chemistry*, 126(4): 1840–1844. doi:10.1016/j.foodchem.2010.11.142.

Boneva Iskra, Yaneva Spaska and Danalev Dancho.2021. "Development and validation of method for determination of organophosphorus pesticides traces in liver sample by GC-MS/MS-ion trap." *Acta Chromatographica*, *33*(2): 188-194.doi: 10.1556/1326. 2020.00778.

Bosak Anita. 2006. "Enzymatic reaction with organophosphorus compounds." *Archives of Industrial Hygiene and Toxicology*, 57(4): 445-457.

Bouaid A., Lourdes Ramos, Gonzalez M. José, Fernandez P. Hernando and Camara, Carmen. 2001. "Solid-phase microextraction method for the determination of atrazine and four organophosphorus pesticides in soil samples by gas chromatography." *Journal of Chromatography A*, 939(1-2):13-21.doi:10.1016/S0021-9673(01)01332-2.

Cáceres Tanya, Ying Guang-Guo, and Kookana Rai.2002. "Sorption of pesticides used in banana production on soils of Ecuador." *Australian Journal of Soil Research*, 40(7): 1085 -1094. doi:10.1071/sr02015.

Cáceres Tanya, Megharaj Mallavarapu, Venkateswarlu Kadiyala, Sethunathan Nambrattil and Naidu Ravi. 2009. "Fenamiphos and Related Organophosphorus Pesticides: Environmental Fate and Toxicology." *Reviews of Environmental Contamination and Toxicology*, 205:117–162. doi:10.1007/978-1-4419-5623-1_3.

Casida E. John, and Quistad B. Gary. 2004. "Organophosphate Toxicology: safety aspects of non-acetylcholinesterase secondary targets." *Chemical Research in Toxicology*, 17(8):983–998. doi:10.1021/tx0499259.

Chambers W. Howard, Meek C. Edward, Chambers E. Janice. 2010. Chapter 64 - *Chemistry of Organophosphorus Insecticides*, Editor(s): Robert Krieger, *Hayes' Handbook of Pesticide Toxicology* (Third Edition), Academic Press, 1395-1398, doi:10.1016/B978-0-12-374367-1.00064-1.

Changlong Wu, and Linden G. Karl. 2010. "Phototransformation of selected organophosphorus pesticides: roles of hydroxyl and carbonate radicals." *Water research*, 44(12): 3585-94. doi:10.1016/ j.watres.2010.04.011.

Clune L. Alison, Ryan P. Barry and Barr D. Boyd. 2012. "Have Regulatory Efforts to Reduce Organophosphorus Insecticide Exposures Been Effective?" *Environmental Health Perspectives*, 120(4): 521–525. doi:10.1289/ehp.1104323.

Collins R. Clare, Ragnarsdóttir K. Vala and Sherman, M. David. 1999. "Effect of inorganic organic ligands on the mechanism of cadmium sorption on goethite." *Geochimicaet Cosmochimica Acta,* 63(19-20): 2989-3002. doi:10.1016/S0016-7037(99)00226-4.

Doong Ruey-An, and Liao Pei-Lin. 2001 "Determination of organochlorine pesticides and their metabolites in soil samples using headspace solid-phase microextraction." *Journal of Chromatography A*, 918(1):177-188.doi:10.1016/S0021-9673(01)00740-3.

Dyguda-Kazimierowicz Edyta, Roszak Szczepanand Sokalski W. Andrzej. 2014. "Alkaline Hydrolysis of Organophosphorus Pesticides: The Dependence of the Reaction Mechanism on the Incoming Group Conformation." *The Journal of Physical Chemistry B*, 118(26): 7277–7289. doi:10.1021/jp503382j.

Edwards L. Felicia, and Tchounwou B. Paul. 2005. "Environmental Toxicology and Health Effects Associated with Methyl Parathion Exposure – A Scientific Review." *International Journal of Environmental Research and Public Health*, 2(3): 430–441. doi: 10.3390/ijerph2005030007.

EFSA (European Food Safety Authority), Medina-Pastor Paula, and TriacchiniGiuseppe. 2020. The 2018 European Union report on pesticide residues in food. *EFSA Journal*, 18(4): 6057. doi:10.2903/j.efsa.2020.6057.

Guitart Raimon, Sachana Magda, Caloni Francesca, Croubels Siska, Vandenbroucke Virginie, and Berny Philipe. 2009. "Animal poisoning in Europe. Part 3: Wildlife." *The Veterinary Journal*, 183(3): 260–265. doi:10.1016/j.tvjl.2009.03.033.

Hernández F. Antonio, Lozano - Paniagua David, González - Alzaga Beatriz, Kavvalakis P. Matthaios, Tzatzarakis N. Manolis, López-Flores Inmaculada, Aguilar-Garduno Clemente, Caparros-Gonzalez A. Rafael, Tsatsakis M. Aristidis and Lacasana Marina. 2019. "Biomonitoring of common organophosphate metabolites in hair and urine of children from an agricultural community." *Environment International*, 131: 104997. doi: 10.1016/j.envint.2019.104997.

Kapeleka A. Jones, Elingarami Sauli, Omowunmi Sadik and Ndakidemi A. Patrick. 2019. "Biomonitoring of Acetylcholinesterase (AChE) Activity among Smallholder Horticultural Farmers Occupationally Exposed to Mixtures of Pesticides in Tanzania." *Journal of Environmental and Public Health*, 2019, 11 pages.doi:10.1155/2019/3084 501.

Kapka-Skrzypczak Lucyna, Cyranka Malgorzata, Skrzypczak Maciej, and Kruszewski Marcin. 2011. "Biomonitoring and biomarkers of organophosphate pesticides exposure - state of the art." *Annals of Agricultural and Environmental Medicine*. 18(2);294-303.

Karpouzas G. Dimitros, and Singh K. Brajesh. 2006. "Microbial Degradation of OrganophosphorusXenobiotics: Metabolic Pathways and Molecular Basis." *Advances in Microbial Physiology*, 52: 119–225. doi:10.1016/s0065-2911(06)51003-3.

Katsikantami Ioanna, Kavvalakisa P. Matthaios, Tzatzarakisa N Manolis, NosyrevbE. Alexander, Vakonakia Elena and Tsatsakis M. Aristidis. 2017. "Advances on Biomonitoring of Organophosphorus and Organochlorine Pesticides." *Journal of Siberian Federal University. Biology*, 10(2): 153-170. doi: 10.17516/1997-1389-0019.

Kaushal Jyoti, Khatri Madhu, and Arya S. Kumaru. 2021. "A treatise on Organophosphate pesticide pollution: Current strategies and advancements in their environmental degradation and elimination." *Ecotoxicology and Environmental Safety*, 207: 111483. DOI:10.1016/j.ecoenv.2020.111483

Kavalci Cemil, Durukan Polat, Ozer Mehmet, Cevik Yunsur and Kavalci Gülsüm. 2009. "Organophosphate Poisoning Due to a Wheat Bagel." *Internal Medicine*, 48(2): 85–88. doi:10.2169/internalmedicine.48.1559

Kumar Shardendu, Kaushik Garima, Dar M. Asratf, Nimesh Surendra, Lopez-Chuken U. Javier, and Villareal-Chiu J. Francisco. 2018. "Microbial Degradation of Organophosphate Pesticides: A Review." *Pedosphere*, 28(2): 190–208. doi:10.1016/s1002-0160(18)60017-7

Koureas Michalis, Tsakalof Andreas, Tzatzarakis Manolis, Vakonaki Elena, Tsatsakis Aristidis, and Hadjichristodoulou Christos. 2014. "Biomonitoring of organophosphate exposure of pesticide sprayers and comparison of exposure levels with other population groups in Thessaly (Greece)." *Occupational Environmental medicine Journal*. 71(2): 126-33. doi: 10.1136/oemed-2013-101490.

Lu Jianhang., Wu Laosheng, Newman Julie, Faber Ben, Merhaut, J. Donald and GanJianying. 2006. "Sorption and Degradation of Pesticides in Nursery Recycling Ponds." *Journal of Environment Quality*, 35(5): 1795. doi:10.2134/jeq2006.0123

Ma Jiping, Xiao Ronghui, Li Jie, Zhao Xiuhua, Shi Benzhang, and Li Shuqing. 2009. "Determination of organophosphorus pesticides in underground water by SPE-GC-MS." *Journal of Chromatographic Science*. 47(2): 110-115. doi: 10.1093/chromsci/47.2.110.

Margariti G. Maria, Tsatsakis M. Aristidis. 2009. "Analysis of dialkyl phosphate metabolites in hair using gas chromatography-mass spectrometry: a biomarker of chronic exposure to organophosphate pesticides." *Biomarkers: Biochemical Indicators of Exposure, Response, and Susceptibility to Chemicals*, 14: 137–147. doi: 10.1080/13547500902792912.

McKelvey Wendy, Bryan J. Jacobson, Kass Daniel, and Barr B. Dana. 2014. "Biomonitoring of Exposure to Organophosphate Pesticides: McKelvey et al. Respond." *Environmental Health Perspectives*, 122(7). doi:10.1289/ehp.1408444R.

Miodovnik Amir. 2019. Prenatal Exposure to Industrial Chemicals and Pesticides and Effects on Neurodevelopment, Editor(s): Jerome Nriagu, *Encyclopedia of Environmental Health* (Second Edition), Elsevier, Pages 342-352, doi: 10.1016/B978-0-12-409548-9.11008-5.

Mohd Ashraf Dar, Garima Kaushik, Juan Francisco Villareal Chiu. 2020. Chapter 2 - Pollution status and biodegradation of organophosphate pesticides in the environment,

Editor(s): Pardeep Singh, Ajay Kumar, Anwesha Borthakur, *Abatement of Environmental Pollutants*, Elsevier, Pages 25-66, doi:10.1016/B978-0-12-818095-2.00002-3.

Mulla I. Sikandar, AmeenFuad, Talwar P. Manjunatha, Eqani Syed Ali Musstjab Akber Shah, Bharagava Ram Naresh, SaxenaGaurav, Tallur N. Preeti, and Ninnekar Z. Harichandra. 2020. Chapter 13 Organophosphate Pesticides: Impact on Environment, Toxicity, and Their Degradation, Editor(S): G. Saxena, R. N. Bharagava, *Bioremediation of Industrial Waste for Environmental Safety*, Springer Singapore, Pages 265-291. doi:10.1007/978-981-13-1891-7_13.

Racke, D. Kenneth. 1992. Degradation of Organophosphorus Insecticides in Environmental Matrices, Editor(s): Janice E. Chambers, Patricia E. Levi, *Chapter 3: Organophosphates Chemistry, Fate, and Effects*, Academic Press, Pages 47-78, doi: 10.1016/B978-0-08-091726-9.50007-0.

Ragnarsdottir K. Vala. 2000. "Environmental fate and toxicology of organophosphate pesticides." *Journal of the Geological Society*, 157(4): 859–876. doi:10.1144/jgs.157.4.859.

Raina-Fulton Renata and XieZhen. 2017. "Sample Preparation Methods for Pesticide Analysis in Food Commodities, Biological and Environment Matricesa." *InTechOpen*. doi: 10.5772/intechopen.69791.

RichesJames. 2015. Chapter 7 - Analysis of Organophosphorus Chemicals Editor(s): Christopher M. Timperley, *Best Synthetic Methods*. Pages 721-752, Academic Press, doi: 10.1016/B978-0-08-098212-0.00007-8.

Ross H. John, and Ginevan E. Michael. 2014. "Biomonitoring of Exposure to Organophosphate Pesticides in New York City." *Environmental Health Perspectives*, 122(7): A178-179.doi:10.1289/ehp.1408444.

Rosenberg M. Norman, Queen M. Robert, and Stamper H. James.1985. "Sweat-patch test for monitoring pesticide absorption by airblast applicators." *Bulletin of Environmental Contamination and Toxicology*, 35(1): 68–72. doi:10.1007/bf01636482.

Schwarzenbach P. Rene, Gschwend M. Philip, and Imboden M. Dieter. 1993. "*Environmental Organic Chemistry*," John Wiley & Sons, Inc., New York.

Shrikant Randhavane, and Khambete A. K. 2015. "PHOTOLYSIS: Case Studies for Organophosphate Pesticides Treatment." *International Research Journal of Engineering and Technology*, 2(3): 1596-1599.

Singh K. Brajesh, and Walker Allan. 2006. "Microbial degradation of organophosphorus compounds." FEMS *Microbiology Reviews*, 30(3): 428–471. doi:10.1111/j.1574-6976.2006.00018.x

Sobhanzadeh Elham, Bakar NorKartiniAbu, AbasMhdRadzi, NematiKeivan. 2009. "Sample Preparation Methods for Pesticides Analysis in Food Matrices and Environmental Samples by Chromatography-Based Techniques: A Review", *Journal of Fundamental*, 5(2): 106-128. doi: 10.11113/mjfas.v5n2.294.

Valcke Mathieu, Samuel Onil, Bouchard Mchele, Dumas Pierre, Belleville Denis and Tremblay Claude. 2006. "Biological monitoring of exposure to organophosphate pesticides in children living in peri-urban areas of the Province of Quebec, Canada." *International Archives of Occupational and Environmental Health*, 79(7): 568–577. doi:10.1007/s00420-006-0085-8.

Van Scoy April, Pennell Ashley, and Zhang Xuyang. 2016. "Environmental Fate and Toxicology of Dimethoate." *Reviews of Environmental Contamination and Toxicology*, 237: 53–70. doi:10.1007/978-3-319-23573-8_3.

Wang Wentao, BingjunMeng, Xiaoxia Lu, Yu Liu, and Shu Tao. 2007. "Extraction of polycyclic aromatic hydrocarbons and organochlorine pesticides from soils: a comparison between Soxhlet extraction, microwave-assisted extraction and accelerated solvent extraction techniques." *Analytica Chimica Acta*, 602(2): 211-222. doi: 10.1016/j.aca.2007.09.023.

Waras MaisarahNasution, BabinaKateryna, Ismail Ismaniza and Lenehan E. Claire. 2020. "*Law and regulations to control pesticide exposure among the general population: Comparing the Australian and the European Union pesticide regulatory system*," Preprints, 2020120117.

Wauchope R. Don, Yeh Simon, Linders B. Jan, Kloskowski Regina, Tanaka Keiji, Rubin Baruch, Katayama Arata, KordelWarmer, Gerstl Zev, Lane Michael and Unsworth B. John. 2002 "Pesticide soil sorption parameters: theory, measurement, uses, limitations and reliability." *Pest Management Science*, 58(5): 419–445. doi:10.1002/ps.489.

Wolfe N. Lee, Zepp Richard, and Paris F. Doris. 1978. "Use of structurere activity relationships to estimate hydrolytic persistence of carbamate pesticides." *Water Research*. 12(8): 561-563.

Yang Zhonghua, Liu Yu, Liu Donghui and Zhou Zhiqiang. 2011. "Determination of Organophosphorus Pesticides in Soil by Dispersive Liquid-Liquid Microextraction and Gas Chromatography." *Journal of Chromatographic Science*, 50(1):15–20. doi:10.1093/chromsci/bmr011.

Zhang Jie, and HianKee Lee. 2006. "Application of liquid-phase microextraction and on-column derivatization combined with gas chromatography–mass spectrometry to the determination of carbamate pesticides." *Journal of Chromatography A*, 1117(1): 31-37.doi: 10.1016/j.chroma.2006.03.102.

Zikankuba V. Lwoga, Mwanyika Gaspary, Julius E. Ntwenya and Armachius James. 2019. "Pesticides Regulations and their Malpractice Implications on Food and Environment Safety." *Cogent Food & Agriculture*, 5(1): 1601544. doi:10.1080/23311932.2019. 1601544.

Chapter 7

Ionic Liquids as a Promising Media in (Pre)analytical Treatments and Degradation of Organophosphate Pesticides

Aleksandra Dimitrijević* **and Ana Jocić**

Department of Physical Chemistry, "Vinča" Institute of Nuclear Sciences - National Institute of the Republic of Serbia, University of Belgrade, Belgrade, Serbia

Abstract

Organophosphorus pesticides (OPs) are highly effective pesticides whose usage accounts for approximately 30% of the pesticides used worldwide. However, overusing OPs in addition to their difficult degradation is threatening environmental and human health even at very low concentrations that are below detection limits of the standard analytical techniques. For that reason, great efforts have been placed in developing rapid and effective preparation methods for OPs prior to analysis, as well as in improving the performance of different analytical techniques for OPs detection from various samples. Much of the recent progress in terms of efficiency and sustainability has come from incorporating ionic liquids (ILs) as solvents or in the creation of IL-supported materials for OPs extraction purposes. ILs are liquid molten salts typically composed of bulky and unsymmetrical organic cations and various anions. ILs are non-flammable and non-volatile, have excellent (electro)chemical stability and solvation ability for a wide range of compounds and materials. Owing to a large number of ion combinations, ILs have been referred to as tunable designer solvents, the feature that overcomes the limited selectivity and efficiency of common organic

* Corresponding Author's Email: daleksandra@vin.bg.ac.rs.

In: Organophosphates
Editor: Tamara Lazarević-Pašti
ISBN: 978-1-68507-652-8

solvents and allows the development of more effective extraction platforms. This review will summarize the main results achieved by the use of IL-based processes in the extraction of OPs as a part of sample preparation (liquid-liquid microextraction and solid-phase micro-extraction) or in designing different nanocomposites to improve electrochemical performance for the detection of different OPs. Moreover, this review will outline and discuss achievements in nucleophilic degradation of OPs where IL can act as both the nucleophile and solvent. The key accomplishments and future challenges to the role of ILs in OPs analysis and degradation will be discussed, with particular emphasis on the more benign ILs while foreseeing the development of more sustainable processes.

Keywords: ionic liquid, extraction, organophosphate pesticides

Introduction

Organophosphorus pesticides (OPs) are widely used worldwide due to their high insecticidal activity and other biological activities (EPA, 2008). Overall, OPs compounds represent more than 38% of total pesticides used globally. The toxicity of these pesticides is attributed to irreversible inhibition of the acetylcholinesterase enzyme, which is essential for the central nervous system (Colovic et al. 2013). On the other hand, the growing use of OPs has increased the risks of environmental contamination of soils and groundwater (Mulla et al. 2020). Therefore, pollution with pesticides represents a widespread global problem, and biomonitoring of the environment is required to assess potential harmful effects. For that reason, great efforts have been placed in developing rapid and effective preparation methods for OPs prior to analysis, as well as in improving the performance of different analytical techniques for OPs detection from various samples (Silva Sousa et al. 2021). In addition, given the adverse effect OPs have, many methods have been developing for their degradation (Sidhu et al. 2019).

Numerous extraction methodologies have been developed as a tool for preanalytical treatments for different compounds. Sample preparation steps, such as extraction, enrichment, and isolation of various analytes, hugely impact the reliability and accuracy of their determination. Extraction techniques are usually complex and present several shortcomings such as low efficiencies, time and energy-consuming, require harsh conditions, and toxic volatile organic solvents (Farajzadeh et al. 2014). Based on these drawbacks, researchers are constantly in the quest for alternative extraction processes with "greener" and more sustainable credentials. Particular efforts are being placed on the development of new time-saving strategies which should also minimize required labor and materials. Therefore, current trends in extraction method development are shifting towards simplification, miniaturization, and process automation. Also of particular interest is the use of solvent-free or environmentally friendly procedures while maintaining good or acceptable extraction efficiencies at the same time. For that reason, ionic liquids (ILs) have received increased attention in the past decade. ILs are considered a more environmentally friendly and safer option than conventional organic solvents. ILs are liquid molten salts at temperatures below 100 °C which are composed of large and unsymmetrical organic cations (imidazolium, phosphonium, ammonium, cholinium, pyrrolidinium, pyridinium piperidinium, morpholinium, etc.) and organic or inorganic anions (halogenides, fluorinated anions, carboxylic acids, amino acids, carbohydrates, etc.) (Rogers and Seddon 2003). However, it should be kept in mind that the 'greenness' of ILs cannot be considered as a 'generic' property for such a broad and diverse class of compounds (Chatel et al. 2014). The history of ILs dates back to 1914, when the first IL was prepared in 1914 by Walden (Walden 1914). It was ethyl ammonium nitrate synthesized by neutralizing ethyl-amine with concentrated nitric acid. Prepared ethyl ammonium nitrate was clear, colorless, viscous, and most importantly, its conductivity was consistent with a composition of purely anions and cations. Afterward, in 1951, Hurley and Wier reported, so-called, the second generation of chloroaluminates ILs, which suffered from extreme sensitivity to moisture, and their acidity/basicity was difficult to regulate (Hurley and Wier 1951). A major step forward was made in 1992 by Wilkes and Zaworotko, who prepared moisture- and air-stable ILs based on imidazolium cation and tetrafluoroborate anion (Wilkes et al. 1992). It was an introduction to the new generation era of ILs. Many advantages of imidazolium ILs enabled them to receive extensive attention from both academia and industry since then. Imidazolium salts represent the most prominent subclass of ILs, and a number of them are commercially available. Their solvent properties such as melting points, solubility, and viscosity can easily be tuned in a wide range by varying the substituents at the nitrogen atoms as well as by varying the counter-ions. Starting with imidazolium cations, various cationic components, including pyridinium, ammonium, phosphonium, thiazolium, and thiazolium species, have been combined with weakly coordinating anions. Common examples include tetrafluoroborate, hexafluorophosphate, triflate, triflimide, and dicyanimide. The list of possible anionic components continues to grow at a rapid rate which makes ILs, in general, real designer solvents. Owing to their low volatility, ILs are considered promising alternative solvents to replace traditional volatile organic compounds, spurred by the green chemistry movement. Moreover, IL chemistry has progressed considerably in the past years. In the last ten years, ILs have become a hot research topic among multidisciplinary areas, including chemistry, physics, biology, and engineering. Various exciting new applications of ILs have been developed continuously, for example, gas adsorbents (Macfarlane et al. 2014), lubricants (Somers et al. 2013), catalysts (Q. Zhang, Zhang, and Deng 2011), extractants (Freire et al. 2012), ionic liquid crystals (Goossens et al. 2016), energy application (Macfarlane et al. 2014), etc. ILs have brought about a revolution in the scientific community, deriving from their unique properties such as extremely low vapor pressure, large liquid range, tunable polarity, nonflammability, wide solubility, and high stability. Despite fascinating findings regarding the IL-based processes, the price and toxicities of some ILs still constrain the large-scale application of these ILs. Several of these IL constituents possess some toxicity and are poorly biodegradable, although biodegradable potential and their toxicity can be tailored by side-chain modifications or by a proper selection of IL anion (J. I. Santos et al. 2015). Thus, the search for safer and cheaper ILs is still an imperative issue given that the ample versatility of ILs permits the synthesis of new formulations with an acceptable environmental footprint and enhanced biocompatibility at the same time (Quental et al. 2015; Gomes, Silva, and Reis 2019). For those reasons, the development of third generation of ILs which includes salts with a room temperature melting points that contain biologically active cations and/or anions in the structure are of high importance (Gomes, Silva, and Reis 2019). Special attention of the scientific community in recent years has been given to ILs with cholinium cation. Recent advances have demonstrated the synthesis of novel cholinium-based ionic liquids combined with different anions (Taha et al. 2015; Khan et al. 2016). These ILs have shown excellent biodegradability and low toxicity to filamentous fungi and the freshwater crustacean Daphnia magna (J. I. Santos et al. 2015). While both biocompatible and biodegradable, cholinium-based ionic liquids share the most relevant ionic liquid properties: a wide liquidus temperature range, non-flammability and negligible vapour pressure at ambient conditions. Moreover, a various studies have showed protein structure and enzyme function can be maintained or even increased in novel cholinium-based ILs (Weaver et al. 2010). The work of Deive et al. has shown the higher capacity of fungi to grow during their exposure to biocompatible cholinium ILs (compared to bacteria), and the authors also analyzed the ILs degradation during cultivation (Deive et al. 2011). Therefore, although most researches have been focused on ILs composed of imidazolium cations, in particular, 1-alkyl-3- methylimidazolium cations, recently, other interesting varieties of ionic liquids are also attracting attention. For example, protic ionic liquids represent a subgroup of ionic liquids that are formed in one-step reactions between a Brønsted acid and base. The ability to transfer protons from the acid to the base is of crucial importance to create hydrogen-bonded networks. The main advantages of PILs comparing to the aprotic ILs include easy synthesis pathway, more benign character, low-price, and capability of being distilled as PILs boiling temperatures are lower than their decomposition temperatures (Greaves and Drummond 2015). Beyond the excellent chemical and thermal stability, non-flammability, and negligible volatility, ILs are also known for their excellent solvation ability for a wide range of compounds, as well as good stabilizing media for many biomolecules (Ventura et al. 2017). As mentioned, ILs have been recognized as tunable designer solvents given a large number of ion combinations and the possibility of designing task-specific fluids. This feature overcomes the limited selectivity of common volatile organic solvents, and it allowed the development of more effective and environmentally friendly extraction platforms. ILs can exhibit interactions with another moiety via cation or anion or both, which makes IL chemistry far more complex than in the case of traditional molecular solvents, which have enabled numerous 'green' applications of ILs. There is an important role for ILs to play in future sustainable processes, so it is very important to explore their outstanding potential in regard to design control by independent modification of cation or/and anion that is one of the greatest assets of this novel class of compounds (Dimitrijević et al. 2020; Rogers et al. 2012). However, understanding and predicting which ions combinations would lead to the specific physical properties and behavior of IL still require further studies. The replacement of volatile organic solvents by nonvolatile ILs eliminates solvent losses to the atmosphere, decreasing both the environmental footprint and the cost of the process. Some water-miscible ILs can induce IL-based aqueous biphasic systems (ABS). These systems were investigated for extraction of a wide range of bioactive compounds, ranging from small organic compounds, such as dyes (Jocić et al. 2020; Dimitrijević et al. 2019), phenolic compounds (Santos et al. 2018), alkaloids (Dimitrijević et al. 2020; Marić et al. 2021), fats, vitamins, amino acids, among others, to more complex molecules, such as nucleic acids, proteins, enzymes, and antibodies (Freire et al. 2012). In addition, imidazolium-based ILs were investigated for simultaneous one-step extraction of different polarity pesticides applying ABS as an extraction method (Dimitrijević et al. 2017). On the other hand, water-immiscible ILs with large, highly charge-delocalized anion can easily form biphasic systems with water and have been used for liquid phase micro-extraction of different compounds (Zhang et al. 2012; Albishri 2016; Pang et al. 2018). These water-immiscible ILs have shown great promise as an organic phase in solvent extraction studies. ILs insoluble in water can be used as diluents for extractants, or task-specific ILs can be prepared in which a functional group is incorporated in the structure of the IL. Different IL-based processes have been developed, ranging from liquid−liquid extractions (carried out with hydrophobic ILs, including microwave-assisted, ultrasound-assisted, among others) (Albishri et al. 2016), solid−liquid extractions, solid-phase extractions, employing IL-modified materials, (Silva Sousa et al. 2021), and induced-precipitation techniques (Pusey et al. 2007). This review will summarize the main achievements regarding IL-based extraction techniques for OPs proposed to date, the improvements in IL-supported materials for OPs extraction or electroanalytical detection as well as degradation of OPs accomplished by incorporating the ILs.

Ionic Liquids in (Pre)analytical Treatments of Organophosphate Pesticides

Extraction of Organophosphorus Pesticides Using Ionic Liquid-Based Liquid-Liquid Extraction Techniques

This section addresses the use of ILs in the extraction of OPs and provides a comprehensive overview of the data hitherto reported on the development of IL-based liquid-liquid extraction (LLE). As mentioned, OPs are widely used for agricultural activities due to their relatively low price and effective ability to control pests, weeds, and diseases. Overusing OPs, in addition to their difficult degradation, is threatening environmental and human health (Colovic et al. 2013) even at very low concentrations. They can be traced in a wide range of surface water, soil, and organisms. Due to the low concentrations, complex matrices, and the protection of instruments, OPs in samples are not directly analyzed with conventional methods such as gas chromatography (GC), high-performance liquid chromatography (HPLC), gas chromatography-mass spectrometry (GC–MS), and liquid chromatography-tandem mass spectrometry (LC-MS-MS). For that reason, great efforts have been placed in developing rapid and effective preparation methods for OPs from various samples before analysis (Foong et al. 2020). The sample preparation step is crucial and most tedious in the analytical process to obtain accurate and sensitive results in analytical and environmental fields. A variety of pretreatment methods had been established for separation and preconcentration of OPs from different matrices, such as liquid-liquid extraction (LLE), solid-phase microextraction (SPME), stir bar sorptive extraction (SBSE), and liquid-phase microextraction (LPME) were typical techniques among them.

Although SPME and SBSE were regarded as alternative methods to LLE, as solvent-free extraction methods, they often suffered from the plugging of cartridges and consumption of an appreciable amount of toxic solvents at the elution step (Farajzadeh et al. 2014). Most sample preparation techniques are based on solvent extraction. LLE is a classical sample preparation to extract analytes from various samples. However, LLE consumes a large amount of hazardous organic solvents, and a large volume of sample is often required for trace analysis. It was gradually substituted by other methods for its disadvantage of time-consuming, large consumption of toxic organic solvents, loss of target analytes, and emulsification. In the last few decades, LPME based on the miniaturization of the traditional LLE technique has been reported as an alternative for sample preparations. LPME achieved extensive attention as a sample pretreatment technique due to its advantages such as simple operation, low cost, low consumption of sample volume and extraction solvent, and rapidity. However, extraction solvent in a single drop or protected in porous hollow fiber is not stable and easy to volatile under stirring conditions. In addition, the extraction procedure in LPME is time-consuming, and extraction equilibrium is not attained during a short time in most cases (Farajzadeh, et al. 2014; Zgoła-Grześkowiak and Grześkowiak 2011).

Dispersive liquid-liquid microextraction (DLLME) is one category of LPME which is based on the dispersion of the extraction solvent into the aqueous sample. Extraction and disperser solvents are rapidly injected into an aqueous sample simultaneously, resulting in the formation of a cloudy solution (Zgoła-Grześkowiak and Grześkowiak 2011). The DLLME is very convenient to operate, having in mind that extraction could be completed in a few seconds. In comparison to LPME, DLLME avoids volatilization, so there is no loss of extractants. DLLME also provides the possibility for the improvement of sensitivity and simplification of the extraction procedure. Assadi and co-workers introduced DLLME techniques, and since, many types of research have been conducted regarding the application of DLLME for the extraction of different analytes (Rezaee et al. 2006). DLLME assures many advantages, including rapidity, simplicity of operation, high recovery, and enrichment factor. In this technique, it is considered important to select an appropriate extraction solvent with a density higher than water and which has a high extraction capability of compounds of interest and good chromatographic behavior. Hydrocarbons are usually selected as extraction solvents (chloroform, chlorobenzene, carbon tetrachloride, tetrachloroethane, n-hexane and hexadecane) (Zgoła-Grześkowiak and Grześkowiak 2011). However, the development of new extraction solvents is an important issue for DLLME further development and its improvement. In that context, ILs made a significant shift in the DLLME evolution. ILs have attracted much attention as solvents in sample preparation thanks to their unique properties and good extraction ability (Rogers and Seddon 2003). In the last decade, a few works have been published addressing the use of ILs as extracting solvent in DLLME for extraction of OPs from different samples. Although IL-based LLE techniques have been receiving increasing attention lately, in general, the role of ILs in the development of alternative extraction media for OPs seems to have been under the radar in comparison to other solvents. Design ILs that possess stronger hydrophobic nature, better extraction efficiency, and appropriate viscosity and density is very important for DLLME. The use of ILs in DLLME avoids using highly toxic volatile organic solvents. In order to be adequate for DLLME, IL as the extracting solvent has to be soluble in the dispersing solvent while its solubility in water has to be very low. Additionally, its density has to differ greatly from the density of water to enable phase separation. These requirements significantly reduce the choice of adequate IL as the most studied hydrophobic ILs are mainly based on $[BF_4]^-$, $[PF_6]^-$, and $[NTf_2]^-$ anions. However, designing both the cation and anion can modify hydrophobicity and polarity, and create new classes of ILs that are both hydrophobic and polar and appropriate for IL-DLLME technique.

In 2008, hydrophobic IL based on $[PF_6]^-$ anion was effectively applied by Zhou and co-workers (Zhou et al. 2008), in temperature-dependent IL-DLLME technique prior to HPLC for the determination of methyl parathion and phoxim in water samples. The authors discussed the factors affecting the extraction efficiency, such as the volume of $[C_6mim][PF_6]$, pH of solutions, extraction time, centrifuging time, temperature, and salt effect. It was concluded that IL-based DLLME increases the stability of extraction without suspending related to single drop microextraction, enhances the sensitivity, and simplifies the extraction procedure. The detection limits achieved were

0.17 μg L^{-1} and 0.29 μg L^{-1}, for methyl parathion and phoxim, respectively, with a linear concentration range of 1–100 μg L^{-1} under the optimal conditions. The proposed method was successfully applied in the analysis of four real environmental water samples, and good spiked recoveries over the range of 88.2–103.6% were obtained. In addition to this pioneering work, He et al. (He et al. 2009) determined OPs (parathion, phoxim, phorate, and chlorpyrifos) in water using DLLME combined with HPLC with [PF_6]-based IL (1-octyl-3-methylimidazolium hexafluorophosphate, [C_8mim][PF_6]) as extracting solvent. The extraction procedure was followed by the formation of a cloudy solution composed of fine drops of [C_8mim][PF_6] dispersed entirely into the sample solution with the help of methanol as a disperser solvent. Up to 200-fold enrichment factor of analytes and extraction recovery higher than 70% were obtained. Authors obtained good linearity in the concentration range of 10.5–1045.0 μg L^{-1} for parathion, 10.2–1020.0 μg L^{-1} for phoxim, 54.5–1089.0 μg L^{-1} for phorate, and 27.2–1089.0 μg L^{-1} for chlorpyrifos, respectively, while the limits of detection were in the range of 0.1–5.0 μg L^{-1}. After the IL-DLLME system optimization, its performance was demonstrated on the analysis of four different sources of water samples (tap, well, rain, and river water) and the relative recoveries of spiked water samples achieved 99.9–115.4%, 101.8–113.7%, and 87.3–117.6% at three different concentration levels of 75, 200 and 1000 μg L^{-1}. The same group of researchers (He et al. 2010) investigated 1,3-dibutylimidazolium hexafluorophosphate IL [C_4C_4im][PF_6] as extraction solvent for extraction and preconcentration of OPs (fenitrothion, parathion, fenthion, and phoxim) from water and fruit samples by DLLME combined with HPLC. An extraction recovery of over 75% and enrichment factor of over 300-fold were obtained under the optimum conditions. They obtained the linearity relationship was also observed in the range of 5–1000 μg L^{-1} with the high correlation coefficients, while limits of detection were in the range of 0.01–0.05 μg L^{-1} for four analytes. Subsequently, the authors analyzed three real samples, including tap water, river water, and pear spiked at three concentration levels and yielded recoveries ranging from 92.7–109.1, 95.0–108.2, and 91.2–108.1%, respectively.

An IL-based vortex-assisted DLLME was developed by Zhang et al. for the analysis of trace amounts of six OPs in apple and pear coupled with HPLC (Lijin Zhang et al. 2012). ILs selected were [C_6C_1im][PF_6] and [C_8C_1im][PF_6].

Table 1. Ionic liquids used in DLLME for the extraction of OPs

Ionic liquid	Pesticides	Publication
$[C_6mim][PF_6]$	Methylparathion, phoxim	(Zhou et al. 2008)
$[C_8mim][PF_6]$	parathion, phoxim, phorate, chlorpyrifos	(He et al. 2009)
$[C_4C_4im][PF_6]$	fenitrothion, parathion, fenthion, phoxim	(He et al. 2010)
$[C_6C_1im][PF_6]$, $[C_8C_1im][PF_6]$	isocarbophos, phtalofos, triazophos, fenthion, phoxim, profenofos	(Zhang et al. 2012)
$[C_4C_1im]PF_6$	azinphos-methyl, chlorpyrifos, parathion-methyl, diazinon, phosalone	(Albishri et al. 2016)
$[Be\text{-}C_1im][NTf_2]$	phoxim, fenitrothion, chlorpyrifos, phorate, parathion	(Wang et al. 2016)
$[C_4C_1im][NTf_2]$, $[C_6C_1im][NTf_2]$, $[C_8C_1im][NTf_2]$, $[C_{10}C_1im][NTf_2]$, $[C_{10}C_1im][NTf_2]$	dimethoate, disulfoton, famphur, parathion, parathion-methyl, phorate, sulfotep, thionazin triethyl thiophosphate	(Cacho et al. 2018)
$[C_8C_1im][NTf_2]$, $[C_6C_6im][NTf_2]$, $[BeC_8im][NTf_2]$	fenitrothion, fenthion, phoxim	(Liu et al. 2018)
$[C_1\text{-}CH_2COOCH_3im]$ $[NTf_2]$ $[C_4C_1im][NTf_2]$, $[C_8C_1im][NTf_2]$, $[He\text{-}C_1im]$ $[NTf_2]$	parathion, parathion-methyl, profenofos, chlorpyrifos, O-ethyl O-4-nitrophenyl phenylpho-sphonothioate	(Lu et al. 2019)
$[BeC_1im]$ $[NTf_2]$, $[C_4C_4im][NTf_2]$, $[C_6C_6im][NTf_2]$, $[C_8C_1im][NTf_2]$	ethoprophos, fenitrothion, and phoxim	(Liu et al. 2020)
$[C_6C_1im][PF_6]$	trichlorfon, chlorpyrifos, phoxim, fenthion, and diazinon	(Zhang et al. 2021)
$[C_4C_1im]Cl$, $[C_4C_1im]Br$, $[C_2C_1im]Cl$	pirimicarb, carbaryl, methiocarb, aldicarb, carbofuran	(Anvar et al. 2020)
cholinium chloride	phosalone, chlorpyrifos	(Heidari et al. 2020)

Under the chosen extraction parameters, the analytes were enriched more than 300-fold, and the limits of detections were greatly dropped to 0.061–0.73 μg L^{-1}. The linearity relationship was observed in the range of 2–100 μg L^{-1} while mean recovery values of the OPs were in the range of 69.8–109.1%. Based on the results obtained, the authors concluded that $[C_8C_1im][PF_6]$ was a good extraction solvent, and the proposed IL-based vortex-assisted DLLME method was suitable for the effective extraction of the OPs in fruits. In 2016, Albishri and co-workers developed a rapid and sensitive ultrasound-assisted temperature-controlled $[C_4mim]PF_6$-based DLLME combined with reversed-phase HPLC for the determination of five OPs (azinphos-methyl, chlorpyrifos, parathion-methyl, diazinon, and phosalone) in water samples (Albishri, Aldawsari, and El-Hady 2016). The authors dealt with the optimization of parameters including IL type, IL volume, ionic strength, sonication time, heating/cooling temperature, centrifugal time, and speed. The extraction procedure was induced by the formation of a cloudy solution, which was composed of IL dispersed entirely into the sample solution with the assistance of ultrasound and higher temperature. Under optimal conditions, the linearity of the five OPs was obtained in the range of 0.09–200 μg L^{-1} with high correlation coefficients. Limits of detection and limits of quantitation ranged from 0.01 to 0.1 μg L^{-1} and from 0.05 to 0.4 μg L^{-1}, respectively. The proposed method exhibited the highest extraction efficiency ranging between 90 and 98% for targeted OPs. The method was successfully applied to different water samples (tap, well, and lake water), showing relative recoveries ranging from 96.9 to 103.2%. Investigation of Ya-Li et al. (Wang et al. 2016) included a synthesis of benzyl functionalized IL, 1-benzyl-3-methylimidazolium bis [(trifluoromethyl) sulfonyl] imide [BeC_1im][NTf_2], as an extraction solvent of DLLME for determination of five OPs (phoxim, fenitrothion, chlorpyrifos, phorate, and parathion) and two aromatic compounds (chloronaphthalene and anthracene) in environmental water samples by HPLC (Wang et al. 2016). [BeC_1im][NTf_2] had higher extraction efficiency than [C_8C_1im][NTf_2] and common organic solvents such as carbon tetrachloride and tetrachloro-ethylene. The extraction was performed using [Be-mim][NTf_2] as respective extraction solvent and methanol as dispersive solvent followed by centrifugation. Under the optimal conditions, the proposed method exhibited good linearity for all analytes. The limits of detection were 0.01–1.0 μg L^{-1} (S/N = 3). The developed method was convenient, rapid, and applicable to the detection of the analytes in real environmental water samples with satisfactory recoveries of 82.7%–118.3%. The authors noted that the introduction of the benzyl group into the imidazolium could considerably elevate the extraction efficiency of the analytes due to enhanced π-π interaction between [Be-mim][NTf_2] and analytes. [Be-mim][NTf_2] was a satisfactory extraction solvent with a high enrichment factor of 339 and extraction efficiency of 81.4%. Cacho et al. (Cacho et al. 2018) determined nine OPs in environmental waters from different origins using in situ DLLME based on different [NTf_2]-based ILs ([C_4C_1im][NTf_2], [C_6C_1im][NTf_2], [C_8C_1im][NTf_2], [$C_{10}C_1$im] [NTf_2], [$C_{10}C_1$im][NTf_2]). The preconcentration technique was coupled to GC-MS using microvial insert thermal desorption, which allowed the introduction of the IL extracts directly into the GC system, facilitating the operation, while increasing sensitivity. The proposed method showed good precision, with recoveries in the 85–118% range, and sensitivity with DLs ranging from 5 to 16 ng L^{-1}.Liu et al. (Liu, Quan, and Hu 2018) proposed the ultrasonic-assisted IL-DLLME method for the extraction of three OPs from wheat combining with HPLC for their analysis. The extraction efficiencies of three ILs with [NTf_2] anion were compared by extracting OPs from wheat samples. It was found that the use of 1-octyl-3-methylimidazolium bis(trifluoromethylsulfonyl)imide ([C_8C_1im][NTf_2]) had both high enrichment efficiency and appropriate extraction recovery. The method was used for the determination of three wheat samples, and the recoveries achieved were 74.8–112.5%, 71.8–104.5%, and 83.8–115.5%, respectively. The results show that the method proposed is simple, fast, and efficient, which can be applied to the extraction of organic matters in wheat samples. Lu et al. (Lu et al. 2019) designed a functionalized IL, 3-methyl-1-(ethoxycarbonylmethyl) imidazolium bis (trifluoromethylsulfonyl)-imide ([C_1-CH_2COOCH_3im] [NTf_2]) and investigated it as an extraction solvent in DLLME combined with ultra-high performance liquid chromatography (UHPLC) for the sensitive determination of OPs. The authors compared results with [C_4C_1im][NTf_2]-, [C_8C_1im][NTf_2]-, and [HeC_1im] [NTf_2]- based DLLME. The proposed IL exhibited a high extraction efficiency which was the outcome of π–π stacking interaction between the parent imidazolium cation core and the OPs. Moreover, during the DLLME process, a cloudy solution containing fine drops of [C_1-CH_2COOCH_3im][NTf_2] allowed for a larger contact area between the OPs and the IL, which accelerated the mass transfer, and therefore the enrichment could be realized quickly. Under the optimal extraction conditions, the developed method was successfully applied in the analysis of OPs in environmental water samples with a high enrichment factor (more than 400) and less volume requirement of both the extraction solvent and the water samples. Liu et al. (Liu and Quan 2020), employed four ILs, [BeC_1im] [NTf_2], [C_4C_4im][NTf_2], [C_6C_6im][NTf_2]) and [C_8C_1im][NTf_2], for extracting three OPs (ethoprophos, fenitrothion, and phoxim) from tea samples by DLLME prior to HPLC with UV detection. Acetonitrile was used as an extracting solvent and as a dispersant solvent in the DLLME process. Among studied ILs, [BeC_1im] [NTf_2] showed the best performance. The authors optimized several parameters affecting the performance of the method, including the selection of extraction solvent type, dispersant type, pH, extraction time, extraction temperature, salt addition, and centrifugation time. Under the optimal conditions, the enrichment factors of the analytes were in the range of 242.8–266.0, the limit of detection was 0.1 μg L^{-1}, while the recoveries obtained were between 78.8 and 102.2%. The developed method proved to be a promising, low-cost, time-saving, and reagent-saving sample technique. Most recently, Zhang and co-workers determined OP residues (trichlorfon, chlorpyrifos, phoxim, fenthion, and diazinon) in coarse cereals using DLLME (Zhang et al. 2021) combined with HPLC to rapidly and efficiently separate and detect. $[C_6C_1im][PF_6]$ was selected as the extraction agent and $[C_6C_1im][BF_4]$ as the dispersant. The results showed that the recoveries are between 85.3 and 107.3% proving that the applied method is simple, rapid, and efficient for extracting and separating OPs from coarse cereals. Recently, Anvar et al. developed a simple and reliable sample pretreatment method for the extraction and preconcentration of carbamate pesticides (pirimicarb, carbaryl, methiocarb, aldicarb, and carbofuran) from juice samples. The proposed method consisted of a two-step procedure in which an elevated temperature homogeneous liquid extraction coupled with ionic liquid-based dispersive liquid–liquid microextraction was developed to extract five carbamate pesticides from commercially packed grape, apple, sour cherry, pomegranate, and apricot juices. Cyclohexylamine, an organic solvent which is water-immiscible at high temperatures and water-miscible at room temperature, was used as an extraction solvent and disperser solvent in LLE and DLLME procedures, respectively. In the first step, the analytes were extracted from a sample solution thermostated at elevated temperature into a water-immiscible organic solvent (cyclohexylamine). Selection of a proper extraction solvent is considered a critical parameter as it should have desirable chromatographic behavior, extraction capability of the analytes, and different density from the aqueous phase. The authors studied three water-miscible ILs as possible extraction solvents, $[C_4C_1im]Cl$, $[C_4C_1im]Br$, and $[C_2C_1im]Cl$. The main parameter in their selection was that their solubility decreased in the presence of cyclohexylamine. To study the kind of extraction solvent in the performance of the developed method, $[C_4C_1im]Cl$, $[C_4C_1im]Br$, and $[C_2C_1im]Cl$ were subjected to the same procedure. Various volumes of the studied ILs to obtain the same volume of the sedimented phase and based on the obtained results 1-butyl-3-methylimidazolium chloride showed the highest analytical signals among the examined ILs. Therefore, it was chosen as the extractant in the next experiments. Therefore, in the second step, the separated organic phase was mixed with 1-butyl-3-methylimidazolium chloride (as the solvent for extraction) and the mixture was spread into deionized water. In this step, cyclohexylamine controlled the solubility of 1-butyl-3-methylimidazolium chloride in the aqueous phase and acted as a dispersive solvent and a phase separation agent, simultaneously. After centrifugation, all the gathered organic phases were removed and analyzed by using high-performance liquid chromatography diode-array detector. Limits of detection and quantification, enrichment factors, extraction recoveries, and relative standard deviations were obtained within ranges 0.4–3.9 and 1.5–13.2 ng mL^{-1}, 147–217, 59–87%, and 3.3–5.5%, respectively. These results indicate that the method is successful in determining the analytes. Heidary et al. (Heidari, Ghanbari-Rad, and Habibi 2020) developed a deep eutectic solvent-based ultrasound-assisted liquid-liquid microextraction (DES-UALLME) followed by quantitative analysis using HPLC-UV for the determination of organophosphorus pesticides (phosalone and chlorpyrifos) in fruit juice samples (red grape juice and sour cherry juice). Three different types of DES were applied composed of choline chloride with urea, ethylene glycol, and phenol. In DES-UALLME, ultrasonication was applied to transfer the aggregated of DES into tiny droplets. Central composite design was used in combination with a desirability function approach to find the optimum value of the parameters that influence on extraction efficiency, including pH, the volume of DES, the volume of the THF, and ultrasonication time. Under the optimized condition, the author obtained wide linear range (1–500 ng mL^{-1}), low detection limits (0.070and 0.096 ng mL^{-1}), high enrichment factors, and good extraction recoveries. The developed extraction/preconcentration technique was successfully applied for the determination of phosalone and chlorpyrifos in red grape juice and sour cherry juice samples with satisfactory relative recoveries (87.3–116.7%).

Summary of publication dealing with IL-DLLME is listed in Table 1. Although numerous ILs were investigated for the extraction of OPs, there is still a huge potential for further improvements.

Extraction of Organophosphorus Pesticides Using Ionic Liquid-Based Solid-liquid Extraction Techniques

Solid-phase microextraction (SPME) is a sensitive, solvent-free sampling and pre-concentration method which integrates several steps into one, namely, sampling, isolation, and pre-concentration (Souza-Silva et al. 2015). After developing by Pawliszyn (Arthur and Pawliszyn 1990) it becomes a very popular method for the detection of various organic compounds from different matrices. SPME is based on partitioning the analytes between a phase immobilized on SPME fiber and the sample matrix. The extraction ability of SPME fiber depends on the properties of fiber coating material (Souza-Silva et al. 2015). Different commercially available coatings such as poly(acrylate), poly(dimethylsiloxane), carbowax-templated resin, etc., express some disadvantages like low chemical and thermal stability and low selectivity of analytes (Nawała et al. 2018). Therefore, this technique can be advanced by developing novel coating materials that have high extraction capacity, good stability and can selectively extract specific groups of target analytes. Considering a great variety of ILs and their tuneable nature reflected in the tailoring of their physical and chemical properties (Rogers and Seddon 2003) by selecting appropriate ions, they seem to be an ideal alternative for the currently used SPME fiber coatings. Summary of publication dealing with different SPME methods is listed in Table 2.

Pang et al. (Pang et al. 2018) used ionogel-based IL fiber coating to determine OPs by headspace SPME (HS-SPME) combined with gas chromatography (GC) coupled to a flame photometric detector (FPD). 1-hexadecyl-3-methylimidazolium bis(trifluoromethylsulfonyl)imide, $[C_{16}C_1im][NTf_2]$, was restricted in the framework of silica-based ionogels, and the system was used as a sorbent in HS-SPME for isolation of four OPs (phorate, parathion, parathion-methyl, and malathion). The authors evaluated the applicability of the proposed method for the determination of OPs from wine and juice. They obtained very good LODs for all analytes extracted from real matrices that ranged from 0.57 – 4.13 ng L^{-1} with recoveries between 54 –118% and 56–109%. The authors concluded that IL-based HS-SPME was comparable to alternative analytical methods such as LLE, SPE, DLLME. Along the same line, GionfriddoIn et al. (Gionfriddo et al. 2018) prepared two polymeric ionic liquid (PIL)-based SPME coatings of different natures for the analysis of OPs (diazinon, methyl parathion, malathion, chlorpyrifos, parathion) in grapes. The PIL coatings contained either poly(1–4-vinylbenzyl-3-hexadecylimidazolium) bis[(trifluoromethyl)sulfonyl] imide, PIL 1, or N,N-didecyl-N-methyl-d-glucaminium poly(2-methyl-acrylic acid 2-[1-(3-{2- [2-(3-trifluoromethanesulfonylamino-propoxy)-ethoxy]-ethoxy}-propylamino)-vinylamino]-ethyl ester), PIL 2. The authors compared extraction efficiencies for these new coatings to that of commercial fibers during analysis by GC-MS. It was observed that PIL 1 coating yielded higher or comparable values with respect to commercial SPME coatings for most of the analytes studied. Obtained LOQs were 0.001 mg L^{-1} for diazinon, 0.025 mg L^{-1} for methyl parathion, and 0.01 mg L^{-1} for malathion, chlorpyrifos, and parathion.

Solid-phase microextraction is available in different configurations, like stir bar solid-phase microextraction (SB-SPME) or hollow fiber solid-phase microextraction (HF-SPME). In SB-SPME and HF-SPME, the adsorptive element is mobile and can turn freely in the analytes solution, increasing the transfer of analyzed molecules from the solution to the adsorbent. In SB-SPME, analytes are adsorbed on a layer of coating on a mobile adsorptive element which is similar to a magnetic stir bar. In HF-SPME, the extraction solvent is placed within a short, hollow fiber membrane sealed at both ends, which is then placed in a stirred aqueous sample solution. After adsorption, the target compounds are usually thermally desorbed or elute with a solvent and analyzed by chromatography (Y. Zhang et al. 2013). Two works of Ebrahimi et al. (Ebrahimi et al. 2011; Ebrahimi et al. 2012) reported the use of IL-mediated sol-gel sorbents for HF-SPME for determination of OP (diazinon, fenitrothion, malathion, fenvalerate, phosalone, and tridemorph) prior to HPLC detection. Novel solid sorbents containing $[C_4C_1im][OH]$ along with different nanoparticles (multi-walled carbon nanotubes, amino-functionalized multi-walled carbon nanotubes, nano SiO(2), nano TiO(2), and nano MgO) were used for pre-concentration, extraction, and determination of pesticides from water and hair samples (Ebrahimi et al. 2011). The authors found that nano MgO provided the best results for simultaneous analysis of all target pesticides. In terms of analytical performance, the LODs obtained were in the range between 0.004 and 0.095 mg L^{-1} for the OPs in the aqueous matrices and 0.003–0.080 mg L^{-1} in the hair matrices (Ebrahimi et al. 2011). The same group of authors evaluated Keggin-based IL ($(PYBS)_3PW_{12}O_{40}$) sol-gel sorbent as an extraction medium for HF-SPME (Ebrahimi et al. 2012). The LOD obtained for all analytes (diazinon, fenitrothion, and malathion) extracted from hair and water samples ranged from 0.0074 to 1.3000 g g^{-1} and 0.00034–0.84 mg L^{-1}, respectively. Excellent extraction efficiencies, and consequently low LODs, obtained applying IL-mediated sol-gel sorbents in solid-phase extraction were the result of IL contribution to the better porosity, higher surface area, and homogeneity of active sites of the sorbents (Ebrahimi et al. 2011; Ebrahimi et al. 2012). Zhang et al. (Zhang et al. 2013) studied IL-based HF-SPME combined with HPLC for the detection and measurement of OPs (parathion-methyl, fenitrothion, parathion, and phoxim) in water samples. The authors optimized various process variables such as types of IL, stirring speed, extraction time, extraction temperature, and salt effect on extraction efficiency. Among the studied IL ($[C_6C_1im][PF_6]$, $[C_8C_1im][PF_6]$, $[C_6C_1im][BF_4]$, and $[C_8C_1imim][BF_4]$) the highest enrichment factor was obtained employing the most hydrophobic $[C_8C_1im][PF_6]$ which also expressed high solubility for the target compounds. The proposed method was successfully applied in the analysis of real environmental water samples under the optimal extraction conditions, and good spiked recoveries of 86.71–

103.7% were obtained with linearity between 1–200 mg L^{-1} and the detection limits in the range 0.015–0.026 mg L^{-1} for each OPs.

Magnetic solid-phase extraction (MSPE) is one of the improved versions of SPME (Souza-Silva et al. 2015). MSPE uses magnetic adsorbents (usually contain iron-oxide nanoparticles modified with different materials) dispersed directly into a sample solution, which can easily separate from bulk solution by application of the external magnetic field. Dispersive extraction mode enhances the contact area between adsorbents and analytes. Furthermore, magnetic adsorbents can be recycled and reused easily, which makes this method cost-effective and environmentally friendly. The selectivity and sensitivity of the MSPE technique mostly depend on the adsorbent material. As ILs can be synthesized with various functional groups in their structure, capable of specific interaction with analytes, IL can be considered as a promising magnetic nanoparticles modification agent (Zheng et al. 2014). Poly(ionic liquid)-modified magnetic nanoparticles were developed as a new adsorbent material via co-polymerization of 1-vinyl-3-hexylimidazolium-based ionic liquid and vinyl-modified magnetic particles (Zheng et al. 2014). Prepared PIL-based material was applied as an adsorbent phase in MSPE coupled with HPLC for detection of parathion, fenthion, phoxim, and temephos. A low LOD of 0.01 μg L^{-1} and high enrichment factors that ranged between 84 to 161 were acquired with the proposed method. Authors successfully applied PIL-based material to determine OPs from three tea drink samples with a quite good recovery of 81.4–112.6% for selected OPs. The reusability of the sorbents was also addressed, and the researchers reported that PIL-based magnetic nanoparticles were mechanically stable and could be recycled 20 times. Recently, Amiri et al. (Amiri et al. 2018) reported a method for the detection of OPs using a magnetic sorbent consisting of a polyoxometalate of Keggin type [$BeW_{12}O_{40}$]-supported on IL functionalized silica-coated cobalt ferrite nanoparticles [ILSCCFNP]. The synthesized nanocomposite [BeW12O40-ILSCCFNPs] was used as an MSPE adsorption material for the extraction of five OPs (diazinon, phosalone, fenthion, fenitrothion, and profenofos) prior to GC–FID detection. The proposed method showed LODs in the range from 0.02 to 0.06 mg L^{-1}. Comparing the extraction efficiencies of ILSCCFNPs and $BeW_{12}O_{40}$-ILSCCFNPs under the same condition revealed the effect of polyoxometalate in the material.

$BeW_{12}O_{40}$-ILSCCFNPs had a higher extraction efficiency than ILSCCFNPs, which explained that the presence of $BeW_{12}O_{40}$ on the ILSCCFNPs surface increased the number of active adsorbent sites. The authors demonstrated that the adsorbent could be reused at least 10 times without significant loss of extraction efficiency. Described MSPE was applied for the determination of OPs in real water samples and fruit juices and provided good recoveries (90.9–99.3%) and acceptable precision (6.2–10.2%). The pretreatment procedure, based on described technique, is very fast, sensitive, and eco-friendly, but on the other hand, the synthesis and preparation of the required material is a complex and long process which is the main disadvantage of the method.

Table 2. Ionic liquids used in various SPME methods for the extraction of OPs

Ionic liquid	Analytes	Ref
$[C_{16}C_1im][NTf_2]$	phorate, parathion, parathion-methyl, and malathion	(Pang et al. 2018)
poly(1–4-vinylbenzyl-3-hexadecylimidazolium) bis[(trifluoromethyl)sulfonyl] imide (poly [VBHDIM][NTf2]), N,N-didecyl-N-methyl-d-glucaminium poly (2-methyl-acrylic acid 2-[1-(3-{2- [2-(3-trifluoromethanesulfonylamino-propoxy)-ethoxy]-ethoxy}-propylamino)-vinylamino]-ethyl ester)	diazinon, methyl parathion, malathion, chlorpyrifos, parathion	(Gionfriddo et al. 2018)
$[C_4C_1im][OH]$	diazinon, fenitrothion, malathion, fenvalerate, phosalone, tridemorph	(Ebrahimi et al. 2011)
$(PY\ BS)_3PW_{12}O_{40}$ *(IL based on Keggin heteropoly acid, $H_3PW_{12}O_{40}$, py -pyridine, BS-1,4 butane sultone)*	diazinon, fenitrothion malathion	(Ebrahimi et al. 2012)
$[C_6C_1im][PF_6]$, $[C_8C_1im][PF_6]$, $[C_6C_1im][BF_4]$, $[C_8C_1imim][BF_4]$	parathion-methyl, fenitrothion, parathion, phoxim	(Y. Zhang et al. 2013)
Poly IL based on [vinyl C_6im] Br	parathion, fenthion, phoxim, temephos	(X. Zheng et al. 2014)
$[BeW_{12}O_{40}]^{6-}$ -IL	diazinon, phosalone, fenthion, fenitrothion, and profenofos	(Amiri et al.2018)
$[C_4C_1im]Br$	methamidophos, malathion, parathion, diazinon	(Chen et al. 2021)

By employing a method known as effervescent reaction-enhanced magnetic solid-phase extraction (ERMSE) could be achieved better dispersion of nanoparticles in microextraction. ERMSE is based on a simple in situ reaction that generates carbon dioxide bubbles, which then accelerates the dispersion of magnetic nanoparticles in solution. Chen et al. (Chen et al. 2021) reported the preparation of effervescent tablets of magnetic nanocomposite based on IL-coated $NiFe_2O_4$-SiO_2-polyaniline. Adsorbent material was prepared using hydrophilic IL, $[C_4C_1im]Br$, and $NiFe_2O_4$-SiO_2-polyaniline, which were mixed with acidic salts (tartaric acid) and alkaline salts (Na_2CO_3) to compose effervescent tablets. The obtained magnetic effervescent tablets were employed prior to HPLC for the detection of four OPs (methamidophos, malathion, parathion, diazinon) from fruit juice samples. Under optimized conditions, the ERMSE method achieved low LOD (0.06–0.17 mg L^{-1}) and high recoveries (80.6–97.3%) for OP quantification in real samples. Good extraction efficiency was a consequence of two effects: enhanced interactions between OPs and the nano adsorbents due to the homogeneous dispersion of adsorbents by CO_2 bubbles and a larger number of active sites, as well as better textural characteristics due to attachment of SiO_2, polyaniline, and ILs onto the surface of $NiFe_2O_4$. Researchers studied the reusability of the nanocomposite and found that it could be reused for at least eight extractions without significant loss.

Overview of IL Role in Electroanalytical Methods for Organophosphate Pesticides Detection

As discussed, the analytical methods usually employed for the detection of OPs (HPLC, GC, GC-MS) require pretreatment (sample preparation and preconcentrating steps), which can be a time-consuming and complex procedure. On the other hand, electrochemical biosensors, commonly based on the inhibition of the acetylcholinesterase enzyme, can offer faster and easier detection of OPs with a significant decrease in cost (Rana et al. 2019; Wei and Wang 2015). More precisely, the AChE immobilized on an electrode surface catalyzes the hydrolysis of acetylthiocholine chloride (ATCl) producing thiocholine, which can induce an irreversible oxidation peak upon a certain applied potential. This peak is directly proportional to the enzyme activity, i.e., inversely proportional to OPs concentration. OPs inhibit the enzyme activity, oxidation of thiocholine decreases, and at the same time, the oxidation peak current decreases (Wei and Wang 2015; Dongfei Chen et al. 2017). Therefore, the selection of materials for enzyme immobilization is a critical step in biosensor design as they need to have a high surface area, fast charge transfer, and simultaneously retain enzyme bioactivity (Rana et al. 2019). Various carbon materials are used in biosensing applications owing to their favorable properties such as surface area, mass transport, conductivity, and chemical stability. However, ILs can significantly improve carbon nanomaterials properties that are important for use in the field of electrochemical biosensors owing to their unique features, such as wide electrochemical windows, excellent conductivity, negligible vapor pressure, and high chemical stability (Wei and Wang 2015). Zhan et al. (Zhan et al. 2015) developed an electrochemical AChE biosensor for detection of OPs (phorate) using Au–Pd bimetallic nanoparticles combined with IL functionalized graphene-chitosan nanocomposites (Au–Pd/IL-GR–CHI). Using hybrid nanocomposites accelerated direct electron transfer from the enzyme to the electrode surface and enhanced the amount of immobilized AChE, which greatly improved the electrochemical properties of the biosensor interface. Under optimal conditions, the obtained linearities were from 5.0 10^{-16} to 2.5 10^{-13} mol L^{-1} and from 4.9 10^{-13} to 9.5 10^{-6} mol L^{-1}, with LOD of 2.5 10^{-16} mol L^{-1}. The biosensor was also tested on apple juice with high recoveries obtained between 91.0% and 110.0% for phorate detection. Wei et al. (Wei and Wang 2015) investigated AChE biosensors based on IL-Au nanoparticles-porous carbon composite modified boron-doped diamond electrode for detection of OP. Honeycomb-like hierarchically porous carbon, AuNPs, and 1-(4-Sulfonic acid) butyl-3-methylimidazolium hydrogensulfate ([BSC_1im] HSO_4) were combined and used to immobilize AChE with retained activity. Prepared composite has few advantages, including the high accessible surface area, which improves AChE adsorption and the number of the reactive site and appropriate conductivity to make the electron transfer easier, enhancing the sensitivity of response. Researchers used dichlorvos as a model molecule to test constructed biosensor and obtained a LOD of 6.61 10^{-11} g L^{-1}. After usage, 91.7% of the activity of AChE/[BSC_1im]HSO_4-AuNPs-porous carbon biosensor was regenerated by the application of pralidoxime iodide. The proposed biosensor was employed to detect dichlorvos in lettuce leaves sample, and recovery was ranged between 80.8% and 93.1%. Zheng et al. (Zheng et al. 2016) modified the glassy carbon electrode with a composite prepared of IL (1-(3-aminopropyl)-3-methylimidazolium bromide) functionalized graphene (IL-GR), Co_3O_4 nanoparticles, and chitosan. Then, AChE was immobilized on obtained composite film and applied for electrochemical detection of OP. The proposed method was successfully applied for the detection of dimethoate with LOD of 1.0 10^{-13} mol L^{-1}. After detection, inhibited biosensor was regenerated 92.3% of original activity by immersion in pralidoxime iodide. Performances of the biosensor were tested for detection of dimethoate in lettuce leaves sample with recovery from 92.5%

to 107.5%, suggesting that the proposed biosensor is adequate for the analysis of real samples. Chen et al. (Chen et al. 2017) constructed a sensitive electrochemical biosensor based on an AChE/multi-walled carbon nanotubes/IL composite modified screen-printed electrode for detection of chlorpyrifos. Octyl pyridinium hexafluorophosphate-based composite used to modify electrode improved biosensor's electrochemical properties, improved microarchitecture of the electrode surface and increased the electron transfer rate. Similar to the above-mentioned publication dealt with IL-based biosensing materials, AChE was also immobilized onto the modified electrode to build OP biosensor. Under optimal conditions, the a wide linearity range (0.05 -1.0 10^5 μg L^{-1}) was obtained and low LOD (0.05 μg L^{-1}) for the determination of chlorpyrifos. Furthermore, determination of chlorpyriphos was performed in rape samples with obtained recovery values ranged between 87% and 101% indicating applicability of studied biosensor for direct analysis of vegetable samples. In order to design efficient electrochemical detector for OPs, Rana et al. (Rana et al. 2019) immobilized the AChE enzyme on reduced graphene oxide which was functionalized by poly(3,4-ethylene dioxythiophene) and IL (1-Butyl-3-methylimidazolium trifluoromethane-sulfonate). Applied IL increased surface area and provided more redox sites on electrode material leading to a higher electron transfer rate and consequently a more sensitive biosensor. The prepared biosensor was utilized for the determination of three OPs with LOD of 0.04 mg L^{-1}, 0.117 mg L^{-1} and 0.108 mg L^{-1} for chlorpyrifos, malathion, and methyl parathion, respectively. Good sensitivity, selectivity, and reproductivity were obtained for determination of OP from real samples (apple juice and aerated drinks). Moreover, the authors showed that the sensor can be 91.7% reactivated after use. Ma et al. (Ma et al. 2016) modified glassy carbon electrode with IL-graphene nanosheets (IL-GNs) and applied for sensitive detection of methyl parathion. In contrary to above described publication, authors constructed an enzymeless electrochemical detector. Synthesized IL-based material enhanceed the redox reaction of OP at the electrode and allows direct detection. The LOD of methyl parathion was 1.1 mg L^{-1}, and it is comparable with previous reported electrochemical sensors. Proposed method was successfully applied for detection of methyl parathion in spiked water sample, with recoveries ranged from 98.5 to 104.4%.

Ionic Liquids as Promising Media for Degradation of Organophosphate Pesticides

Organophosphate pesticides are used to save crops from pests. However, most of their unused portion and by-products remain as a contaminant in the soil, causing loss of fertility, acidification of soil, nitrate leaching, increased resistance of weed species, and loss of biodiversity. Therefore, due to the high toxicity and bioaccumulation shown by OPs, many methods have been developed for their degradation, such as homogeneous and heterogeneous hydrolysis, photolysis, photochemical degradation, and biodegradation, as well as chemical treatments based on the use of strong oxidants or nucleophiles (Sidhu et al. 2019). Different pathways of degradation depend primarily on their alkylation or arylation state and the nature of the nucleophile. Nevertheless, some of these treatments may not be very efficient or are unfriendly to the environment due to the formation of products that have mild or acute toxicity. Therefore, the efficiency of nucleophilic degradation of organophosphorus pesticides depends on the nucleophile, the nature of the phosphoryl center of the OP, and the type of solvent used. Chemical degradation of organophosphorus pesticides through different approaches in green solvents such as ionic liquids has been studied in recent years. These studies helped to understand the mechanism of the reaction of OP, which can be used to create efficient and sustainable degradation methods.The first demonstration of the effects of various ILs on the efficiency of chloroperoxidase oxidation of methyl-parathion was carried out by Boškin et al. (Boškin, Tran, and Franko 2009). The effects of ILs on the activity and function of various enzymatic systems are of significance for bioanalytical methods. These techniques rely on enzymes in media, affecting the activity of the enzyme and other parameters contributing to higher sensitivity and/or selectivity of an analytical technique. For example, thio-forms of OPs (e.g., methyl-parathion) exhibit low in vitro inhibition potency towards AChE, and in order to achieve a satisfactory limit of detection of thio-OPs, such compounds need to be oxidized to their oxo-forms (e.g., methyl-paraoxon). Enhancements in the sensitivity of analytical measurements can be achieved in ILs as a solvent, and Boškin et al. investigated the most suitable IL and experimental conditions required to achieve efficient oxidation of thio-OPs compounds and the highest possible sensitivity of a flow injection system by thermal lens spectrometry assay. In general, the activity of chloroperoxidase decreases with the increased concentration of ILs. ILs with PF_6anion exhibit the strongest inhibition toward chloroperoxidase. It was further demonstrated that even for buffer solutions containing up to 30% concentration of ILs such as ethyl ammonium nitrate, 1-butyl-3-methylimidazolium methyl- sulfate, and N-ethyl pyridinium trifluoroacetate, chloroperoxidase still can retain high activity (70–100% compared to that in citrate buffer) for oxidation of methyl-parathion. Photothermal enhancement factors of 3.5 times and corresponding improvements of sensitivity in the determination of OPs by the thermal lens spectrometry method were predicted in 30% ionic liquids. In 2013 Pavez et al. studied mechanisms of degradation of paraoxon in different ILs (Pavez et al. 2013). The effects of ILs as solvents on rate and product distribution of the nucleophilic substitution reactions of phosphoester compounds was studied using the reaction of paraoxon (**1**, O,O-diethyl 4-nitrophenyl phosphate trimester) with piperidine in ten ILs ([C_4C_1im]BF_4, [C_4C_1im]DCA, [C_4C_1im]NTf_2, [C_4C_1im]PF_6, [C_2C_1im]NTf_2, [$C_4C_1C_1$im]BF_4, [$C_2C_1C_3$am] NTf_2, [C_4C_1pyrr]NTf_2, [C_4C_1py]DCA, [C_4C_1pyrr]DCA) and in three conventional organic solvents (dioxane, ACN and DMSO). The reactivity and selectivity of the reaction of OP with piperidine in different solvents were studied employing ^{31}P NMR, UV−VIS, and GC/MS methods. The authors reported that three phosphorylated products were identified: O,O-diethyl piperidinophosphate diester (**2**), O,O-diethyl phosphate (**3**), and O-ethyl 4-nitrophenyl phosphate diester (**4**). Compound 4 also reacted with piperidine to yield O-ethyl piperidinophosphate monoester (**5**). The authors found that both selectivity and rate constant depend on the nature of ILs' anion and cation. Correlation was established between hydrogen-bond acceptor ability of the anion and rate constants values. Regarding IL cation effect, the highest rate constant values are obtained when aromatic ILs are used suggesting that the highest π stacking interaction between the aromatic cation of the IL and the transition state of the reaction, in comparison with the π stacking interaction between aromatic cation and reactants. Both hydrogen-bond acceptor ability and aromaticity of the ILs' cations must be taken into account to establish the effect of the cations' nature rate constant. A better understanding of the ILs' effect could be explained, taking into account the three-dimensional structure of ILs, where solute−solvent interactions are much more complex. The overall conclusion of the authors was that paraoxon degradation was obtained with high efficiency, yielding phosphate products less lipophilic and therefore less toxic to the human being. Following a similar line of research, in 2016, the same research group investigated and compared the effect of twelve ILs in comparison to six bio-based solvents on the degradation of two OPs, fenitrothion and paraoxon, through a nucleophilic substitution reaction (Pavez, Oliva, and Millán 2016). The kinetic results showed that when ILs and a bio-based solvent are used as reaction media for the degradation of fenitrothion, the process is more efficient, showing half-life values in the range of 4−11 min. Although ILs were generally more efficient in comparison to bio-based solvents for the degradation of OPs, the authors gave an advantage to bio-based solvents having in mind their lower cost and toxicity. The results showed that the use of bio-based solvents as media for degradation of studied OPs not only leads to a more efficient degradation but also assures a more sustainable process. After this study, the authors continued to search for more efficient ILs and enlarged the number of ILs suitable for degradation processes providing important insights for task-specific IL design. In 2018, a series of 1-butyl 3-methylimidazolium-based ILs were synthesized with different amino acids as anions, namely, L-cysteine, L-alanine, L-proline, L-histidine, L-phenylalanine, L-serine, L-glycine, L-methionine, and L-asparagine (Morales et al. 2018). These ILs were used for the first time as reaction media to achieve more eco-friendly paraoxon degradation. Kinetics and product distribution results showed that paraoxon degradation was reached more efficiently without an extra nucleophilic agent in comparison to conventional ILs based on inorganic anion where the addition of a nucleophile was necessary. Therefore, the authors proposed that all the ILs used in this study have a dual role in the outcome of this reaction; as a nucleophile and a solvent to carry out degradation of the OPs. Both kinetics and product distribution results found in this study for paraoxon degradation turned out to be promising as this process is achieved in a reaction medium with a better environmental profile. Recently, following a similar analogy that some ILs play a nucleophilic role, Millan et al. (Millán et al. 2020) investigated the influence of microwave heating on the degradation of OP with amino acid-based ILs and piperidine as nucleophiles using propylene carbonate as the solvent. The pesticides malathion, fenitrothion, paraoxon, diazinon and parathion were chosen based on their electrophilic center phosphate (P=O) or phosphorothioate (P=S). The authors concluded that fenitrothion and malathion do not need MW irradiation to achieve complete degradation with either nucleophile used in this study. Fenitrothion degradation was faster with piperidine as the nucleophile, while malathion degradation was carried out at a similar rate with both nucleophiles. On the other hand, the degradation of paraoxon with ILs was accomplished at room temperature in a few hours, which is an important result because the degradation with piperidine was not successful. MW heating is a promising technique that accelerates the degradation of the OP paraoxon, diazinon, and parathion with piperidine which is effective in only 30 minutes at above 90 °C. The temperature does not affect the reaction pathways of the degradation reactions studied. For the reaction of paraoxon, only one phosphorylic compound was found in contrast to the three phosphorylic compounds observed in the nucleophilic degradation of paraoxon in other solvents. This study highlights the efficiency of the microwave-assisted degradation of OP in propylene carbonate that proceeded at a significantly higher rate compared to conventional heating and silent conditions. In 2020, a novel strategy for the degradation of OPs was proposed by Pavez et al. using bio-based ILs with choline cation (Pavez et al. 2020). ILs composed of choline cation and different amino acids as anions and their respective aqueous mixtures were prepared in a range of 0.4−46 mol % IL. These solvents were used for the first time to achieve paraoxon degradation with improved green credentials. The results showed that IL/water mixtures were an effective reaction medium to degrade paraoxon, even when the IL content in the mixture is low (0.4 mol % IL) and without the need for an extra nucleophile. Both the kinetics and the degradation pathways of pesticides depend on the nature of the IL anion and the amount of an IL present in the mixture. The authors demonstrated that in those mixtures with a low amount of [Ch][AA], the hydrolysis reaction is the main pathway for paraoxon degradation, showing a catalytic effect of the IL. However, as the percentage of [Ch][AA] increases in the mixture, the nucleophilic attack of IL was evident. Finally, the aim of this study was to provide evidence of a promising and biocompatible methodology to degrade a toxic compound using a minimal quantity of an IL designed totally from natural resources.

Conclusion

This work aimed to provide an overview of the application of ILs for the extraction, electrochemical detection, and degradation of different OPs. The ILs were either applied as solvents or as support materials to tune the extraction or adsorption capacity of different materials. In general, it was demonstrated that, if properly selected, IL-based solvents and materials are able to afford higher extraction efficiency compared to traditional systems. However, regardless of the efforts made by different authors to test different ILs and to investigate various process conditions, most studies are based on ILs comprising imidazolium cations and fluorinated anions, such as $[NTf_2]^-$, $[PF_6]^-$, and $[BF_4]^-$. These ILs are more expensive and also moderately toxic, have low biodegradability, and some of them, such as $[PF_6]$ and $[BF_4]$-based ILs, are water-unstable. After screening the literature dealing with IL-based processes for the determination of different OPs, it is clear that there is a lot of room for further improvement in this area. For example, using hydrophilic ILs seems to be a more promising option due to their more prominent chemical structure diversity, which can allow more task-specific applications. The potential application of this class of ILs has been demonstrated in the extraction of different polarity pesticides. Moreover, hydrophilic ILs are a particularly good option given that their aqueous solutions allow overcoming the high viscosity of some ILs. At the same time, they would contribute to an increase in the biocompatible nature of the solvent and to a decrease in the overall solvent cost. ILs have been shown to be successful in the extraction of small extractable compounds by favoring their solubility in aqueous solutions. In more recent years, research has been directed to the use of more benign IL alternatives, such as those composed of cholinium cations. Further work needs to be carried out toward the development of cost-effective and more sustainable IL-based processes as they can take their place in the determination and degradation of OPs.

References

Ahmadzadeh Anvar, Sanaz, Mohammadali Torbati, Mir Ali Farajzadeh, and Mohammad Reza Afshar Mogaddam. 2020. "Elevated Temperature Homogeneous Liquid Phase Extraction Coupled to Ionic Liquid–Based Dispersive Liquid–Liquid Microextraction Followed by High-Performance Liquid Chromatography: Application of Water-Miscible Ionic Liquids as Extraction Solvent in Determ." *Food Analytical Methods* 13 (6): 1282–91. https://doi.org/10.1007/s12161-020-01742-2.

Albishri, Hassan M., Naflaa A. M. Aldawsari, and Deia Abd El-Hady. 2016. "Ultrasound-Assisted Temperature-Controlled Ionic Liquid Dispersive Liquid-Phase Microextraction Combined with Reversed-Phase Liquid Chromatography for Determination of Organophosphorus Pesticides in Water Samples." *Electrophoresis* 37 (19): 2462–69. https://doi.org/10.1002/elps.201600107.

Amiri, Amirhassan, Reza Hamid Saadati-Moshtaghin, and Farokhzad Mohammadi Zonoz. 2018. "A Hybrid Material Composed of a Polyoxometalate of Type BeW12O40 and an Ionic Liquid Immobilized onto Magnetic Nanoparticles as a Sorbent for the Extraction of Organophosphorus Pesticides Prior to Their Determination by Gas Chromatography." *Microchimica Acta* 185 (3). https://doi.org/10.1007/s00604-018-2713-x.

Arthur, C. L., and J. Pawliszyn. 1990. "Solid Phase Microextraction with Thermal Desorption Using Fused Silica Optical Fibers." *Analytical Chemistry*, 2145–48.

Boškin, Aleš, Chieu D. Tran, and Mladen Franko. 2009. "Oxidation of Organophosphorus Pesticides with Chloroperoxidase Enzyme in the Presence of an Ionic Liquid as Co-

Solvent." *Environmental Chemistry Letters* 7 (3): 267–70. https://doi.org/10.1007/s10311-008-0161-2.

Cacho, J. I., N. Campillo, P. Viñas, and M. Hernández-Córdoba. 2018. "In Situ Ionic Liquid Dispersive Liquid-Liquid Microextraction Coupled to Gas Chromatography-Mass Spectrometry for the Determination of Organophosphorus Pesticides." *Journal of Chromatography A* 1559: 95–101. https://doi.org/10.1016/j.chroma.2017.12.059.

Chatel, Gregory, Jorge F. B. Pereira, Varun Debbeti, Hui Wang, and Robin D. Rogers. 2014. "Mixing Ionic Liquids-"simple Mixtures" or 'Double Salts'?" *Green Chemistry* 16 (4): 2051–83. https://doi.org/10.1039/c3gc41389f.

Chen, Dechao, Sai Ma, Xiaofan Zhang, Xuedong Wang, Ming Gao, Jieyi Li, and Huili Wang. 2021. "Enhanced Extraction of Organophosphorus Pesticides from Fruit Juices Using Magnetic Effervescent Tablets Composed of the NiFe2O4@SiO2@PANI-IL Nanocomposites." *RSC Advances* 11 (3): 1668–78. https://doi.org/10.1039/d0ra09100f.

Chen, Dongfei, Jiayun Fu, Zengning Liu, Yemin Guo, Xia Sun, Xiangyou Wang, and Zhiqiang Wang. 2017. "A Simple Acetylcholinesterase Biosensor Based on Ionic Liquid/Multiwalled Carbon Nanotubes-Modified Screen-Printed Electrode for Rapid Detecting Chlorpyrifos." *International Journal of Electrochemical Science* 12 (10): 9465–77. https://doi.org/10.20964/2017.10.12.

Colovic, Mirjana B., Danijela Z. Krstic, Tamara D. Lazarevic-Pasti, Aleksandra M. Bondzic, and Vesna M. Vasic. 2013. "Acetylcholinesterase Inhibitors: Pharmacology and Toxicology." *Current Neuropharmacology* 11 (3): 315–35. https://doi.org/10.2174/1570159x11311030006.

Deive, F. J., a. Rodríguez, a. B. Pereiro, J. M. M. Araújo, M. a. Longo, M. a. Z. Coelho, J. N. Canongia Lopes, J. M. S. S. Esperança, L. P. N. Rebelo, and I. M. Marrucho. 2011. "Ionic Liquid-Based Aqueous Biphasic System for Lipase Extraction." *Green Chemistry* 13 (2): 390. https://doi.org/10.1039/c0gc00075b.

Dimitrijević, Aleksandra, Ljubiša Ignjatović, Aleksandar Tot, Milan Vraneš, Nebojša Zec, Slobodan Gadžurić, and Tatjana Trtić-Petrović. 2017. "Simultaneous Extraction of Pesticides of Different Polarity Applying Aqueous Biphasic Systems Based on Ionic Liquids." *Journal of Molecular Liquids* 243: 646–53. https://doi.org/10.1016/j.molliq.2017.08.077.

Dimitrijević, Aleksandra, Ana Jocić, Nebojša Zec, Aleksandar Tot, Snežana Papović, Slobodan Gadžurić, Milan Vraneš, and Tatjana Trtić-Petrović. 2019. "Improved Single-Step Extraction Performance of Aqueous Biphasic Systems Using Novel Symmetric Ionic Liquids for the Decolorisation of Toxic Dye Effluents." *Journal of Industrial and Engineering Chemistry* 76: 500–507. https://doi.org/10.1016/j.jiec.2019.04.017.

Dimitrijević, Aleksandra, Ana P. M. Tavares, Mafalda R. Almeida, Milan Vraneš, Ana C. A. Sousa, Ana C. Cristóvão, Tatjana Trtić-Petrović, Slobodan Gadžurić, and Mara G. Freire. 2020. "Valorization of Expired Energy Drinks by Designed and Integrated Ionic Liquid-Based Aqueous Biphasic Systems." *ACS Sustainable Chemistry and Engineering* 8 (14): 5683–92. https://doi.org/10.1021/acssuschemeng.0c00429.

Dimitrijević, Aleksandra, Jelena Milićević, Ana Jocić, Slađana Marić, Tatjana Trtić-Petrović, Snežana Papović, Aleksandar Tot, Slobodan Gadžurić, and Milan Vraneš.

2020. "Further Insight into the Influence of Functionalization and Positional Isomerism of Pyridinium Ionic Liquids on the Aqueous Two-Phase System Equilibria." *Fluid Phase Equilibria* 512. https://doi.org/10.1016/j.fluid.2020.112520.

Ebrahimi, Mahmoud, Zarrin Es'haghi, Fatemeh Samadi, Fatemeh Farrash Bamoharram, and Mohammad Saeid Hosseini. 2012. "Rational Design of Heteropolyacid-Based Nanosorbent for Hollow Fiber Solid Phase Microextraction of Organophosphorus Residues in Hair Samples." *Journal of Chromatography A* 1225: 37–44. https://doi.org/10.1016/j.chroma.2011.12.077.

Ebrahimi, Mahmoud, Zarrin Es'haghi, Fatemeh Samadi, and Mohammad Saeid Hosseini. 2011. "Ionic Liquid Mediated Sol-Gel Sorbents for Hollow Fiber Solid-Phase Microextraction of Pesticide Residues in Water and Hair Samples." *Journal of Chromatography A* 1218 (46): 8313–21. https://doi.org/10.1016/j.chroma.2011.09.058.

Epa, Ocspp, and Opp. 2008. "*US EPA - Pesticides Industry Sales and Usage 2008 - 2012*."

Farajzadeh, M. A., L. Khoshmaram, and S. Sheykhizadeh. 2014. "A Review on Application of Microextraction Techniques for Analysis of Chemical Compounds and Metal Ions in Foodstuffs." *Analytical and Bioanalytical Chemistry Research* 1 (1): 1–19. https://doi.org/10.22036/abcr.2014.4807.

Foong, Shin Ying, Nyuk Ling Ma, Su Shiung Lam, Wanxi Peng, Felicia Low, Bernard H. K. Lee, Aage K. O. Alstrup, and Christian Sonne. 2020. "A Recent Global Review of Hazardous Chlorpyrifos Pesticide in Fruit and Vegetables: Prevalence, Remediation and Actions Needed." *Journal of Hazardous Materials* 400 (April): 123006. https://doi.org/10.1016/j.jhazmat.2020.123006.

Freire, Mara G., Ana Filipa M. Cláudio, João M. M. Araújo, João A. P. Coutinho, Isabel M. Marrucho, José N. Canongia Lopes, and Luís Paulo N. Rebelo. 2012. "Aqueous Biphasic Systems: A Boost Brought about by Using Ionic Liquids." *Chemical Society Reviews* 41 (14): 4966. https://doi.org/10.1039/c2cs35151j.

Gionfriddo, Emanuela, Érica A. Souza-Silva, Tien D. Ho, Jared L. Anderson, and Janusz Pawliszyn. 2018. "Exploiting the Tunable Selectivity Features of Polymeric Ionic Liquid-Based SPME Sorbents in Food Analysis." *Talanta* 188 (April): 522–30. https://doi.org/10.1016/j.talanta.2018.06.011.

Gomes, Joana M., Simone S. Silva, and Rui L. Reis. 2019. "Biocompatible Ionic Liquids: Fundamental Behaviours and Applications." *Chemical Society Reviews* 48 (15): 4317–35. https://doi.org/10.1039/c9cs00016j.

Goossens, Karel, Kathleen Lava, Christopher W. Bielawski, and Koen Binnemans. 2016. "Ionic Liquid Crystals: Versatile Materials." *Chemical Reviews*. American Chemical Society. https://doi.org/10.1021/cr400334b.

Greaves, Tamar L., and Calum J. Drummond. 2015. "Protic Ionic Liquids: Evolving Structure-Property Relationships and Expanding Applications." *Chemical Reviews* 115 (20): 11379–448. https://doi.org/10.1021/acs.chemrev.5b00158.

He, Lijun, Xianli Luo, Xiuming Jiang, and Lingbo Qu. 2010. "A New 1,3-Dibutylimidazolium Hexafluorophosphate Ionic Liquid-Based Dispersive Liquid-Liquid Microextraction to Determine Organophosphorus Pesticides in Water and Fruit Samples by High-Performance Liquid Chromatography." *Journal of Chromatography A* 1217 (31): 5013–20. https://doi.org/10.1016/j.chroma.2010.05.057.

He, Lijun, Xianli Luo, Hongxue Xie, Chunjian Wang, Xiuming Jiang, and Kui Lu. 2009. "Ionic Liquid-Based Dispersive Liquid-Liquid Microextraction Followed High-Performance Liquid Chromatography for the Determination of Organophosphorus Pesticides in Water Sample." *Analytica Chimica Acta* 655 (1–2): 52–59. https://doi.org/10.1016/j.aca.2009.09.044.

Heidari, Hassan, Samira Ghanbari-Rad, and Esmaeil Habibi. 2020. "Optimization Deep Eutectic Solvent-Based Ultrasound-Assisted Liquid-Liquid Microextraction by Using the Desirability Function Approach for Extraction and Preconcentration of Organophosphorus Pesticides from Fruit Juice Samples." *Journal of Food Composition and Analysis* 87 (April): 103389. https://doi.org/10.1016/j.jfca.2019.103389.

Hurley, Frank H., and Thomas P. Wier. 1951. "Electrodeposition of Metals from Fused Quaternary Ammonium Salts." *Journal of The Electrochemical Society* 98 (5): 203. https://doi.org/10.1149/1.2778132.

Jocić, Ana, Slađana Marić, Aleksandra Dimitrijević, Aleksandar Tot, Slobodan Gadžurić, Milan Vraneš, and Tatjana Trtić-Petrović. 2020. "Protic Ionic Liquids as Adjuvants to Enhance Extraction and Separation Performance of Diverse Polarity Compounds in PEG-Salt Based Aqueous Biphasic System." *Journal of Molecular Liquids* 303. https://doi.org/10.1016/j.molliq.2020.112484.

Khan, Imran, Reddicherla Umapathi, Márcia C. Neves, João A. P. P. Coutinho, and Pannuru Venkatesu. 2016. "Structural Insights into the Effect of Cholinium-Based Ionic Liquids on the Critical Micellization Temperature of Aqueous Triblock Copolymers." *Physical Chemistry Chemical Physics* 18 (12): 8342–51. https://doi.org/10.1039/C5CP07079A.

Liu, Wei, and Ji Quan. 2020. "A Novel Ionic Liquid of [BeMIM] [Tf2N] for Extracting Pesticides Residues in Tea Sample by Dispersive Liquid–Liquid Microextraction." *Chromatographia* 83 (1): 41–51. https://doi.org/10.1007/s10337-019-03819-5.

Liu, Wei, Ji Quan, and Zeshu Hu. 2018. "Detection of Organophosphorus Pesticides in Wheat by Ionic Liquid-Based Dispersive Liquid-Liquid Microextraction Combined with HPLC." *Journal of Analytical Methods in Chemistry* 2018. https://doi.org/10.1155/2018/8916393.

Lu, Dingkun, Chang Liu, Jingjing Deng, Xinguang Zhou, Guoyue Shi, and Tianshu Zhou. 2019. "Rational Design of an Ionic Liquid Dispersive Liquid-Liquid Micro-Extraction Method for the Detection of Organophosphorus Pesticides." *Analyst* 144 (6): 2166–72. https://doi.org/10.1039/c9an00123a.

Ma, Huili, Li Wang, Zhiguang Liu, and Yujing Guo. 2016. "Ionic Liquid–Graphene Hybrid Nanosheets-Based Electrochemical Sensor for Sensitive Detection of Methyl Parathion." *International Journal of Environmental Analytical Chemistry* 96 (2): 161–72. https://doi.org/10.1080/03067319.2015.1114111.

Macfarlane, Douglas R., Naoki Tachikawa, Maria Forsyth, Jennifer M. Pringle, Patrick C. Howlett, Gloria D. Elliott, James H. Davis, Masayoshi Watanabe, Patrice Simon, and C. Austen Angell. 2014. "Energy Applications of Ionic Liquids." *Energy and Environmental Science*. Royal Society of Chemistry. https://doi.org/10.1039/c3ee42099j.

Marić, Slađana, Ana Jocić, Aleksandar Krstić, Miloš Momčilović, Ljubiša Ignjatović, and Aleksandra Dimitrijević. 2021. "Poloxamer-Based Aqueous Biphasic Systems in Designing an Integrated Extraction Platform for the Valorization of Pharmaceutical Waste." *Separation and Purification Technology* 275 (June). https://doi.org/10.1016/j.seppur.2021.119101.

Millán, Daniela, Mabel Rojas, Ricardo A. Tapia, and Paulina Pavez. 2020. "Microwave-Assisted Nucleophilic Degradation of Organophosphorus Pesticides in Propylene Carbonate." *Organic and Biomolecular Chemistry* 18 (39): 7868–75. https://doi.org/10.1039/d0ob01620a.

Morales, Javiera I., Roberto Figueroa, Mabel Rojas, Daniela Millán, Ricardo A. Tapia, and Paulina Pavez. 2018. "Dual Function of Amino Acid Ionic Liquids (Bmim[AA]) on the Degradation of the Organophosphorus Pesticide, Paraoxon®." *Organic and Biomolecular Chemistry* 16 (40): 7446–53. https://doi.org/10.1039/c8ob01928b.

Mulla, Sikandar I., Fuad Ameen, Manjunatha P. Talwar, Syed Ali Musstjab Akber Shah Eqani, Ram Naresh Bharagava, Gaurav Saxena, Preeti N. Tallur, and Harichandra Z. Ninnekar. 2020. "Organophosphate Pesticides: Impact on Environment, Toxicity, and Their Degradation." *Bioremediation of Industrial Waste for Environmental Safety*, 265–90. https://doi.org/10.1007/978-981-13-1891-7_13.

Nawała, Jakub, Barbara Dawidziuk, Daniel Dziedzic, Diana Gordon, and Stanisław Popiel. 2018. "Applications of Ionic Liquids in Analytical Chemistry with a Particular Emphasis on Their Use in Solid-Phase Microextraction." *TrAC - Trends in Analytical Chemistry* 105: 18–36. https://doi.org/10.1016/j.trac.2018.04.010.

Pang, Long, Peijie Yang, Rong Pang, Xiaoling Lu, Jingwen Xiao, Shunyi Li, Hongzhong Zhang, and Jihong Zhao. 2018. "Ionogel-Based Ionic Liquid Coating for Solid-Phase Microextraction of Organophosphorus Pesticides from Wine and Juice Samples." *Food Analytical Methods* 11 (1): 270–81. https://doi.org/10.1007/s12161-017-0997-9.

Pavez, Paulina, Roberto Figueroa, Mayte Medina, Daniela Millán, R. Darió Falcone, and Ricardo A. Tapia. 2020. "Choline [Amino Acid] Ionic Liquid/Water Mixtures: A Triple Effect for the Degradation of an Organophosphorus Pesticide." *ACS Omega* 5 (41): 26562–72. https://doi.org/10.1021/acsomega.0c03305.

Pavez, Paulina, Daniela Millán, Javiera I. Morales, Enrique A. Castro, Claudio López A., and José G. Santos. 2013. "Mechanisms of Degradation of Paraoxon in Different Ionic Liquids." *Journal of Organic Chemistry* 78 (19): 9670–76. https://doi.org/10.1021/jo401351v.

Pavez, Paulina, Guillermo Oliva, and Daniela Millán. 2016. "Green Solvents as a Promising Approach to Degradation of Organophosphorate Pesticides." *ACS Sustainable Chemistry and Engineering* 4 (12): 7023–31. https://doi.org/10.1021/acssuschemeng.6b01923.

Pusey, Marc L., Mark Steve Paley, Megan B. Turner, and Robin D. Rogers. 2007. "Protein Crystallization Using Room Temperature Ionic Liquids." *Crystal Growth and Design* 7 (4): 787–93. https://doi.org/10.1021/cg060696t.

Quental, Maria V., Magda Caban, Matheus M. Pereira, Piotr Stepnowski, João A. P. P. Coutinho, and Mara G. Freire. 2015. "Enhanced Extraction of Proteins Using Cholinium-Based Ionic Liquids as Phase-Forming Components of Aqueous Biphasic

Systems." *Biotechnology Journal* 10 (9): 1457–66. https://doi.org/10.1002/biot.201500003.

Rana, Shweta, Ranjeet Kaur, Roshni Jain, and Nirmal Prabhakar. 2019. "Ionic Liquid Assisted Growth of Poly(3,4-Ethylenedioxythiophene)/Reduced Graphene Oxide Based Electrode: An Improved Electro-Catalytic Performance for the Detection of Organophosphorus Pesticides in Beverages." *Arabian Journal of Chemistry* 12 (7): 1121–33. https://doi.org/10.1016/j.arabjc.2018.08.008.

Rezaee, Mohammad, Yaghoub Assadi, Mohammad Reza Milani Hosseini, Elham Aghaee, Fardin Ahmadi, and Sana Berijani. 2006. "Determination of Organic Compounds in Water Using Dispersive Liquid-Liquid Microextraction." *Journal of Chromatography A* 1116 (1–2): 1–9. https://doi.org/10.1016/j.chroma.2006.03.007.

Rogers, Robin D., and Kenneth R. Seddon. 2003. "Ionic Liquids - Solvents of the Future?" *Science* 302 (5646): 792–93. https://doi.org/10.1126/science.1090313.

Rogers, Robin D., Kenneth Richard Seddon, and Sergei Volkov. 2012. *Green Industrial Applications of Ionic Liquids*. Vol. 92. Springer Science & Business Media.

Santos, J. I., A. M. M. Gonçalves, J. L. Pereira, B. F. H. T. Figueiredo, F. A. E. Silva, J. A. P. Coutinho, S. P. M. Ventura, and F. Gonçalves. 2015. "Environmental Safety of Cholinium-Based Ionic Liquids: Assessing Structure–Ecotoxicity Relationships." *Green Chemistry* 17: 4657. https://doi.org/10.1039/c5gc01129a.

Santos, João H. P. M. P. M., Mafalda R. Almeida, Cláudia I. R. R. Martins, Ana C. R. V. R. V. Dias, Mara G. Freire, João A. P. P. Coutinho, and Sónia P. M. M. Ventura. 2018. "Separation of Phenolic Compounds by Centrifugal Partition Chromatography." *Green Chemistry* 20 (8): 1906–16. https://doi.org/10.1039/C8GC00179K.

Sidhu, Gurpreet Kaur, Simranjeet Singh, Vijay Kumar, Daljeet Singh Dhanjal, Shivika Datta, and Joginder Singh. 2019. "Toxicity, Monitoring and Biodegradation of Organophosphate Pesticides: A Review." *Critical Reviews in Environmental Science and Technology* 49 (13): 1135–87. https://doi.org/10.1080/10643389.2019.1565554.

Silva Sousa, Jonas da, Hélio Oliveira do Nascimento, Hiago de Oliveira Gomes, and Ronaldo Ferreira do Nascimento. 2021. "Pesticide Residues in Groundwater and Surface Water: Recent Advances in Solid-Phase Extraction and Solid-Phase Microextraction Sample Preparation Methods for Multiclass Analysis by Gas Chromatography-Mass Spectrometry." *Microchemical Journal*. Elsevier Inc. https://doi.org/10.1016/j.microc.2021.106359.

Somers, Anthony, Patrick Howlett, Douglas MacFarlane, and Maria Forsyth. 2013. "A Review of Ionic Liquid Lubricants." *Lubricants* 1 (1): 3–21. https://doi.org/10.3390/lubricants1010003.

Souza-Silva, Érica A., Ruifen Jiang, Angel Rodríguez-lafuente, and Emanuela Gionfriddo. 2015. "A Critical Review of the State of the Art of Solid-Phase Microextraction of Complex Matrices I . Environmental Analysis." *Trends in Analytical Chemistry* 71: 224–35. https://doi.org/10.1016/j.trac.2015.04.016.

Taha, Mohamed, Mafalda R. Almeida, Francisca A. e. Silva, Pedro Domingues, Sónia P. M. Ventura, João A. P. Coutinho, and Mara G. Freire. 2015. "Novel Biocompatible and Self-Buffering Ionic Liquids for Biopharmaceutical Applications." *Chemistry (Weinheim an Der Bergstrasse, Germany)* 21 (12): 4781–88. https://doi.org/10.1002/chem.201405693.

Ventura, Sónia P. M., Francisca A. e Silva, Maria V. Quental, Dibyendu Mondal, Mara G. Freire, and João A. P. Coutinho. 2017. "Ionic-Liquid-Mediated Extraction and Separation Processes for Bioactive Compounds: Past, Present, and Future Trends." *Chemical Reviews*, acs.chemrev.6b00550. https://doi.org/10.1021/acs.chemrev.6b00550.

Walden, P. 1914. "Molecular Weights and Electrical Conductivity of Several Fused Salts." *Bulletin de l'Academie Imperiale Des Sciences de St.-Petersbourg* 8: 405–22. https://doi.org/citeulike-article-id:12856962.

Wang, Ya Li, Li Qin You, Yu Wen Mei, Jian Ping Liu, and Li Jun He. 2016. "Benzyl Functionalized Ionic Liquid as New Extraction Solvent of Dispersive Liquid-Liquid Microextraction for Enrichment of Organophosphorus Pesticides and Aromatic Compounds." *Chinese Journal of Analytical Chemistry* 44 (6): 942–49. https://doi.org/10.1016/S1872-2040(16)60937-4.

Weaver, Katherine D., Hye Jin Kim, Jiazeng Sun, Douglas R. MacFarlane, and Gloria D. Elliott. 2010. "Cyto-Toxicity and Biocompatibility of a Family of Choline Phosphate Ionic Liquids Designed for Pharmaceutical Applications." *Green Chemistry* 12 (3): 507–51. https://doi.org/10.1039/b918726j.

Wei, Min, and Jingjing Wang. 2015. "A Novel Acetylcholinesterase Biosensor Based on Ionic Liquids-AuNPs-Porous Carbon Composite Matrix for Detection of Organophosphate Pesticides." *Sensors and Actuators, B: Chemical* 211: 290–96. https://doi.org/10.1016/j.snb.2015.01.112.

Wilkes, John S., Michael J. Zaworotko, F. H. Hurley, T. P. Wier, John S. Wilkes, J. A. Levisky, R. A. Wilson, et al. 1992. "Air and Water Stable 1-Ethyl-3-Methylimidazolium Based Ionic Liquids." *Journal of the Chemical Society, Chemical Communications*, no. 13: 965. https://doi.org/10.1039/c39920000965.

Zgoła-Grześkowiak, Agnieszka, and Tomasz Grześkowiak. 2011. "Dispersive Liquid-Liquid Microextraction." *TrAC - Trends in Analytical Chemistry* 30 (9): 1382–99. https://doi.org/10.1016/j.trac.2011.04.014.

Zhan, Haijun, Jie Li, Zhimin Liu, Yingying Zheng, and Yanfeng Jing. 2015. "A Highly Sensitive Electrochemical OP Biosensor Based on Electrodeposition of Au-Pd Bimetallic Nanoparticles onto a Functionalized Graphene Modified Glassy Carbon Electrode." *Analytical Methods* 7 (9): 3903–11. https://doi.org/10.1039/c5ay00702j.

Zhang, Lijin, Fang Chen, Shaowen Liu, Biyu Chen, and Canping Pan. 2012. "Ionic Liquid-Based Vortex-Assisted Dispersive Liquid-Liquid Microextraction of Organophosphorus Pesticides in Apple and Pear." *Journal of Separation Science* 35 (18): 2514–19. https://doi.org/10.1002/jssc.201101060.

Zhang, Liyuan, Runzhong Yu, Xue Zhang, and Dongjie Zhang. 2021. "Ionic Liquid–Based Dispersive Liquid–Liquid Micro-Extraction of Five Organophosphorus Pesticides in Coarse Cereals." *Food Analytical Methods* 14 (1): 10–17. https://doi.org/10.1007/s12161-020-01851-y.

Zhang, Qinghua, Shiguo Zhang, and Youquan Deng. 2011. "Recent Advances in Ionic Liquid Catalysis." *Green Chemistry* 13 (10): 2619–37. https://doi.org/10.1039/ c1gc15334j.

Zhang, Yao, Ruoyu Wang, Ping Su, and Yi Yang. 2013. "Ionic Liquid-Based Solvent Bar Microextraction for Determination of Organophosphorus Pesticides in Water Samples." *Analytical Methods* 5 (19): 5074–78. https://doi.org/10.1039/c3ay40803e.

Zheng, Xiaoyan, Lijun He, Yajing Duan, Xiuming Jiang, Guoqiang Xiang, Wenjie Zhao, and Shusheng Zhang. 2014. "Poly(Ionic Liquid) Immobilized Magnetic Nanoparticles as New Adsorbent for Extraction and Enrichment of Organophosphorus Pesticides from Tea Drinks." *Journal of Chromatography A* 1358: 39–45. https://doi.org/10.1016/j.chroma.2014.06.078.

Zheng, Yingying, Zhimin Liu, Haijun Zhan, Jie Li, and Chengcheng Zhang. 2016. "Studies on Electrochemical Organophosphate Pesticides (OPs) Biosensor Design Based on Ionic Liquid Functionalized Graphene and Co3O4 Nanoparticles Modified Electrode." *Analytical Methods*. https://doi.org/10.1039/C6AY01346E.

Zhou, Qingxiang, Huahua Bai, Guohong Xie, and Junping Xiao. 2008. "Trace Determination of Organophosphorus Pesticides in Environmental Samples by Temperature-Controlled Ionic Liquid Dispersive Liquid-Phase Microextraction." *Journal of Chromatography A* 1188 (2): 148–53. https://doi.org/10.1016/ j.chroma. 2008.02.094.

Chapter 8

Electrooxidative Removal of Organophosphates - A Combined Experimental and Theoretical Approach

Danka D. Aćimović and Dragana D. Vasić Anićijević*

Department of Physical Chemistry, "Vinča" Institute of Nuclear Sciences - National Institute of the Republic of Serbia, University of Belgrade, Belgrade, Serbia

Abstract

Electrochemical oxidation is in general considered to be among the best methods for removing organic pesticides from wastewater. The oxidation of organophosphates induces the formation of a great number of various more or less toxic short-time living intermediates and finally leads to their complete degradation. This chapter presents an overview of the literature data concerning the electrochemical oxidation of several widely used organophosphate pesticides (malathion, diazinon, chlorpyrifos, dimethoate, methyl parathion). The influence of the electrode materials on selected organophosphates, the choice of an appropriate electrode, supporting electrolyte, and knowing the electrolysis parameters such as current density, temperature, oxidation rate, were elucidated. The theoretical calculation enabled us to identify the phenomena engaged near the electrode and to completely determine the structures of the products of electrochemical oxidation formed during the degradation. The comparison of the results relating to the various electrodes indicates that these materials have the power to reduce the quantity of the organic matter in the electrolyzed solution.

* Corresponding Author's Email: draganav@vin.bg.ac.rs.

In: Organophosphates
Editor: Tamara Lazarević-Pašti
ISBN: 978-1-68507-652-8

Keywords: electrooxidation, organophosphates, anode materials, DFT calculations, electronic descriptors

Introduction

In recent decades, along with population growth, demand for food and its production has also been rising. Increased agricultural productivity is enabled by the use of mechanization, selection of better plant varieties, and application of advanced agricultural practices. Likewise, in order to obtain high yields of agricultural crops, pesticides are increasingly used.

In the last century, the development of organochlorines and organophosphates went simultaneously, but the successful application of the insecticide dichlorodiphenyltrichloroethane (DDT) favored the widespread use of organochlorines. The recognition and better understanding of the toxicity and accumulative properties of the organochlorines resulted in the increased use of the organophosphorus pesticides (diazinon, methyl parathion, parathion, malathion, chlorpyrifos, etc.). Depending on the function, they can be classified as fungicides, bactericides, herbicides, insecticides, etc. Numerous organophosphates have been recognized as persistent organic pollutants (POPs). Nowadays, organophosphates are one of the most widespread pollutants, and they are present in all ecosystems. Also, their retention in the environment is directly influenced by physical and chemical parameters, such as temperature, pH, oxygen levels, vapor pressure, water solubility, adsorption coefficients, the presence of other organic pollutants or heavy metals, soil type and structure, sunlight, etc. (Büyüksönmez et al. 1999; V. Anićijević et al. 2019).

A number of organophosphates are persistent (remain in the environment for a long time and have been difficult to degrade) and bioaccumulative (accumulate in the organism) and, as such, pose a danger to the environment and living organisms. Their toxic effect is reflected in the irreversible inhibition of acetylcholinesterase, an enzyme that assists in the transmission of nerve signals. Hence, it induces the systemic loss of function of cholinergic neurons with numerous consequences in almost all organs, including loss of muscular coordination, convulsions, and ultimately death. Acetylcholinesterase is an essential component of the nervous system of both insects and mammals, and therefore the toxic effect of the inhibition of the enzyme is the same in both (Soltaninejad and Abdollahi 2009; Casida and Quistad 2004).

The extensive use of pesticides in agriculture and lack of attention to environmental issues causes pesticides found their way in surface water, soil and atmosphere. Also, pesticides enter the food chain and create a significant impact on agricultural ecosystems, groundwater, as well as drinking water. The European Parliament and the council of the European Union set out general provisions for the protection and conservation of groundwater (The European Parliament and Council 2006). A directive concerning the quality of the water intended for human consumption has also been adopted (The Council of European Union 1998). These directives set limits of 100 ng L^{-1} and 500 ng L^{-1} as quality standards for the presence of individual and total pesticides (including their relevant metabolites, degradation, and reaction products), respectively, in groundwater and drinking water (Postigo et al. 2010).

In order to remove the resulting pollution, more attention is paid to the development of decontamination techniques.

The best-known techniques for organophosphates removal from the environment are:

- Biodegradation-bioremediation (by bacteria, fungi, algae, and plants),
- Physico-chemical methods
    - Photooxidation
    - Electrochemical oxidation
    - Ozonation
    - Chemical oxidation
    - Volatilization
    - Leaching
    - Adsorption.

Each of the listed techniques of removing organophosphates degrades individual organophosphate at different rates, depending on their structure, and thus their physical and chemical characteristics (Büyüksönmez et al. 1999).

Treatment of persistent organic pollutants by conventional techniques (chemical oxidation, adsorption, leaching, volatilization) does not always lead to their complete decomposition. The application of these techniques can lead to incomplete degradation and the formation of more ecotoxic products (Moreira et al. 2017). Therefore, in recent decades, significant attention has been paid to the development of more effective techniques for removing organic pollutants. Advanced Oxidation Processes (AOPs) is a common name for a group of de-pollution methods that use hydroxyl radicals - the strongest known oxidant for the oxidation of organic pollutants. Hydroxyl radicals can be generated by chemical or electrochemical methods.

Hence, electrochemical oxidation of organic pollutants, which is the central subject of this chapter, belongs to the AOPs. Electrochemical removal of organic pollutants from water has remarkable potential due to its unique mechanism of action in the oxidation of all organic matter - including organophosphates. Also, it is clean, environmentally friendly (since there are no chemicals involved, and there is no toxic waste after the treatment is completed), and it requires simple apparatus and experimental conditions. On the other hand, the whole concept of electrochemical oxidation is based on the high expected efficiency of the process, finally resulting in the full mineralization of the pollutants up to carbon dioxide, phosphate, and water. This is, however, an idealized picture, overlooking the impact of external anions, electrode corrosion, and incomplete efficiency of the process with various more or less harmful byproducts. To attain this goal of "ideal electrooxidation," numerous researchers in recent years have been pursuing novel, advanced, high-surface electrode materials, highly efficient and highly stable in oxidative conditions, and for the complex essential principles beyond the process.

In this chapter, main achievements related to the electrochemical (anodic, electrooxidative) removal of organophosphates from the aqueous solutions are presented. Electrochemical degradation was compared to other types of advanced oxidation processes (H_2O_2 with UV radiation, ozone-based processes, and Fenton reaction). The main principles of electrochemical oxidation of organophosphates are presented, with the focus on the nature of electrode material (active and non-active electrodes) and the role of external anions in the outcome of the process. Finally, we have presented possible approaches to use *ab initio* calculations (in particular, density functional theory) in the investigation of basic principles of electrooxidation of organic compounds.

Advanced Oxidation Processes (AOPs)

Electrochemical oxidation of organophosphates belongs to the group of de-pollution methods, which is called advanced oxidation processes. AOPs are based on the *in situ* generation of highly reactive species, such as hydroxyl radicals (•OH), in sufficient quantity to interact with organic pollutants. These powerful oxidizing agents are capable of efficiently (10^8-10^{10} M^{-1} s^{-1}) degrading organophosphorus pesticides, as well as a wide range of other persistent organic pollutants such as polycyclic aromatic hydrocarbons, azo dyes, pharmaceuticals, phenolic compounds, etc. (J. L. Wang and Xu 2012). The ultimate aim of chemical oxidation is the mineralization of pollutants by oxidizing agents to carbon dioxide, water, and inorganic products - phosphates, or their transformation into non-toxic products (Figure 1).

The following techniques are included in AOPs, and the difference between them is in the way in which hydroxyl radicals are generated (Aćimović 2020):

- H_2O_2 with UVC radiation,
- zone-based processes - O_3, O_3/UVC, O_3/H_2O_2/UVC,
- Photocatalytic processes based on semiconductor metals and their oxides - (TiO_2, ZnO, WO_3, NiO, $BiTiO_3$)/UV, (TiO_2, ZnO, WO_3, NiO, $BiTiO_3$)/H_2O_2/UV,
- Processes based on the Fenton method - Fenton - Fe^{2+}/H_2O_2, photo Fenton - Fe^{2+}/H_2O_2/UV,
- Electrochemical advanced oxidation processes (electrochemical oxidation) - anodic oxidation (AO), AO with electrogenerated H_2O_2, electro-Fenton, photoelectro-Fenton, solar photoelectro-Fenton.

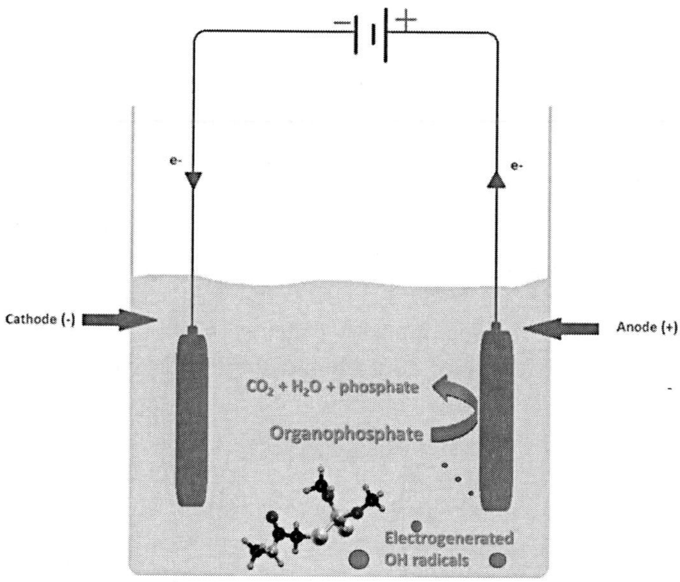

Figure 1. Concept of electrooxidative mineralization.

AOPs are technologies compatible with the environment because they don't produce additional waste, and generated hydroxyl radicals don't remain in the environment due to their short lifetimes. During advanced oxidation processes, equipment for the production of hydroxyl radicals is simple to manipulate.

The main positive characteristics of AOPs are:

1. The potential for complete mineralization of organic pollutants to carbon dioxide, water, and inorganic salts (phosphates, chlorides, nitrates, sulfates, etc.);
2. Non-selective reactivity with organic compounds- primary pollutants or byproducts formed during incomplete degradation are equally oxidized by the same principles.

The main limitations of the AOP processes are:

1. High energy consumption (UV radiation and O_3 generation), the high price of used reagents (Fenton), and expensive electrodes (BDD electrode).
2. Complex analytical procedures for tracking degradation byproducts require extensive equipment.

Comparison of AOPs with Other Methods

Most of the AOP methods are based on chemical generation of OH radicals (H_2O_2/UV) based process, ozone/UV based process, and methods based on Fenton reaction. Besides OH radical, other oxidizing agents such as sodium hypochlorite, chlorine dioxide (ClO_2), chlorine gas (Cl_2) permanganates, persulfates, hydrogen peroxide can also be directly used. Application of the chemical oxidation is determined by the specific conditions which are necessary for its performance, for example, the pH of the environment. Wu et al. investigated the degradation of common organophosphates by ozonization. By-products of organophosphate ozonation, such as phosphoric acid, phosphate, carbonate, nitrate, and sulfate, have been identified. The biggest detected problem during organophosphate ozonation has been the production of oxones, which are even more toxic than their parental form. Diazoxon, methyl paraoxon, and paraoxon were identified as ozonation intermediates of diazinon, methyl parathion, and parathion, respectively. The potential accumulation of oxons could be overcome by an excessive ozone dose and proper ozonation conditions (Wu et al. 2009). Along with the possible formation of toxic oxons, the main disadvantage of the method is the use of chemicals, so it cannot be said it is convenient for the environment (Druzina and Stegu 2007).

Contrary to chemical oxidation or electrooxidation, photooxidation in natural condition rarely leads to complete decomposition of organophosphates (to carbon dioxide, water, and inorganic residues). Photolysis causes the excitation of molecules via absorption of UV-light, and the excited molecules undergo various chemical and physical transformations (Büyüksönmez et al. 1999). Wan and co-workers studied photolysis of 16 organophosphates in aqueous solutions, and the results indicate that complete mineralization did not occur, but phenolic compounds were formed (Wan, Wong, and Mok 1994). Photolysis in the presence of TiO_2 and ZnO shows better degradation, which sometimes even leads to complete mineralization (Kanan et al. 2019).

Compared to AOPs, biodegradation of organophosphates is a slow process. Organophosphates need to be dissolved in water or in the gas phase so that they are available to live organisms for ingestion and metabolic degradation. Further adjustment of environmental conditions (availability of oxygen, humidity, temperature, sunlight, media characteristic, nutrient content, and surfactants) optimizes the biodegradation of organophosphates (Krishna and Philip 2009; Singh, Kaur, and Singh 2014). A large number of bacteria, fungi, and algae are capable of decomposing organophosphates. These processes are mainly conditioned by the presence of oxygen, although today, anaerobic degradation conditions are also being examined (Sidhu et al. 2019; Kumar et al. 2019; Büyüksönmez et al. 2000).

Electrochemical (Anodic) Oxidation of Organophosphates

Because of simplicity and universal principle, electrochemical oxidation has been applied for the degradation of various organic pollutants - organic dyes, polycyclic aromatic hydrocarbons, polychlorinated biphenyls, phenolic compounds, including organophosphates. In previous research, electrochemical oxidation has proven to be an effective technique for treating complex contaminated media (Aćimović et al. 2017; Savić et al. 2020).

Due to the different physico-chemical characteristics of pollutants, there is no unique strategy for their removal. Electrochemical (anodic) oxidation stands out as an alternative technique, which is still being developed and improved. It can be used as a separate method or in combination with other processes, such as anodic oxidation with electrogenerated hydrogen peroxide (AO-H_2O_2), electro-Fenton (EF), photoelectro-Fenton (PEF) and solar-photo-electro-Fenton process-Table 1 (Moradi et al. 2020; Savic 2021; Rodrigo, Oturan, and Oturan 2014).

The main advantage of electrochemical methods for degradation is that the electron is used as a "pure agent," and there is no additional threat to the environment. The electrochemical methods are often mentioned as environmentally friendly.

In recent decades, the efforts of researchers have focused on the improvement of electrode materials, analysis of the oxidation efficiency of various contaminants on electrodes made of existing and newly synthesized materials, as well as studying the mechanism and kinetics of pollutants degradation (Martı and Ferro 2006).

Electrochemical oxidation is a method by which even the most persistent organic pollutants are completely mineralized (conversion to CO_2, water and inorganic ions) or only partially degraded to forms that are not toxic to the environment. Anodic (electrochemical) oxidation of organic molecules can take place in the following ways:

Table 1. Review of electrochemical advanced oxidation processes

Anodic oxidation (AO)	**Oxidation of water on the anode surface** $\mathbf{M + H_2O \rightarrow M(^{\cdot}OH) + H^+ + e^-}$
AO with electrogenerated H_2O_2 (AO-H_2O_2)	AO reaction + Electrogenerated H_2O_2 on the cathode surface $O_{2(g)} + 2H^+ + 2e^- \rightarrow H_2O_2$
Electro-Fenton (EF)	AO and AO - H_2O_2 reaction + Fenton reaction: $Fe^{2+} + H_2O_2 \rightarrow Fe^{3+} + \cdot OH + OH^-$ Reaction on cathode: $Fe^{3+} + e^- \rightarrow Fe^{2+}$
Photoelectro-Fenton (PEF) Solar-photo-electro-Fenton (SPEF)	AO, AO - H_2O_2, EF reactions + Photoreduction: $FeOH^{2+} + h\nu \rightarrow Fe^{2+} + \cdot OH$ or Direct photolysis of complexes formed between Fe^{3+} and organic compounds: $Fe^{3+}(L)_n + h\nu \rightarrow Fe^{2+}(L)_{n-1} + L^{\cdot}_{ox}$

1. direct oxidation at the anode (by direct electron transfer to the anode)
2. indirect oxidation through reactive oxygen species formed at the anode.

Direct anodic oxidation is performed on the anode by direct transfer of charge between the anode surfaces and organic pollutants. In processes of direct anodic oxidation reaction participate only electrons that oxidize organic pollutants on the more negative potentials than the oxygen evolution potentials. Prior to this process, organic compounds have to adsorb on the anode surface. After electrochemical oxidation at the anode, it is necessary for the product to be desorbed from the electrode and transported to the solution. This limits the rate of decomposition of organic molecules and reduces the efficiency of the process.

During indirect electrochemical oxidation of organic compounds, the first reaction is the oxidation of water molecules leading to the formation of adsorbed hydroxyl radicals (Equation 1).

$$M + H_2O \rightarrow M(\cdot OH) + H^+ + e^- \qquad (1)$$

The following steps have been shown to depend on the nature of the electrode materials. It has been established in many studies, the electrochemical activity (overpotential for oxygen evolution) and chemical reactivity (rate of organics oxidation with electrogenerated hydroxyl radicals) of adsorbed hydroxyl radicals strongly depends on the nature of the used electrode material (Comninellis et al. 2008; Simond, Schaller, and Comninellis 1997; Kapałka, Fóti, and Comninellis 2008). Comninellis and co-workers defined two classes of electrodes regarding the binding of OH radical on their surface:

1. active electrodes and
2. non-active electrodes.

The active anodes easily bind oxygen on the surface and interact stronger with the adsorbed hydroxyl radicals (forming chemisorbed hydroxyl radicals). Consequently, the interaction between adsorbed hydroxyl radicals and anode may form a so-called higher oxide, MO (Equation 2).

$$M(\cdot OH) \rightarrow MO + H^+ + e^- \qquad (2)$$

The surface redox couple MO/M acts as a mediator in the conversion or partial oxidation of organic compounds (R) on active anodes (Equation 3).

$$MO + R \rightarrow M + RO \qquad (3)$$

Due to the chemical decomposition of the higher oxide, this reaction is in competition with the side reaction of oxygen evolution (OER) (Equation 4).

$$MO \rightarrow M + \frac{1}{2}O_2 \tag{4}$$

Different from active electrodes, non-active electrodes have weak interactions between their surface and hydroxyl radicals, forming physisorbed hydroxyl radicals. Physisorbed hydroxyl radical is among the strongest known oxidants that can completely mineralize the organic compound (R) to carbon dioxide and water (Equation 5).

$$M(\cdot OH) + R \rightarrow M + mCO_2 + H_2O + H^{+} + e^{-} \tag{5}$$

This reaction is in competition with the oxygen evolution side reaction (Equation 6).

$$M(\cdot OH) \rightarrow M + \frac{1}{2}O_2 + H^{+} + e^{-} \tag{6}$$

Electrode Materials

When observing the nature of anode materials, the following rule applies: the higher the potential for oxygen evolution of the anode material, the lower the M(·OH) interaction with the anode surface, and the higher the chemical reactivity. The most important electrode materials in use include various carbons, boron-doped diamond, and mixed metal/metal oxide electrodes. Non-active anodes exhibit OER overpotential lower than 1.7 V/SHE, and active anodes exhibit OER overpotential between 1.7 V/SHE and 2.6 V/SHE). The most frequently used anodes are listed in Table 2.

Table 2. Potential of oxygen evolution of different anodes materials

Anode material	Potential for O_2 evolution, V vs. SHE
RuO_2	1.4-1.7
IrO_2	1.5-1.8
Pt	1.6-1.9
Graphite	1.7
Titanium oxides	1.7-1.8
PbO_2	1.8-2.0
SnO_2	1.9-2.2
BDD	2.2-2.6

Non-Active Electrodes

Conductive diamond electrodes, especially the boron-doped diamond (BDD) electrodes, some metallic oxides, and mixed metal oxides (those containing PbO_2 and/or SnO_2), are characteristic representatives of non-active anodes. Their oxygen evolution potentials are in the range of 1.7 to 2.6 V/SHE (Table 2). These anodes are highly efficient electrodes for the oxidation of organic compounds and prefer the complete mineralization of organic pollutants. The main objection to these electrodes is the limited mass transport which slows down the whole mineralization process.

Particularly attractive results regarding the degradation of diverse organic pollutants have been obtained with electrodes comprising boron-doped diamond (BDD) films. In comparison with other electrode materials, the conductive diamond appears to offer a number of advantages, including a high overpotential for water electrolysis and enhanced stability (Alves et al. 2013). In particular, the hydroxyl radicals that appear to be directly involved in oxidation mechanisms occurring at BDD surfaces are produced with high efficiency within a large potential window (Marselli et al. 2003). Alves et al. investigated electrochemical degradation of methyl parathion in aqueous Na_2SO_4 solution using a boron-doped diamond film electrode on Ti support. Mineralization efficiency (ME is defined as the ratio of final to initial TOC values expressed as a percentage) attained 67.6% at 100 mA cm^{-2}. The percentage of mineralization was lower than the percentage reduction in the concentration of methyl parathion (81.2%), indicating that electrochemical degradation of the insecticide was accompanied by the formation of intermediates and/or byproducts. The toxicity of the treated solution was investigated against the gram-negative bacteria *Vibrio fischeri*. At current density 5 mA cm^{-2} and 100 mA cm^{-2} the relative toxicity (initial EC50/final EC50) was 82% and 74% compared to the initial solution. At a glance, the observed toxicity seems too high regarding the high extent of mineralization. As will be discussed later, themain problem of electrooxidation of organophosphate pesticides on BDD (and most other electrode materials) represent the chlorine species which are present in the electrolyte solution – although their presence normally reduces the energy demand of degradation, it also results in the formation of toxic chlorinated species and even induce electrode breakdown in case of BDD thin film.

Samet et al. investigated electrochemical degradation of chlorpyrifos pesticide on BDD electrode on Si support. It was found that the removal rate increases with applied current density until 20 mA cm^{-2} due to the increase of the mass transport caused by oxygen evolution reaction but decreases for higher values due to the predominance of OER reaction on the electrode. After 6 hours at optimized conditions of electrooxidation (j_{app} = 20 mA cm^{-2}, pH = 2 and at 70°C), COD removal was very close to 100% (Samet, Agengui, and Abdelhédi 2010).

A comparative study of electrochemical degradation efficiency on BDD and SnO_2 electrodes was performed by Hatchami et al. in 2% NaCl solution, as a supporting electrolyte using the same conditions for these two electrodes (60 mA cm^{-2} and 25°C) (Hachami et al. 2015). The results indicated that both electrode materials have the power to reduce the quantity of organophosphate - methidathion. The speed of oxidation of methidathion is different according to the materials of the electrodes used. Results showed that the BDD electrode is more effective than SnO_2 for the electrochemical degradation of methidathion. For BDD, the achieved reduction of chemical oxygen demand (COD) was about 85%, while for SnO_2, it was about 73%. Better energetic efficiency of the electrochemical process on BDD compared to SnO_2 was also confirmed through Instantaneous Current Efficiency (ICE) measurements – which represent a variation of COD with time. The mechanism of degradation was proposed based on theoretical calculations of frontier electron density and partial charges of isolated methidation molecule. The OH radical is predicted to attack the C-S bond first, and subsequently, the P-S bond, independently on the electrode used. However, the predicted degradation products still were not confirmed experimentally.

PbO_2 based electrodes represent a matter of compromise since their use is closely connected to the possible release of toxic Pb^{2+} ions into the treated sample, particularly in acidic media. On the other hand, low cost and high OER onset potential, which results in the efficient formation of OH radicals, make PbO_2 permanently investigated in the field of organophosphate removal.

Tran et al. investigated the ability of electrochemical oxidation for the degradation and mineralization of glyphosate herbicide using Ti/PbO_2 anode in Na_2SO_4 solution as electrolyte (Aquino Neto and de Andrade 2009). Degradation efficiency and TOC removal were most efficient in acidic conditions and achieved 95% and 85%, respectively. In neutral and alkaline environments, efficiency and TOC removal are, however, reduced to 70-80%.

Martinez-Huitle and co-workers have investigated electrochemical removal of pesticide methamidophos from aqueous solutions using Pb/PbO_2, Ti/SnO_2, and Si-supported BDD electrodes (Martínez-Huitle et al. 2008). The study represents illustrative proof that the degradation mechanism of organophosphates does not only depend on current density and electrochemical conditions but also on the nature of the electrode material itself. Two previously established products of organophosphate degradation, formaldehyde, and phosphate ion, were tracked to determine the mineralization extent (Tanaka, Harada, and Hisanaga 1990). In the case of the Pb/PbO_2 electrode, almost complete elimination of initial pesticide methamidophos was achieved near the electrode surface, while the faradaic efficiency of the electrochemical process is moderately low at low current densities (10-20 mA cm^{-2}). Overall mineralization extent achieves 85%. At higher current densities, the speed of the oxidation process is limited by the mass transfer step.

It was found that methamidophos degradation using Pb/PbO_2 generates formaldehyde, subsequently being degraded to formic acid, as found from FTIR analysis. Moreover, it is observed that on Pb/PbO_2, formaldehyde production is favored at low pH values, and it disappears within 60 minutes of electrolysis at a pH higher than 6. This indicates the mechanism of indirect mineralization. Compared to the Pb/PbO_2 electrode, under the same conditions, Ti/SnO_2 showed poor formaldehyde production. Furthermore, BDD electrodes showed the formation of phosphate, ammonia, and CO_2 as main reaction products, with no detected formaldehyde suggesting complete, direct methamidophos mineralization. The removal efficiency of methamidophos follows the order BDD $>Pb/PbO_2>Ti/SnO_2$ (Martínez-Huitle et al. 2008).

Electrochemical degradation of organophosphate chlorpyrifos using Nb/PbO_2 anodes was demonstrated by Samet and co-workers. The mineralization efficiency was evaluated by means of the COD measurement. It was concluded that complete mineralization of chlorpyrifos was not reached, and the best COD removal was 76% after 10 h of electrolysis (Samet, Agengui, and Abdelhédi 2010a).

Pure graphite, according to its "boundary" value of OER overpotential (1.7 V vs. SHE), stands between active and inactive anodes. Rueffer et al. provided a detailed analysis of the behavior of OH radical on the graphite anode surface (Rueffer, Bejan, and Bunce 2011). Although positive polarization of graphite anode results in the formation of oxygenated functional groups, finally resulting in the formation and deliberation of carboxyl ions (i.e., mineralization of the anode), they found no evidence that OH radical acts as an intermediate in the anodic oxidation of organic matter. In other words, in the presented experiment, graphite did not act as the inactive anode towards oxidation of organic substrates – at potentials below OER direct electron transfer mechanism has been occurring. In conclusion, it was found that graphite works as a modified active anode. The oxygen atom in the generation of OH radicals is not bonded directly to carbon but to the surface functional groups. The generation of OH radicals is competitive with the OER.

Active Electrodes

Metal electrodes (Pt, Ti/Pt), metal oxide electrodes (IrO_2, RuO_2), mixed metal oxides electrodes (Ti/SnO_2, Ti/IrO_2/SnO_2, SnO_2-Sb2O_5, Ti/IrO_2), graphite electrodes (with sp^2 hybridized carbon) are considered as active anodes and have O_2 evolution below 1.8 V/SHE (Table 2), which is in competition with the formation of higher oxides. As a result, the generation of free and loosely bound hydroxyl radicals at the electrode surface is considerably less efficient compared to the non-active electrodes. As a result, the mineralization of organic compounds is also less efficient, and intermediate byproducts are generated in greater amounts (Feng et al. 2016; Panizza and Cerisola 2009). However, this class of anodes, also called mixed metal-oxides (MMO), is still extensively investigated for the degradation of organic pollutants and is considered a promising and cheaper alternative to the BDD electrodes.

Vlyssides and co-workers studied the electrooxidation of various organophosphorus pesticides (monocrotophos, phosphamidon, phosalone, azinphos-methyl, methadathion, methyl parathion, demeton-S-methyl, methamidophos, fenthion, and diazinon) using Ti/Pt anodes in an aqueous solution of sodium chloride. The COD and the biochemical oxygen demand (BOD) were measured to monitor the efficiency of degradation. COD values are in the range from 28% for monocrotophos and fospamidone to 93% for methyl parathion. The efficiency of electrolysis, among other things, is influenced by pH, organophosphate concentration, temperature, etc. The electrolysis of these organophosphorus pesticides was more effective in terms of % COD reduction when the pH was in the acid range (A. Vlyssides et al. 2004; 2005b; Arapoglou et al. 2003; A. G. Vlyssides et al. 2004; A. Vlyssides et al. 2005a)

Babu et al. investigated the degradation of methyl parathion by continuous electro-oxidation using a Ti/RuO_2 anode and a stainless steel cathode in NaCl aqueous solution as electrolyte (Babu et al. 2013). A COD reduction of 98.5% has been observed at a current density 150 mA cm^{-2}. Better efficiency of COD reduction in acidic medium was ascribed to the presence of oxidizing hypochlorite anions. The secondary products such as p-nitrophenol, dimethyl phosphoric acid, benzoquinone, and hydroquinone are confirmed by the HPLC study.

Aquino Neto and Andrade investigated the degradation of glyphosate on mixed RuO_2, SnO_2, and IrO_2 based electrodes (in variable ratio) on Ti support (Aquino Neto and de Andrade 2009). $Ti/Ir_3Sn_7O_{20}$ was found to be the most efficient among the investigated MMO electrodes. Phosphate anion concentration, TOC analysis, and UV-Vis spectrometry of glyphosate were all used as the measure of the degree of glyphosate degradation. In chloride medium at low operating current (30 $mAcm^{-2}$ for 4 hours), the release of phosphate has achieved 91%. However, in the absence of chloride, the mineralization extent does not exceed 35%. Obtained poor efficiency was ascribed to the competition between OER and organic compound oxidation, occurring in the same potential range at the investigated electrodes. Hence, controlled oxidation at low currents in the presence of chloride was recommended by the authors.

Role of Supporting Electrolyte and External Anions in Indirect Electrooxidation

In practice, it is impossible to provide enough conductivity of electrolytic solution to maintain electrooxidation of organophosphate solutions using only distilled water and electric current. That is the first reason why electrooxidation is always conducted in the presence of a supporting electrolyte – chloride, sulfate, phosphate, etc. On the other hand, it should be emphasized here that the anions from a supporting electrolyte are not only "spectators" but are also reduced at the anode and generate additional oxidizing species in the electrolyzed solution.

These novel oxidizing species also significantly contribute to the overall depletion of the initial contaminant. One-electron water discharge process (Equation 1), which results in the formation of hydroxyl radical, has a standard electrode potential of 2.80 V vs. SHE – which is much higher than competing oxygen evolution reaction (OER) standard potential (1.23 V vs. SHE) (Henke et al. 2018). As a consequence, this crucial process for electrochemical oxidation of organophosphates has rather slow kinetics. At the same time, electrode processes involving supporting electrolyte anions will proceed faster due to their lower reduction potentials (Table 3). Hypothetically, in a true wastewater matrix with various present anions (chloride, sulfate, nitrate), organophosphates would be reacting not only with hydroxyl radical but also by a whole sequence of electrogenerated oxidants (chlorine, hypochlorite, peroxydisulfate, nitrite, etc.), whose impact would be determined by concentrations of parent anions and operating electrode potential, i.e., applied current density in galvanostatic conditions. In other words, the hydroxyl radical generation will have much slower kinetics than the generation of many other oxidizing species generated by the reduction of external anions. So, in an electrochemical system with chloride as a supporting electrolyte, oxidation of organic compounds by hypochlorite will dominate over oxidation by hydroxyl radical.

Table 3 represents standard electrode potentials in reference to a standard hydrogen electrode (SHE) of some possible electrode processes that involve supporting electrolyte anions during the electrooxidation process.

Most organophosphate degradation studies include the evaluation of the impact of external species on the efficiency of the process. Electrochemical (anodic) oxidation processes are improved in controlled conditions by the activity of oxidants, such as activated chlorine species, persulfates, perphosphates, percarbonates, and hydrogen peroxide that are electrochemically generated from compounds present in the bulk solution like chloride, sulfate, carbonate, phosphate, and oxygen, respectively. The final result is a matter of compromise between the acceleration of kinetics and the remnant byproducts.

Due to the natural abundance of chloride ions in most waters, active chlorine- chlorine (Cl_2), hypochlorite (HOCl), and hypochlorite ion (OCl^-) can be *in situ* electrochemically generated for oxidation of organic pollutants (Equation (7-9)).

$$2Cl^- \rightarrow Cl_2 + 2e^- \quad (7)$$

$$Cl_2 + H_2O \rightarrow HClO + Cl^- + H^+ \quad (8)$$

$$HClO \leftrightarrow ClO^- + H^+ \quad (9)$$

The predominance of different chemical species in active chlorine is pH-dependent. Cl_2 is the predominant species at pH <3. When the pH value of medium is between 3 and 8, HClO and ClO^- are predominant species, and for pH values greater than 7.5, only ClO^- is present (Feng et al. 2016).

The high anodic potential typical of non-active electrodes, including BDD, SnO_2, and PbO_2, can lead to oxidation of Cl^- to produce chlorate (ClO_3^-) and perchlorate (ClO_4^-) according to the following equations:

$$Cl^- + \cdot OH \rightarrow ClO^- + H^+ + e^- \quad (10)$$

$$ClO^- + \cdot OH \rightarrow ClO_2^- + H^+ + e^- \quad (11)$$

$$ClO_2^- + \cdot OH \rightarrow ClO_3^- + H^+ + e^- \quad (12)$$

$$ClO_3^- + \cdot OH \rightarrow ClO_4^- + H^+ + e^- \quad (13)$$

Table 3. Formation potentials of the most common oxidants

Oxidant		Standard Reduction Potential, E^0 (V)/SHE
Hydroxyl radical	($H_2O/OH^{\cdot}$)	2.80
Ozone	(O_2/O_3)	2.07
Peroxodisulphate	($SO_4^{2-}/S_2O_8^{2-}$)	2.01
Hydrogen peroxide	(H_2O/H_2O_2)	1.77
Permanganate ion	(MnO_2/MnO_4^-)	1.67
Hypochlorous acid	($Cl^-/HOCl$)	1.48
Chlorine	(Cl^-/Cl_2)	1.36
Dichromate	($Cr^{3+}/Cr_2O_7^{2+}$)	1.23
Oxygen	(H_2O/O_2)	1.23
Hypochlorite	(Cl^-/ClO^-)	0.81

The study performed by Muff and co-workers has shown that the organophosphorus pesticides are oxidized due to the chloride-mediated indirect electrochemical oxidation process (by hypochlorous/hypochlorite pair as the main oxidizing agent). The organophosphate pesticides, parathion, methyl-parathion, malathion, and degradation products ethyl-amino-parathion, paraoxon, and several triesters were all degraded during the electrolysis except O, O, O-tri-ethyl-phosphoric. This study indicates that is the practical use of electrochemical chloride mediated indirect oxidation is significant and imposes itself as a potential possible solution for the remediation of organophosphorus pesticide polluted natural waters (Muff et al. 2009).

Strong oxidants, persulfate ($S_2O_8^{2-}$), perphosphate ($P_2O_8^{4-}$), and percarbonate ($C_2O_6^{2-}$), can be used as mediators in the indirect anode oxidation and can be generated according to equation (14)–(16), respectively. In order to reach the formation potentials of these active species, non-active anodes such as BDD and PbO_2 are used (Moradi et al. 2020).

$$2SO_4^{2-} \rightarrow S_2O_8^{2-} + 2e^- \quad (14)$$

$$2PO_4^{3-} \rightarrow P_2O_4^{4-} + 2e^- \quad (15)$$

$$2CO_3^{2-} \rightarrow C_2O_6^{2-} + 2e^- \quad (16)$$

In summary, external anions have an active and significant role in the electrochemical degradation process. Their concentration is easily controllable in model solutions, while in real samples, it could be more difficult to define the electrooxidation potential and predict the outcome of electrooxidation.

Tracking Degradation Efficiency

During the degradation of organic pollutants (pesticides, aromatic hydrocarbons, phenolic compounds, antibiotics, etc.), it is necessary to constantly monitor these compounds in order to see whether this leads to their degradation. For monitoring the mineralization of organic compounds, the determination of total organic carbon (TOC) and/or the measurement of the chemical oxygen demand (COD) or the biological oxygen demand (BOD) is used. However, for better insight into the process of degradation, it is imperative to identify and quantify organic pollutants and their by-products in various media. These methods usually include analytical techniques such as electrospray ionization-mass spectroscopy (ESI-MS), liquid chromatography-mass spectroscopy (LC-MS), gas chromatography-mass spectroscopy (GC-MS), ultra-performance liquid chromatography (UPLC) with photodiode array (PDA) detector, matrix-assisted laser desorption ionization-time of flight (MALDI-TOF), etc. UV-Vis spectroscopy and FTIR spectroscopy play an important role during the monitoring of degradation. Different NMR spectroscopic assays have been developed for the identification of transformation products.

In recent decades, electrochemical methods often have been used for the detection of organic compounds, including organophosphates. By applying different materials to the electrodes, better sensitivity and detection of organophosphates are achieved. There are numerous examples of organophosphate detection by electrochemical sensors in the literature (Aćimović et al. 2017; Sidhu et al. 2019; Musameh et al. 2012).

Ab Initio Modeling of Organophosphate Electrooxidation

A fast expansion of computers in the past few decades has enabled the progressive development of computational chemistry methods, further resulting in the increase of the size of chemical systems that can be successfully studied. Today, *ab initio* calculations can be performed on systems up to several thousands of atoms in size. This enables the conduction of *in silico* experiments, being applied equally to the experimental investigations in almost every aspect of science, including materials science and electrochemical processes (D. Anićijević Vasić 2015).

In this section, we summarized the common methods in *ab initio* modeling that are useful in the investigation of organophosphate electrooxidation. Calculated electronic properties of isolated molecules (ability to release the electron and the susceptibility to the radical attack) can be used to predict the degradation efficiency and mechanism. Calculated adsorption properties of organophosphate and their degradation byproducts can provide deeper insight into the kinetics and mechanism of the degradation process. Ionization energy, along with semiempirical hydratation energies of the organophosphate pollutant, can be used to predict the Gibbs free energy of oxidation, which is in a fundamental relationship with oxidation potential (in direct oxidation process).

Ab Initio Modeling of Isolated Organophosphate Molecules – Electronic Descriptors of Indirect Oxidation

Knowledge of the structural and electronic properties of organophosphate molecules is necessary for further modeling of their interactions in real electrochemical systems. Modeling of isolated organic molecules and their electronic distribution is usually performed using the meta-hybrid GGA functionals, which are implemented within the majority of quantum chemistry software (Lewars 2010). The organic pollutant molecule is first optimized in the gas phase, and then the vibrational analysis is performed to check that the true energy minimum is obtained. The solvent effect can be included subsequently using the implicit (parametrized) solvent models (SMD, PCM), normally implemented within the software (Marenich, Cramer, and Truhlar 2009; Mennucci et al. 2002). Analysis of the charge population is also usually implemented in the software (for example, Mulliken or Lowdin charges, or can be performed subsequently using a separate code (AIM in Bader code) (Mulliken 1955; Löwdin 1970; Henkelman, Arnaldsson, and Jónsson 2006).

Ab initio calculations are commonly used to determine structural properties of isolated organophosphate molecules (i.e., bond lengths, bond angles), which is not always trivial due to a large number of atoms in the molecule. As the organophosphate pesticides have multiple conformational isomers, conformational analysis is desirable prior to any further investigation of their interactions in electrochemical systems. The number of conformers increases with the complexity of the organophosphate alkyl chains. In practice, energy separation between the conformers is usually considered low and thus neglected. The study of Ford et al., which confirms this preposition, deals with the conformational analysis of parathion carried out at the density functional theory (DFT) and Møller–Plesset second-order perturbation (MP2) level of theory (Ford-green, Majumdar, and Leszczynski 2006). Hydration effects were encountered through the PCM – COSMO model (Klamt and Schuurmann 1993). Six low-energy conformers of parathion have been identified in the gas phase and in an aqueous medium with very low energy separation.

Besides the determination of isolated organophosphate molecule structural properties, one can also extract the electronic structure data from *ab initio* calculations, which can be further applied for defining various reactivity descriptors. These reactivity descriptors can be of interest when predicting organophosphate degradation pathways upon electrooxidation. Although not dealing with organophosphorus pesticides, an illustrative example comes from the paper by Wang et al., where the atrazine degradation pathway on BDD anode was proposed using reactivity descriptors obtained from DFT electronic structure analysis. Generally, the highest occupied molecular orbital (HOMO) suggests the easy escape of electrons, which is of the highest possibility to be attacked by •OH (T. Wang et al. 2021). The regioselectivity of generated radicals in the electrochemical oxidation of atrazine molecules was assessed by the Fukui function to provide more information on the degradation process. The natural population analysis (NPA) charge distribution and Fukui index representing radical attack (f^0) of carbon, nitrogen, and chlorine atoms in atrazine were also calculated. The results revealed that the bond cleavage upon radical attack occurs at amine C – N and C – Cl bond on the ring.

As shown in the example, the Fukui index of susceptibility to radical attack (f^0) can be used for the prediction of the reactive site of organophosphate molecule in the indirect electrooxidation (•OH radical attack).

$$f_A^0 = \frac{1}{2}\left(P_A(N-1) - P_A(N+1)\right) \tag{17}$$

Here $P_A(N)$ stands for the electronic population of atom A in the molecule with N electrons. The atoms with the highest f^0 index will probably be first attacked by OH radical in indirect electrooxidation, thus enabling prediction of its mechanism and kinetics.

Fukui index of nucleophilicity (Padmanabhan, 2004) can be calculated using the equation:

$$f_A^- = P_A(N) - P_A(N-1) \tag{18}$$

for each atom in a considered molecule. The higher nucleophilicity index – the higher ability of the atom to be directly oxidized on the anode. In other words, one can expect that the atom with the highest nucleophilicity is expected to release the electron in the direct oxidation process.

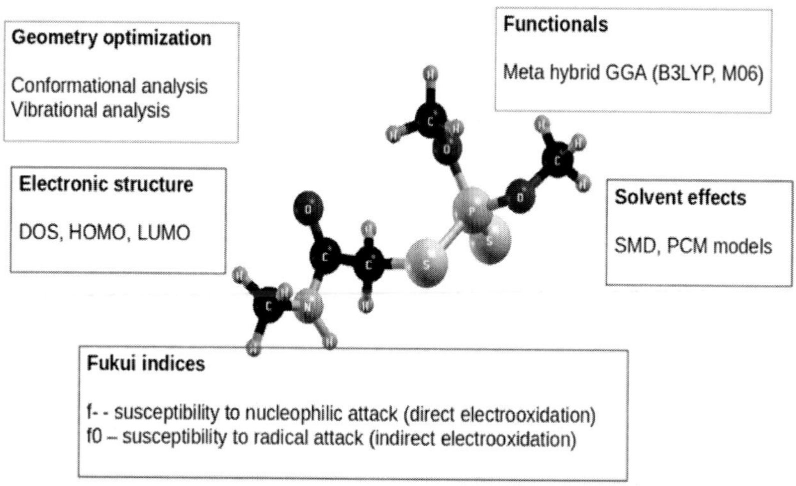

Figure 2. DFT calculations on isolated organophosphate molecules – a summary.

Computational study of solvolysis of toxic organophosphorus pesticide, parathion by anionic nucleophiles (NH_2O^-, HOO- and CH_3S^-) has been reported in both gas and aqueous phase. Fukui function was used as the reactivity descriptor, and the COSMO model was used for accounting solvent effects (Sahu and Das 2017). The solvolysis reaction involves multistep addition-elimination pathways which pass through the trigonal bipyramidal intermediates at the phosphorus center. Obtained results, illustrating the use of Fukui nucleophilicity index as a reactivity descriptor, also point to the phosphorus center as the most probable site of parathion to be oxidized on the anode.

Figure 2 represents a graphical summary of the application of DFT calculations on isolated organophosphate molecules.

Ab Initio Modeling of Adsorption - Interaction between Organophosphates and Electrode Surface

The anodic oxidation process is strongly dependent on the thermodynamic properties of the electrode material and its interaction with the oxidized compound on the fundamental level. Although many studies have determined that the process of direct oxidation is not the primary mechanism of organics electrooxidation, rather it occurs only when operating in the potential electrical regime of water stability (i.e., potentials below the onset of water oxidation), it is important to understand any immediate interaction between the electrode and degraded chemicals in order to prevent unwanted side effects, such as electrode poisoning or damage (Mcbeath, Wilkinson, and Graham 2019). For example, too strong adsorption would mean rapid electrode poisoning and subsequent inability of the electrode to generate hydroxyl radicals, while too weak adsorption means that direct charge transfer between the electrode and organic pollutant is impossible at all, so the indirect oxidation will be prevalent, and its kinetics will be probably limited by mass flow. Also, modeling of adsorption becomes useful when different electrode materials are compared as electrodes for the electrochemical oxidation of a particular pollutant – *in silico* screening is helpful when selecting the most appropriate anodes for the experiment.

Modeling of adsorption requires the inclusion of the electrode surface model and the size of the system increases compared to the modelling of isolated organophosphate molecules. Although a cluster of atoms can also be used as a surface model, and the surface–adsorbate interaction can be calculated in non-periodic conditions, these calculations are generally much more expensive than for isolated molecules, and this approach is rarely used. Usually, the electrode surface is represented as a crystalline slab with periodically adsorbed organophosphate molecules (Vasić Anićijević, 2020).

Adsorption is described using GGA-PBE and a plane-vawe[1] basis set, implemented in the periodic DFT codes.Commonly, a pseudopotental approach is used (Perdew, Burke, and Ernzerhof 1997; Kresse and Furthmüller 1996). During the past decade, models of solvent effect have been developed for periodic systems, and/or directly calculated solvent effect can be taken from literature and encountered (Can, Dario, and Frédéric 2021).

Adsorption energy represents a measure of the interaction strength between adsorbate and electrode surface. The adsorption energy is usually calculated as:

$$E_{ads} = E_{mol/surf} - \left(E_{mol} + E_{surf}\right) \tag{19}$$

Where $E_{mol/surf}$ is the total energy of the bound system and E_{mol} and E_{surf} are the respective energies of the isolated molecule and the clean surface.

Some representative studies of organophosphate adsorption on the most frequent electrode materials are summarized in the following text.

Le et al. investigated adsorption geometry and energetics of nerve agent sarin and simulant dimethyl methyl-phosphonate (DMMP) on TiO_2 (anatase) (Le et al. 2018). Periodic DFT calculations on GGA-PBE level were employed in combination with ultrasoft pseudopotentials. Kinetic energy and charge density cutoffs were set to 60 Ry and 600 Ry, respectively. Rutile slabs consisted of 192 atoms (4x4x1 or 4x2x2 unit cells), with 20 angstroms of vacuum in the z-direction. Pseudohydrogens of +2/3 and +4/3 nuclear charge was initially placed 1 Å below two-coordinated O atoms and five-coordinated Ti atoms, respectively. Limited by the computational expense of these calculations, initial geometries were constrained to those in which the P=O group is directed toward the anatase surface, to be preferred for the adsorption of organophosphates on metal oxides. Within this constraint, optimizations were tested based on initial geometries with the adsorbate molecule above either a surface Ti or O site and with the phosphoryl O from 2.7 Å to 3.7 Å above the site. For all surfaces, the strongest interaction occurs by the formation of a dative bond between the phosphoryl O, which carries a lone pair of electrons and acts as a donor, and a Ti in the surface that acts as an electron acceptor. Calculated binding energies of sarin on anatase were -18 - 35 kJ mol^{-1}, while for DMMP they go from -20 - 59 kJ mol^{-1}, depending on

[1] In a plane – wave basis sets, wave functions are expressed as a linear combination of plane waves. Accuracy of the basis is defined by the energy cutoff – upper limit of energy of the plane vawes

the orientation of the crystallographic plane of the slab. Electronic density of state calculations revealed that the highest occupied states of sarin hybridize with energetically close TiO_2 states. Obtained adsorption energies are within a range of common values for molecular physisorption. Strong and dissociative chemisorption, which would represent a stumbling block in anodic degradation, was not observed.

Using a similar setup based on pseudopotentials, plane-wave basis, and GGA-PBE approximation, Anicijevic et al. investigated the adsorption of dimethoate on four types of graphene surface models: pristine graphene, monovacancy defect graphene, oxy-defect graphene, and Stone-Wales defect graphene (V. J. Anićijević et al. 2021). Dimethoate adsorption geometry was optimized in two sizes of slab cell (4x4 – 32-atom cell and 6x6 – 128-atom cell). Two initial adsorption geometries (S-terminated and O-terminated) were investigated. Interestingly, final optimized geometries, as well as density of state electronic structure calculations, pointed that Van-der-Waals forces represent the main driving force of adsorption of dimethoate on plain graphene. Although the adsorption of dimethoate on the investigated surfaces was found to be thermodynamically favorable, no evidence for the formation of a real chemical bond between graphene and dimethoate was found – except in the case of graphene with monovacancy defect, where dissociation of the adsorbate was observed. Obtained results are in good agreement with the preceding experimental and theoretical study, as well as general trends in reactivity of investigated graphene model surfaces towards atomic and molecular adsorbates (Lazarević-Pašti et al. 2018; D. D. Anićijević Vasić et al. 2016; Anićijević et al. 2021). From the point of view of electrochemical performance of graphene-based anodes, the absence of chemical interaction between dimethoate and electrode surface implies that the electrode surface remains free for generation of hydroxyl radicals and corroborates indirect oxidation of the dimethoate molecules.

Another DFT study of adsorption of glyphosate pesticide on Pt-Cu decorated pyridine-like nitrogen-doped graphene by Kakkar et al. represents one more example where organophosphate adsorption on graphene is investigated by DFT calculations (Mandeep, Gulati, and Kakkar 2020). GGA-PBE approximation was used with numerical grid - DNP basis set and DFT semi-core-pseudo potentials (DSPP). A Γ-centered Monkhorst-Pack scheme with a $4 \times 4 \times 1$ k-point grid was employed (Monkhorst and Pack 1976). Prior to adsorption calculations, the glyphosate molecule was structurally optimized. The most stable conformer was found, and HOMO/LUMO distribution and Hirshfeld charges were calculated for every atom. The large positive charge of the phosphorus atom was confirmed.

The optimized structure of glyphosate was placed on the optimized structure of pyridine-like nitrogen-doped graphene. The combined system of glyphosate and graphene was then further put to geometry optimization to obtain the lowest energy structure and to study the adsorption. Similarly, as in the two studies presented above, it was observed that the glyphosate interaction with PNG is exothermic physisorption. This was confirmed by the low magnitude of the adsorption energy (−0.31 eV), the absence of changes in structural parameters after adsorption, and also the absence of overlap between glyphosate and nitrogen-dopped graphene in the electron density difference plot.

Theoretical Correlation between Organophosphate Oxidation Potential and *Ab Initio* Calculated Quantities

Ab initio calculations are used to calculate relative total internal energies of the neutral and cationic organophosphate forms, which are in close correlation with all other thermodynamical quantities, including Gibbs free energy and standard oxidation potential (Vasić Anićijević et al. 2013).

The standard electrode potential of the organophosphate molecule oxidation in aqueous solution:

$$OP_{(aq)} \rightarrow OP^{+}_{(aq)} \tag{20}$$

is calculated from the Gibbs free energy of the oxidation process through the following equation :

$$E^{0}_{\mathrm{OP/OP+}} = -\frac{\Delta G_{ox}}{nF} \tag{21}$$

Here, OP denotes organophosphate molecule, and n is the number of electrons transferred to the electrode in the process (n = 1). F is a Faraday's constant.

In order to reduce the quantities that can be calculated by theoretical chemistry methods, the oxidation process (20) can be imagined as the combination of three independent processes:

1. the neutral organophosphate molecule passes from aqueous to the gaseous state, and the used amount of energy equals the hydratation energy.

$$OP_{(aq)} \rightarrow OP_{(g)} \tag{22}$$

$\Delta E_1 = - \Delta E_{hyd, OP}$
$\Delta G_1 = -\Delta G_{hyd, OP}$

2. the molecule is ionized in the gaseous state, and the ionization energy is deliberated:

$$OP_{(g)} \rightarrow OP^+_{(g)} + e^- \tag{23}$$

$\Delta E_2 = \Delta E_{ion}$
$\Delta G_2 = \Delta G_{ion} = \Delta E_{ion} + \Delta ZPE - T\Delta S$
Here, ΔZPE and TΔS stand for representing the energetic contribution of entropy and zero-point energy changes upon ionization, respectively (Boyer 1970).

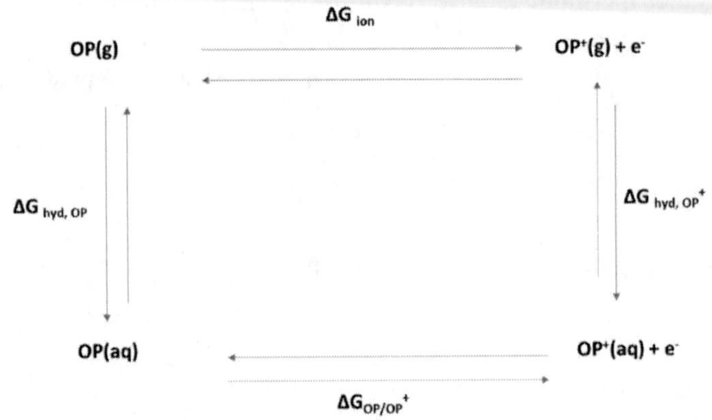

Figure 3. The Born-Haber cycle of direct electrode oxidation of organophosphate molecule.

3. the gaseous ion is dissoluted, and the hydratation energy of the ion is deliberated:

$$OP^+{}_{(g)} \rightarrow OP^+_{(aq)} \tag{24}$$

$\Delta E_3 = \Delta E_{hyd,\,OP^+}$

$\Delta G_3 = \Delta G_{hyd,\,OP^+}$

The Born-Haber cycle of the process is represented in Figure 3 (Laing 2003).

Finally, from Hess law, we have (Stevenson 1976):

$$\Delta G_{OP/OP^+} = \Delta G_{ion} - \Delta G_{hyd,OP} + \Delta G_{hyd,OP^+} \tag{25}$$

As obvious from Figure 3 and equations 22-25, to determine Gibbs free energy of the oxidation process, it is necessary to calculate ionization energy of the organophosphate molecule in the gas phase, as well as the solvation energy of both oxidized (OP^+) and non-oxidized (OP) molecule. Ionization energy is calculated from the difference of *ab initio* total energies of ionized and non-ionized organophosphate molecules:

$$\Delta E_{ion} = E_{tot,OP} - E_{tot,OP^+} \tag{26}$$

To calculate total energies, both OP and OP^+ structures, $E_{hyd,OP}$ and E_{hyd,OP^+} have to be geometrically optimized. The vibrational analysis is performed to confirm that the resulting geometry represents a true energy minimum and to determine zero-point energy (ZPE) and entropy of the system, which are further used to calculate Gibbs's free energy. Replacement of ΔG_2 from (23) into (26) gives (Vansteenkiste et al. 2003):

$$\Delta G_{OP/OP^+} = \left(E_{tot,OP^+} - E_{tot,OP}\right) - \Delta G_{hyd,OP} + \Delta G_{hyd,OP^+} - T\Delta S + \Delta ZPE \tag{27}$$

Hydratation free energies of OP and OP^+ species, $\Delta G_{hyd,OP}$ and $\Delta G_{hyd,OP^+}$ can be calculated semiempirically, by applying the solvent model to the optimized structures, or directly, from *ab initio* calculated interaction energy between organophosphate and water molecules (Mejías and Lago 2000). Replacement of (27) into (21) provides a link between *ab initio* calculated ionization energy, semiempirically calculated hydration properties and standard oxidation potential of a certain organophosphate molecule and can be used for prediction and explanation of its behavior in the potential window applied for electrooxidation.

Conclusion

The removal of organophosphates from aqueous solutions by electrooxidation represents a sophisticated technology with promising high efficiency. The concept comprises complete de-pollution of water using only electrical current, with no additional chemicals involved, and without any remnant waste. Within the process, the pollutant is fully mineralized into carbon dioxide, water, and phosphate anions. Simple electrochemical apparatus requires careful selection of highly efficient and non-corrosive electrode materials, and precision in the determination of experimental parameters, enabling maximal concentration of OH radicals – the most important oxidizing agent for the desired outcome of the process. Generally, a large number of electrode materials provide considerable degradation of initial contaminant, and the leading among them until now is excessively expensive boron-dopped diamond (BDD). Nevertheless, precise control of the composition of the depolluted solution is as important as the choice of the electrode in order to assure that the sample is fully mineralized.

The electrolyte composition remains among the main challenges. Side reactions involving supporting electrolyte anions that also actively participate in oxidation should be minimized, or at least precisely controlled, in order to manage undesired toxic byproducts. However, as electrochemical generation of OH radicals in pure water is kinetically too slow, the presence of supporting electrolytes is inevitable, and the concentration of external ions is a matter of compromise between sustainable kinetics and mineralization efficiency. The problem increases in real wastewater samples, where the possibility to control the presence of anions without additional chemicals is reduced.

Ab initio modeling of organophosphate degradation represents an interesting challenge since it can provide knowledge on the fundamental processes beyond electrooxidation. It is possible to define electronic descriptors that predict reactive sites of organophosphate molecules with OH radical and electrode surface. *Ab initio* modeling of organophosphate adsorption on the anode surface can be used to predict direct organophosphate oxidation or the origin of electrode poisoning. Finally, *ab initio* methods can be used to calculate thermodynamical quantities significant for the electrooxidation process.

Acknowledgments

The research was funded by the Ministry of Education, Science, and Technological Development of the Republic of Serbia, Grant No. 451-03-9/2021-14/200017.

References

Acimovic, Danka D. 2020. "*Electrochemical Oxidation of Polycyclic Aromatic Hydrocarbons in Concrete and Toxicity Estimation of Their Degradation Products.*" Faculty of Technology and Metallurgy, University of Belgrade.

Aćimović, Danka D., Slavko D. Karić, Željka M. Nikolić, Tanja P. Brdarić, Gvozden S. Tasić, Milica P. Marčeta Kaninski, and Vladimir M. Nikolić. 2017. "Electrochemical Oxidation of the Polycyclic Aromatic Hydrocarbons in Polluted Concrete of the Residential Buildings." *Environmental Pollution*, 220 (January): 393–99. https://doi.org/10.1016/J.ENVPOL.2016.09.075.

Acimovic, Danka D., Zeljka M. Nikolic, Milos S. Tosic, Dubravka S. Milovanovic, Vladimir M. Nikolic, Tanja P. Brdaric, and Milica P. Marceta-Kaninski. 2017. "Validation and Uncertainty Estimation of UPLC-PDA Method for the Analysis of Polycyclic Aromatic Hydrocarbons in Concrete." *Journal of Hazardous Materials*, 325 (March): 271–78. https://doi.org/10.1016/j.jhazmat.2016.12.008.

Alves, Suellen A., Tanare C. R. Ferreira, Fernanda L. Migliorini, Maurício R. Baldan, Neidênei G. Ferreira, and Marcos R. V. Lanza. 2013. "Electrochemical Degradation of the Insecticide Methyl Parathion Using a Boron-Doped Diamond Film Anode." *Journal of Electroanalytical Chemistry*, 702: 1–7. https://doi.org/10.1016/j.jelechem.2013.05.001.

Anicijevic Vasic, Dragana. 2015. "*Theoretical Analysis of Tungsten Carbide Properties as an Electrocatalyst Support for Hydrogen Electrode Reactions.*" University of Belgrade Faculty of Physical Chemistry.

Anicijevic Vasic, Dragana D., Ivana M. Perovic, Sladana L. Maslovara, Snezana M. Brkovic, Dragana L. Zugic, Zoran V. Lausevic, and Milica P. Marceta Kaninski. 2016. "AB Initio Study of Graphene Interaction with O2, O, and O-." *Macedonian Journal of Chemistry and Chemical Engineering*, 35 (2): 271–74. https://doi.org/10.20450/mjcce.2016.1038.

Anićijević, Vladan J, Tamara D Lazarević-Pašti, Vesna M Vasić, and Dragana D Vasić Anićijević. 2021. "An Insight into the Efficient Dimethoate Adsorption on Graphene-Based Materials—A Combined Experimental and DFT Study." *Applied Sciences*, 11 (9). https://doi.org/10.3390/app11094014.

Anicijevic, Vladan, Marko Jelic, Aleksandar Jovanovic, Nebojsa Potkonjak, Igor Pasti, and Tamara Lazarevic-Pasti. 2021. "Organophosphorous Pesticide Removal from Water by Graphene-Based Materials - Only Adsorption or Something Else as Well?" *Journal of the Serbian Chemical Society*, 86 (7–8): 699–710. https://doi.org/10.2298/jsc210108012a.

Anićijević, Vladan, Tamara Lazarević-Pašti, Dragana Vasić-Anićijević, and Radovan Karkalić. 2019. "Esters of Organophosphorus Acids: Toxicity, Application and Removal from the Environment." *Scientific Technical Review*, 69 (3): 15–29. https://doi.org/10.5937/str1903015a.

Aquino Neto, S., and A. R. de Andrade. 2009. "Electrooxidation of Glyphosate Herbicide at Different DSA® Compositions: PH, Concentration and Supporting Electrolyte Effect." *Electrochimica Acta*, 54 (7): 2039–45. https://doi.org/10.1016/j.electacta.2008.07.019.

Arapoglou, D, A Vlyssides, C Israilides, A. Zorpas, and P. Karlis. 2003. "Detoxification of Methyl-Parathion Pesticide in Aqueous Solutions by Electrochemical Oxidation." *Journal of Hazardous Materials*, 98 (1–3): 191–99. https://doi.org/10.1016/S0304-3894(02)00318-7.

Babu, B. Ramesh, R. Kanimozhi, P. Venkatesan, and K. Seeni Meera. 2013. "Electrochemical Degradation of Methyl Parathion." *International Journal of Environmental Engineering*, 5 (3): 311. https://doi.org/10.1504/ijee.2013.054702.

Boyer, Timothy H. 1970. "Quantum Zero-Point Energy and Long-Range Forces." *Annals of Physics*, 56 (2): 474–503. https://doi.org/10.1016/0003-4916(70)90027-8.

Büyüksönmez, Fatih, Robert Rynk, Thomas F. Hess, and Edward Bechinski. 1999. "Occurrence, Degradation and Fate of Pesticides during Composting: Part i: Composting, Pesticides, and Pesticide Degradation." *Compost Science and Utilization*, 7 (4): 66–82. https://doi.org/10.1080/1065657X.1999.10701986.

Büyüksönmez, Fatih, Robert Rynk, Thomas F. Hess, and Edward Bechinski. 2000. "Literature Review: Occurrence, Degradation and Fate of Pesticides During Composting: Part II: Occurrence and Fate of Pesticides in Compost and Composting Systems." *Compost Science & Utilization*, 8 (1): 61–81. doi:10.1080/1065657X.2000.10701751.

Can, Ismail, Oğuz Dario, and Vassetti Frédéric. 2021. "Assessing the Performances of Different Continuum Solvation Models for the Calculation of Hydration Energies of Molecules, Polymers and Surfaces : A Comparison between the SMD, VASPsol and FDPB Models." *Theoretical Chemistry Accounts*, 1–13. https://doi.org/10.1007/s00214-021-02799-w.

Casida, John E., and Gary B. Quistad. 2004. "Organophosphate Toxicology: Safety Aspects of Nonacetylcholinesterase Secondary Targets." *Chemical Research in Toxicology*, 17 (8): 983–98. https://doi.org/10.1021/tx0499259.

Comninellis, Christos, Agnieszka Kapalka, Sixto Malato, Simon A Parsons, Ioannis Poulios, and Dionissios Mantzavinos. 2008. "Advanced Oxidation Processes for Water Treatment: Advances and Trends for R&D." *Journal of Chemical Technology & Biotechnology*, 83 (6): 769–76. https://doi.org/10.1002/jctb.1873.

Druzina, Branko, and Matej Stegu. 2007. "Degradation Study of Selected Organophosphorus Insecticides in Natural Waters." *International Journal of Environmental Analytical Chemistry*, 87 (15): 1079–93. https://doi.org/10.1080/03067310701485341.

Feng, Yujie, Lisha Yang, Junfeng Liu, and Bruce E. Logan. 2016. "Electrochemical Technologies for Wastewater Treatment and Resource Reclamation." *Environmental*

Science: Water Research and Technology, 2 (5): 800–831. https://doi.org/10.1039/c5ew00289c.

Ford-green, Jason, D Majumdar, and Jerzy Leszczynski. 2006. "*Conformational Studies on Parathion*", 106 (January): 2356–65. https://doi.org/10.1002/qua.

Hachami, Fatima, Mohamed Errami, Lahcen Bazzi, Mustapha Hilali, Rachid Salghi, Shehdeh Jodeh, Belkheir Hammouti, and Othman A. Hamed. 2015. "A Comparative Study of Electrochemical Oxidation of Methidation Organophosphorous Pesticide on SnO2 and Boron-Doped Diamond Anodes." *Chemistry Central Journal*, 9 (1): 1–7. https://doi.org/10.1186/s13065-015-0136-x.

Henke, Austin, Timothy P Saunders, Joel A Pedersen, and Robert J Hamers. 2018. "*Enhancing Electrochemical Efficiency of Hydroxyl Radical Formation on Diamond Electrodes by Functionalization with Hydrophobic Monolayers Enhancing Electrochemical Efficiency of Hydroxyl Radical Formation on Diamond Electrodes by Functionalization with Hy*." https://doi.org/10.1021/acs.langmuir.8b04030.

Henkelman, Graeme, Andri Arnaldsson, and Hannes Jónsson. 2006. "A Fast and Robust Algorithm for Bader Decomposition of Charge Density." *Computational Materials Science*, 36 (3): 354–60. https://doi.org/https://doi.org/10.1016/j.commatsci.2005.04.010.

Kanan, Sofian, Matthew A Moyet, Robert B Arthur, Howard H Patterson, Sofian Kanan, Matthew A Moyet, Robert B Arthur, and H Howard. 2019. "Recent Advances on TiO 2 -Based Photocatalysts toward the Degradation of Pesticides and Major Organic Pollutants from Water Bodies." *Catalysis Reviews*, 0 (0): 1–65. https://doi.org/10.1080/01614940.2019.1613323.

Kapałka, Agnieszka, György Fóti, and Christos Comninellis. 2008. "Kinetic Modelling of the Electrochemical Mineralization of Organic Pollutants for Wastewater Treatment." *Journal of Applied Electrochemistry*, 38 (1): 7–16. https://doi.org/10.1007/s10800-007-9365-6.

Klamt, A, and G Schuurmann. 1993. "*COSMO : A New Approach to Dielectric Screening in Solvents with Explicit Expressions for the Screening Energy and Its Gradient*."

Kresse, G, and J Furthmüller. 1996. "Efficiency of Ab-Initio Total Energy Calculations for Metals and Semiconductors Using a Plane-Wave Basis Set." *Computational Materials Science*, 6 (1): 15–50. https://doi.org/https://doi.org/10.1016/0927-0256(96)00008-0.

Krishna, K. Rama, and Ligy Philip. 2009. "Biodegradation of Mixed Pesticides by Mixed Pesticide Enriched Cultures." *Journal of Environmental Science and Health - Part B Pesticides, Food Contaminants, and Agricultural Wastes*, 44 (1): 18–30. https://doi.org/10.1080/03601230802519520.

Kumar, Smita S., Pooja Ghosh, Sandeep K. Malyan, Jyoti Sharma, and Vivek Kumar. 2019. "A Comprehensive Review on Enzymatic Degradation of the Organophosphate Pesticide Malathion in the Environment." *Journal of Environmental Science and Health - Part C Environmental Carcinogenesis and Ecotoxicology Reviews*, 37 (4): 288–329. https://doi.org/10.1080/10590501.2019.1654809.

Laing, Michael. 2003. "A Graphical Presentation of the Born-Haber Cycle for Estimating the Electrode Potentials of Metals." *Journal of Chemical Education*, 80 (9): 1057. https://doi.org/10.1021/ed080p1057.

Lazarević-Pašti, Tamara, Vladan Anićijević, Miloš Baljozović, Dragana Vasić Anićijević, Sanjin Gutić, Vesna Vasić, Natalia V. Skorodumova, and Igor A. Pašti. 2018. "The Impact of the Structure of Graphene-Based Materials on the Removal of Organophosphorus Pesticides from Water." *Environmental Science: Nano*, 5 (6): 1482–94. https://doi.org/10.1039/C8EN00171E.

Le, Nam Q, Chinedu E Ekuma, Brett I Dunlap, and Daniel Gunlycke. 2018. "First-Principles Calculations of Sarin Adsorption on Anatase Surfaces." *The Journal of Physical Chemistry C* 122 (5): 2832–39. https://doi.org/10.1021/acs.jpcc.7b11509.

Lewars, Errol G. 2010. *Computational Chemistry: Introduction to the Theory and Applications*. 2nd editio. Springer. https://doi.org/10.1007/978-90-481-3862-3.

Löwdin, Per-Olov. 1970. "On the Nonorthogonality Problem**The Work Reported in This Paper Has Been Sponsored in Part by the Swedish Natural Science Research Council, in Part by the Air Force Office of Scientific Research (OSR) through the European Office of Aerospace Research (OA." In, edited by Per-Olov Löwdin, 5:185–99. *Advances in Quantum Chemistry*. Academic Press. https://doi.org/https://doi.org/10.1016/S0065-3276(08)60339-1.

Mandeep, Archa Gulati, and Rita Kakkar. 2020. "DFT Study of Adsorption of Glyphosate Pesticide on Pt-Cu Decorated Pyridine-like Nitrogen-Doped Graphene." *Journal of Nanoparticle Research*, 22 (1). https://doi.org/10.1007/s11051-019-4730-z.

Marenich, Aleksandr V, Christopher J Cramer, and Donald G Truhlar. 2009. "*Universal Solvation Model Based on Solute Electron Density and on a Continuum Model of the Solvent Defined by the Bulk Dielectric Constant and Atomic Surface Tensions*," 6378–96.

Marselli, B., J. Garcia-Gomez, P.-A. Michaud, M. A. Rodrigo, and Ch. Comninellis. 2003. "Electrogeneration of Hydroxyl Radicals on Boron-Doped Diamond Electrodes." *Journal of The Electrochemical Society*, 150 (3): D79. https://doi.org/10.1149/1.1553790.

Martínez-Huitle, Carlos A., Achille De Battisti, Sergio Ferro, Silvia Reyna, Mónica Cerro-López, and Marco A. Quiro. 2008. "Removal of the Pesticide Methamidophos from Aqueous Solutions by Electrooxidation Using Pb/PbO2, Ti/SnO2, and Si/BDD Electrodes." *Environmental Science and Technology*, 42 (18): 6929–35. https://doi.org/10.1021/es8008419.

Martı, Carlos A, and Sergio Ferro. 2006. "*Electrochemical Oxidation of Organic Pollutants for the Wastewater Treatment : Direct and Indirect Processes*," 1324–40. https://doi.org/10.1039/b517632h.

Mcbeath, Sean T, P Wilkinson, and Nigel J D Graham. 2019. "Application of Boron-Doped Diamond Electrodes for the Anodic Oxidation of Pesticide Micropollutants in a Water Treatment Process: A Critical Review," 2090–2107. https://doi.org/10.1039/c9ew00589g.

Mejías, J. A., and S. Lago. 2000. "Calculation of the Absolute Hydration Enthalpy and Free Energy of H+ and OH−." *The Journal of Chemical Physics*, 113 (17): 7306–16. https://doi.org/10.1063/1.1313793.

Mennucci, B., J. Tomasi, R. Cammi, J. R. Cheeseman, M. J. Frisch, F. J. Devlin, S. Gabriel, and P. J. Stephens. 2002. "Polarizable Continuum Model (PCM) Calculations of

Solvent Effects on Optical Rotations of Chiral Molecules." *The Journal of Physical Chemistry A*, 106 (25): 6102–13. https://doi.org/10.1021/jp020124t.

Monkhorst, Hendrik J., and James D. Pack. 1976. "Special Points for Brillouin-Zone Integrations." *Physical Review B*, 13 (12): 5188–92. https://doi.org/10.1103/PhysRevB.13.5188.

Moradi, Masoud, Yasser Vasseghian, Alireza Khataee, Mehmet Kobya, Hossein Arabzade, and Elena-niculina Dragoi. 2020. "Journal of Industrial and Engineering Chemistry Service Life and Stability of Electrodes Applied in Electrochemical Advanced Oxidation Processes : A Comprehensive Review." *Journal of Industrial and Engineering Chemistry*, 87: 18–39. https://doi.org/10.1016/j.jiec.2020.03.038.

Moreira, Francisca C, Rui A R Boaventura, Enric Brillas, and Vítor J P Vilar. 2017. "Applied Catalysis B : Environmental Electrochemical Advanced Oxidation Processes : A Review on Their Application to Synthetic and Real Wastewaters." *"Applied Catalysis B, Environmental"*, 202: 217–61. https://doi.org/10.1016/j.apcatb.2016.08.037.

Muff, Jens, Christian D Andersen, Rasmus Erichsen, and Erik Gydesen Soegaard. 2009. "*Electrochimica Acta Electrochemical Treatment of Drainage Water from Toxic Dump of Pesticides and Degradation Products*", 54: 2062–68. https://doi.org/10.1016/j.electacta.2008.09.032.

Mulliken, R. S. 1955. "Electronic Population Analysis on LCAO–MO Molecular Wave Functions. II. Overlap Populations, Bond Orders, and Covalent Bond Energies." *The Journal of Chemical Physics*, 23 (10): 1841–46. https://doi.org/10.1063/1.1740589.

Musameh, Mustafa M., Yuan Gao, Mark Hickey, and Ilias Louis Kyratzis. 2012. "Application of Carbon Nanotubes in the Extraction and Electrochemical Detection of Organophosphate Pesticides: A Review." *Analytical Letters*, 45 (8): 783–803. https://doi.org/10.1080/00032719.2012.655678.

Padmanabhan, J., R. Parthasarathi, U. Sarkar, V. Subramanian, and P. K. Chattaraj, 2004. "Effect of solvation on the condensed Fukui function and the generalized philicity index" *Chemical Physics Letters*, 383 (1-2):122-128. https://doi.org/10.1016/j.cplett.2003.11.013.

Panizza, Marco, and Giacomo Cerisola. 2009. "Direct And Mediated Anodic Oxidation of Organic Pollutants." *Chemical Reviews*, 109 (12): 6541–69. https://doi.org/10.1021/cr9001319.

Perdew, John P, Kieron Burke, and Matthias Ernzerhof. 1997. "Generalized Gradient Approximation Made Simple [Phys. Rev. Lett. 77, 3865 (1996)]." *Phys. Rev. Lett.*, 78 (7): 1396. https://doi.org/10.1103/PhysRevLett.78.1396.

Postigo, Cristina, Maria José, López De Alda, Damià Barceló, Antoni Ginebreda, and Teresa Garrido. 2010. "Analysis and Occurrence of Selected Medium to Highly Polar Pesticides in Groundwater of Catalonia (NE Spain): An Approach Based on on-Line Solid Phase Extraction – Liquid Chromatography – Electrospray-Tandem Mass Spectrometry Detection." *Journal of Hydrology*, 383 (1–2): 83–92. https://doi.org/10.1016/j.jhydrol.2009.07.036.

Rodrigo, M. A., M. A. Oturan, and N. Oturan. 2014. "Electrochemically Assisted Remediation of Pesticides in Soils and Water: A Review." *Chemical Reviews*, 114 (17): 8720–45. https://doi.org/10.1021/cr500077e.

Rueffer, Matthew, Dorin Bejan, and Nigel J. Bunce. 2011. "Graphite: An Active or an Inactive Anode?" *Electrochimica Acta*, 56 (5): 2246–53. https://doi.org/10.1016/j.electacta.2010.11.071.

Sahu, Chandan, and Abhijit K Das. 2017. "Solvolysis of Organophosphorus Pesticide Parathion with Simple and α Nucleophiles : A Theoretical Study." *Journal of Chemical Sciences*. https://doi.org/10.1007/s12039-017-1322-2.

Samet, Youssef, Lamia Agengui, and Ridha Abdelhédi. 2010a. "Anodic Oxidation of Chlorpyrifos in Aqueous Solution at Lead Dioxide Electrodes." *Journal of Electroanalytical Chemistry*, 650 (1): 152–58. https://doi.org/10.1016/j.jelechem.2010.08.008.

Samet, Youssef, Lamia Agengui, and Ridha Abdelhédi. 2010b. "Electrochemical Degradation of Chlorpyrifos Pesticide in Aqueous Solutions by Anodic Oxidation at Boron-Doped Diamond Electrodes." *Chemical Engineering Journal*, 161 (1–2): 167–172. doi:10.1016/j.cej.2010.04.060.

Savic, Branislava. 2021. "*Application of Electrochemical Treatment for Decontamination of Wastewater Contaminated with Phenolic Compounds.*" Faculty of Technical Sciences, University of Novi Sad.

Savić, Branislava G., Dalibor M. Stanković, Sanja M. Živković, Miloš R. Ognjanović, Gvozden S. Tasić, Ivana J. Mihajlović, and Tanja P. Brdarić. 2020. "Electrochemical Oxidation of a Complex Mixture of Phenolic Compounds in the Base Media Using PbO2-GNRs Anodes." *Applied Surface Science*, 529 (July): 147120. https://doi.org/10.1016/j.apsusc.2020.147120.

Sidhu, Gurpreet Kaur, Simranjeet Singh, Vijay Kumar, Daljeet Singh Dhanjal, Shivika Datta, and Joginder Singh. 2019. "Toxicity, Monitoring and Biodegradation of Organophosphate Pesticides: A Review." *Critical Reviews in Environmental Science and Technology*, 49 (13): 1135–87. https://doi.org/10.1080/10643389.2019.1565554.

Simond, O., V. Schaller, and Ch Comninellis. 1997. "Theoretical Model for the Anodic Oxidation of Organics on Metal Oxide Electrodes." *Electrochimica Acta*, 42 (13–14): 2009–12. https://doi.org/10.1016/S0013-4686(97)85475-8.

Singh, Baljinder, Jagdeep Kaur, and Kashmir Singh. 2014. "Microbial Degradation of an Organophosphate Pesticide, Malathion." *Critical Reviews in Microbiology*, 40 (2): 146–54. https://doi.org/10.3109/1040841X.2013.763222.

Soltaninejad, Kambiz, and Mohammad Abdollahi. 2009. "Current Opinion on the Science of Organophosphate Pesticides and Toxic Stress: A Systematic Review." *Medical Science Monitor : International Medical Journal of Experimental and Clinical Research*, 15 (3): RA75-90.

Stevenson, Kenneth L. 1976. "Oxidation-Reduction Potentials and Hess' Law." *Journal of Chemical Education*, 53 (7): 453. https://doi.org/10.1021/ed053p453.

Tanaka, Keiichi, Kenji Harada, and Teruaki Hisanaga. 1990. "Photocatalytic Degradation of Organophosphorous Insecticides in Aqueous Semiconductor Suspensions." *Water Research*, 24 (11): 1415–17.

The Council of European Union. 1998. "Directive 98/83/EC, 1998. Council Directive of 3 November 1998 on the Quality of Water Intended for Human Consumption." *Official Journal of the European Communities*, 41. https://eur-lex.europa.eu/legal-content/EN/TXT/?uri=CELEX%3A01998L0083-20151027.

The European Parliament and Council. 2006. "DIRECTIVE 2006/118/EC OF THE EUROPEAN PARLIAMENT AND OF THE COUNCIL on the Protection of Groundwater against Pollution and Deterioration." *Official Journal of the European Communities*, 49. https://eur-lex.europa.eu/legal-content/EN/TXT/?uri=celex%3A32006L0118.

Vansteenkiste, P., V. Van Speybroeck, G. B. Marin, and M. Waroquier. 2003. "Ab Initio Calculation of Entropy and Heat Capacity of Gas-Phase n -Alkanes Using Internal Rotations." *The Journal of Physical Chemistry A*, 107 (17): 3139–45. https://doi.org/10.1021/jp027132u.

Vasić Anićijević, Dragana D., Vladimir M. Nikolić, Milica P. Marčeta-Kaninski, and Igor A. Pašti. 2013. "Is Platinum Necessary for Efficient Hydrogen Evolution? – DFT Study of Metal Monolayers on Tungsten Carbide." *International Journal of Hydrogen Energy*, 38 (36): 16071–79. https://doi.org/10.1016/j.ijhydene.2013.09.079.

Vasić Anićijević, Dragana D. 2020. *Computational modelling of organophosphorous pesticides–density functional theory calculations In: Organophosphate Pesticides*, Nova Science Publishers: 75-97.

Vlyssides, A. G., D. G. Arapoglou, C .J. Israilides, E. M. P. Barampouti, and S. T. Mai. 2004. "Electrochemical Treatment of Methyl Parathion Based on the Implementation of a Factorial Design." *Journal of Applied Electrochemistry*, 34 (12): 1265–69. https://doi.org/10.1007/s10800-004-1698-9.

Vlyssides, Apostolos, Dimitris Arapoglou, Cleanthes Israilides, and Panagiotis Karlis. 2004. "Electrochemical Oxidation of Three Obsolete Organophosphorous Pesticides Stocks." *Journal of Pesticide Science*, 29 (2): 105–9. https://doi.org/10.1584/jpestics.29.105.

Vlyssides, Apostolos, Dimitris Arapoglou, Sofia Mai, and Elli Maria Barampouti. 2005a. "Electrochemical Oxidation of Two Organophosphoric Obsolete Pesticide Stocks." *International Journal of Environment and Pollution*, 23 (3): 289. https://doi.org/10.1504/IJEP.2005.006868.

Vlyssides, Apostolos, Dimitris Arapoglou, Sofia Mai, and Elli Maria Barampouti. 2005b. "Electrochemical Detoxification of Four Phosphorothioate Obsolete Pesticides Stocks." *Chemosphere*, 58 (4): 439–447. doi:10.1016/j.chemosphere.2004.09.037.

Wan, Hai Bin, Ming Keong Wong, and Chup Yew Mok. 1994. "Comparative Study on the Quantum Yields of Direct Photolysis of Organophosphorus Pesticides in Aqueous Solution." *Journal of Agricultural and Food Chemistry*, 42 (11): 2625–30. https://doi.org/10.1021/jf00047a046.

Wang, Jian Long, and LE Jin Xu. 2012. "Advanced Oxidation Processes for Wastewater Treatment: Formation of Hydroxyl Radical and Application." *Critical Reviews in Environmental Science and Technology*, 42 (3): 251–325. https://doi.org/10.1080/10643389.2010.507698.

Wang, Ting, Taobo Huang, Huan Jiang, and Ruoqi Ma. 2021. "Electrochemical Degradation of Atrazine by BDD Anode : Evidence from Compound-Speci Fi c Stable Isotope Analysis and DFT Simulations." *Chemosphere*, 273: 129754. https://doi.org/10.1016/j.chemosphere.2021.129754.

Wu, Jiguo, Chongyu Lan, Gilbert Yuk, and Sing Chan. 2009. "Chemosphere Organophosphorus Pesticide Ozonation and Formation of Oxon Intermediates." *Chemosphere*, 76 (9): 1308–14. https://doi.org/10.1016/j.chemosphere.2009.04.060.

Chapter 9

Carbon Materials as Adsorbents for Organophosphate Pesticides in Aqueous Media – Critical Overview

Igor Pašti[1,*], Stefan Breitenbach[2,3], Christoph Unterweger[2] and Christian Fürst[2]
[1]University of Belgrade – Faculty of Physical Chemistry, Belgrade, Serbia
[2]Wood K plus – KompetenzzentrumHolz GmbH, Linz, Austria
[3]Institute for Chemical Technology of Inorganic Materials (TIM), Johannes Kepler University Linz, Linz, Austria

Abstract

Organophosphates are considered some of the most toxic substances synthesized by man due to their high neurotoxicity. Their wide use in modern agriculture and other areas results in an urgent need for their efficient removal from the environment. Among numerous removal strategies, adsorption is considered a promising approach for organophosphate pesticide removal because it is simple, economical, and environmentally friendly. Carbon-based materials are a very good choice for this purpose because of their properties, including high specific surface areas, various pore structures, and rich surface chemistry. Furthermore, carbon materials not only have a high adsorption capacity towards organophosphates, but they are generally cheap, and their properties could be easily tuned. When analyzing the adsorption of organophosphates on carbon materials, the key is to understand the mechanisms of these interactions. However, available literature shows a great diversity in used approaches and sometimes diverges from basic

* Corresponding Author's Email: igor@ffh.bg.ac.rs.

In: Organophosphates
Editor: Tamara Lazarević-Pašti
ISBN: 978-1-68507-652-8

theoretical principles. In this chapter, we critically address the analysis of organophosphates adsorption on carbon materials. First, we present the properties of organophosphate pesticides, followed by the classification of carbon materials and an overview of techniques used in their characterization. This is essential to understand how organophosphates can interact with carbon surfaces and what are the key factors determining adsorption performance. Next, we summarize thermodynamic and kinetic descriptions of adsorption processes and outline critical points in the analysis of adsorption thermodynamics. This is followed by an overview of organophosphate-carbon interactions and experimentally obtained adsorption capabilities of different carbons reported in the literature. We conclude with the suggestions to standardize adsorption measurements, making them easier to compare and suitable for extracting the key carbon properties for developing advanced superior adsorbents for organophosphate pesticides.

Keywords: organophosphate pesticides, removal, remediation, carbon materials, adsorption

Introduction

Pesticides are synthetic substances developed to and extensively used to repel, kill, or control damaging or interfering pests during the manufacture, processing, transport, or marketing of foodstuffs, agricultural commodities, animal feed, or products that may be administered to animals to control insects, bacteria or other pests (Gruden, Buchholz, and Kanoun 2014, Kanoun et al. 2021). Their role in increasing agricultural yields is crucial, and they are used in practically every country in the world. Increased demands in food followed the growth of the world population, so the use of pesticides followed this trend and also increased. Interestingly, 45% of the usage comes from European countries, which, together with the USA, contributes 60% of the worldwide usage of pesticides, and 40% comes from the remaining countries (Hanrahan, Patil, and Wang 2004). These data show a certain misbalance in using pesticides considering the percentage of the world population in Europe and the USA.

Besides the direct effects on target species, the extensive use of pesticides results in their accumulation in food, water, and the environment, in general. This accumulation of pesticides presents a severe hazard for non-target organisms, including humans (Hossain and Mansour 2019), as many of the used pesticides also exert toxic effects on humans. Furthermore, chronic or acute exposures of humans to certain pesticides cause a wide range of pathological conditions, starting from anxiety, depression, convulsions, severe neurological disorders, or even cancer (Panjan et al. 2017). For all these reasons, it is necessary to precisely monitor the concentration of pesticides in the environment, but even more to use different remediation strategies to remove them from the environment and minimize the risk of unintended exposure of non-target species and humans to pesticide residues.

Based on the chemical structure, pesticides are classified as organophosphates, neonicotinoids, pyrethroids, carbamates, organochlorines (Panjan et al. 2017), substituted phenols and triazines (Panjan, Virtanen, and Sesay 2017). Based on their target groups, pesticides are divided into herbicides, fungicides, insecticides and bactericides. Organophosphates and carbamates are the main groups of pesticides in use today, with a primary role as insecticides. Their toxicity relies on inhibiting acetylcholinesterase (AChE) (Liu et al. 2019), one of the key enzymes in the nervous system responsible for neurotransmission. One AChE gets inhibited, acetylcholine accumulates, and neurotransmission gets disrupted (Ayyalusamy, Mishra, and Suryanarayanan 2018).

The removal of pesticides *via* adsorption is sought as a simple and low-risk strategy, as there are no chemical reactions with pesticides and possible conversion to more toxic compounds, which could be particularly dangerous in the case of organophosphate pesticides (OPs). In a traditional view, an ideal adsorbent would possess a large specific surface area and suitable surface chemistry for removing OPs. However, carbon materials have been present in chemical laboratories for a long time and possess some of the desired properties of good adsorbents like large specific surface areas and rich surface chemistry. For these reasons, carbons are used to remove OPs and other classes of pesticides, particularly from aqueous media. However, the literature survey suggests that different approaches are used to test carbons as adsorbents for OPs, which ultimately results in a very difficult comparison of experimental results from different research groups. In this contribution, we summarize the properties of OPs and carbon materials, crucial to understanding the mechanisms of their interactions, and critically address available experimental data regarding carbon adsorbents for OPs.

Organophosphate Pesticides – Physical and Chemical Properties

From the chemical point of view, OPs are esters of phosphoric acid, and their common feature is the existence of phosphorus atoms double-bonded to sulfur or oxygen (Figure 1). The existence of P=S or P=O bond in, essentially, the same molecule discriminates between thio- and oxo-form (oxon). In general, thio-forms are considered parental forms of corresponding oxons, and these simple chemical changes, caused by oxidation agents (Lazarevic-Pasti et al. 2013), light, or microbes (Van Scoy, Pennell and Zhang 2016), has tremendous effects on the toxicity of a given molecule. Namely, conversion of the OP thio-form to its oxo-form increases the acute toxicity by several orders of magnitude (Colovic et al. 2013). From the point of OPs remediation, this fact is extremely important as one has to apply methods that will not cause the oxidation of OPs but their physical removal from the environment or decomposition into non-toxic products.

Due to the extreme toxicity of oxo-forms, practically only thio-forms are in use as OPs. Besides the central P=S or P=O bond, the P atom forms three more single bonds with different chemical groups of different complexity, which, in turn, determine the toxicity of the final molecule. Apparently, there is no precise rule on how these side groups determine the toxicity (i.e., inhibitory power towards AChE), but only a slight change of the chemical structure can significantly alter the toxicity of the OP molecule. For example, chlorpyrifos is classified as "moderately hazardous," while chlorpyrifos-methyl falls in the "unlikely to present an acute hazard" class. A great diversity of the side chains bonded to the central P atom in an OP molecule resulted in more than one hundred commercially available OPs, which are generally well described and used with different frequencies in different countries worldwide (Table 1).

In OP molecules, the electron-deficient P atom presents the sites of nucleophile attack, which makes the P=S moiety susceptible to oxidation. Besides the toxicity, side chains determine the physical properties of these molecules and the mechanisms of OPs interactions with different materials. For six OPs presented in Figure 1, some basic physicochemical properties are assembled in Table 2. From the point of OPs remediation, solubility is the most important, and one can see that OPs with aromatic side chains have the lowest solubility, followed by phorate with aliphatic low-polarity side chains.

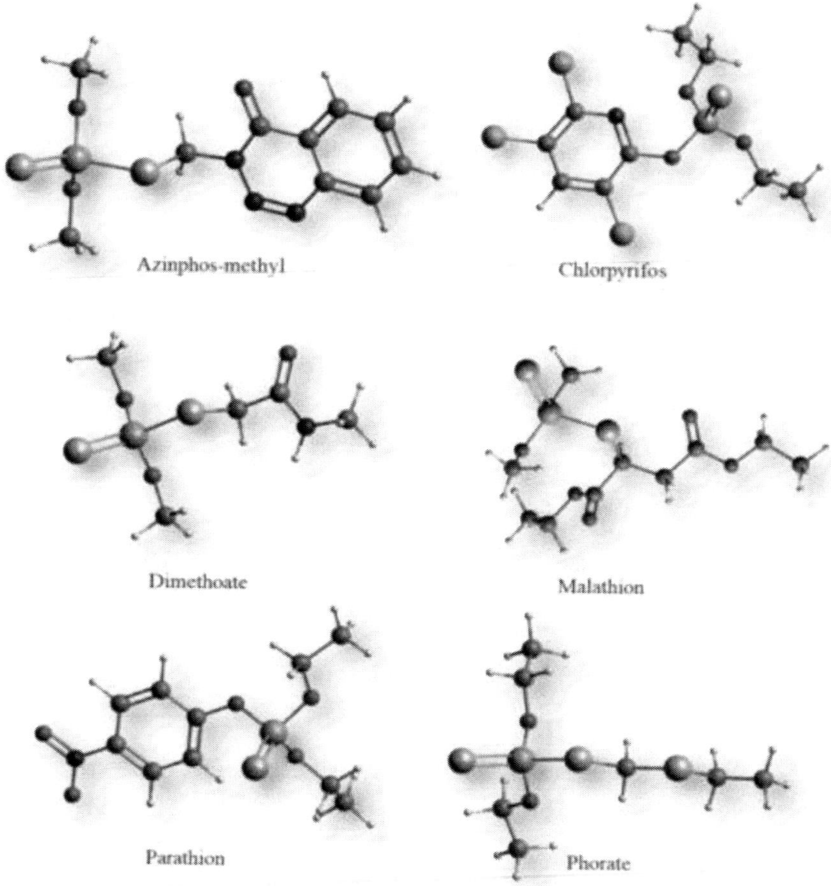

Figure 1. 3D chemical structures of selected, widely used organophosphate pesticides (white – H, grey – C, red – O, orange – P, blue – N, green – Cl, yellow – S).

The nature of the side chains of phosphate moiety in an OP molecule will define how a given molecule will interact with adsorbents. One can immediately conclude that weakly soluble compounds, characterized by hydrophobic chains, will prefer hydrophobic adsorbents. At the same time, a molecule with polar groups or H atoms bonded to O, N, or S could also interact with an adsorbent surface via dipole interactions or the formation of hydrogen bonds. However, the side chains also determine the chemical stability of OP molecules under different environmental pH.

Table 1. Commercially available OPs with CAS identification numbers (data compiled from https://inchem.org/documents/pims/chemical/pimg001.htm#SectionTitle:1.1%20%20Substance)

Name	CAS No.	Name	CAS No.
Azinphos methyl	86-50-0	Formothion	2540-82-1
Bromophos	2104-96-3	Heptenophos	23560-59-0
Chlorphoxim	14816-20-7	Isothioate	36614-38-71
Chlorpyrifos	2921-88-2	Isoxathion	18854-01-8
Chlorpyrifos-methyl	5598-13-0	Jodfenphos	18181-70-9
Crotoxyphos	7700-17-6	Leptophos	21609-90-5
Crufomate	299-86-5	Malathion	121-75-5
Cyanofenphos	13067-93-1	Mephosfolan	950-10-7
Cyanophos	2636-26-2	Mevinphos	7786-34-7
Demeton	8065-48-3	Monocrotophos	6923-22-4
Demeton-O	298-03-3	Naled	300-76-5
Demeton-S	126-75-0	Omethoate	1113-02-6
Diazinon	333-41-5	Oxydemeton-methyl	301-12-2
Dichlofenthion	97-17-6	Parathion	56-38-2
Dichlorvos	62-73-7	Parathion-methyl	298-00-0
Dicrotophos	141-66-2	Phorate	298-02-2
Dimefox	115-26-4	Phospholan	947-02-4
Dimethoate	60-51-5	Phoxim	14816-18-3
Dioxabenzophos	3811-49-2	Pirimiphos-ethyl	23505-41-1
Dioxathion	78-34-2	Prothiofos	34643-46-4
Edifenphos	17109-49-8	Quinlphos	13593-03-8
EPBP	3792-59-4	Sulfotep	3689-24-51
Ethion	563-12-2	Sulprofos	35400-43-2
Ethopropos	13194-48-4	Temephos	3383-96-8
Etrimfos	38260-54-7	Tetrachlorvinphos	22248-79-9
Fenamiphos	22224-92-6	Thiometon	640-15-3
Fenchlorphos	299-84-3	Thionazin	297-97-2
Fenitrothion	122-14-5	Triazophos	24017-47-8
Fenthion	55-38-9	Trichlorfon	52-68-6
Fonofos	944-22-9	Vamidothion	2275-23-2

Table 2. Physicochemical properties of selected OPs

Organophosphate	Solubility in water/mg dm^{-1}	Vapor pressure /mmHg	Melting point/°C	Density/g cm^{-3}
Azinphos-methyl	28	8×10^{-9}	73	1.44
Chlorpyrifos	2.0	1.9×10^{-5}	43	1.40
Dimethoate	2.5×10^{4}	8.5×10^{-6}	43 to 45	1.30
Malathion	145	4.0×10^{-5}	3	1.23
Parathion	24	3.8×10^{-5}	6	1.26
Phorate	50	6.4×10^{-4}	−43	1.16

Fundamental studies of pH stability of OPs are available, but they are rarely systematic. Recently, Lockridge et al. (Lockridge, Verdier, and Schopfer 2019) performed a systematic analysis of the half-life of chlorpyrifos oxon and other organophosphorus esters in an aqueous solution. From this work, it can be unambiguously concluded that OPs decompose rapidly in highly alkaline media. However, for some OPs, the same decomposition is also seen in acids. For example, both chlorpyrifos and chlorpyrifos-oxon decompose at pH 12–13 with the first-order rate constant of the order of 10^{-2} min^{-1}, resulting in half-lives of 18 to 13 min. The same is found for dichlorvos, diazoxon, parathion-ethyl, and parathion-methyl (Lockridge, Verdier, and Schopfer 2019, Macalady and Wolfe 1983). However, diazinon and its oxo-analog diazoxon also rapidly decompose in acidic media (pH below 3) (Gomaa, Suffet, and Faust 1969). It should be emphasized that most of the data on pH stability of OPs is rather old and scattered to a great extent. The summary of Lockridge et al. (Lockridge, Verdier, and Schopfer 2019) confirms this and points out that the experiments are done in different solutions. For example, degradation analyses were done in sterile water with pH adjusted to a given value or buffers with different buffer capacities like carbonate, tris or phosphate (Broomfield, Lenz, and MacIver 1986, Macalady and Wolfe 1983). Such diversity in experimental setups found in the literature causes problems with the data comparison and verification and induces uncertainties in the provided data. For example, the kinetics of OPs degradation could be different if the measurements are done in water with adjusted pH and in a buffer with a high capacity, even if the starting pH is the same. Namely, in the latter case, the concentration of H^+ or OH^- is held constant during the entire experiment, which can take days or months, while in the former case, this is not true. For this reason, the analysis of the OPs decomposition kinetics could be influenced by the change of the pH during OPs decomposition, while for the experiments made in buffers, usually pseudo-first-order kinetics is observed (Anićijević et al. 2021, Lockridge, Verdier, and Schopfer 2019).

Carbon Materials – Underlying Properties Enabling Their Use as Adsorbents

Different Forms of Carbons Present in Chemical Laboratories

Carbon materials are present with a great diversity in chemical laboratories. The classical view of two carbon allotropes, diamond, and graphite, has been extended in the last years with carbon structures of different dimensionality (Figure 2). This classification is closely related to the dimensions of carbon structures at the nanoscale and the definition of nanomaterials itself. Depending on the lateral dimension of carbon structures, the materials are classified as 0D, 1D, 2D, or 3D materials (Belenkov and Greshnyakov 2013).

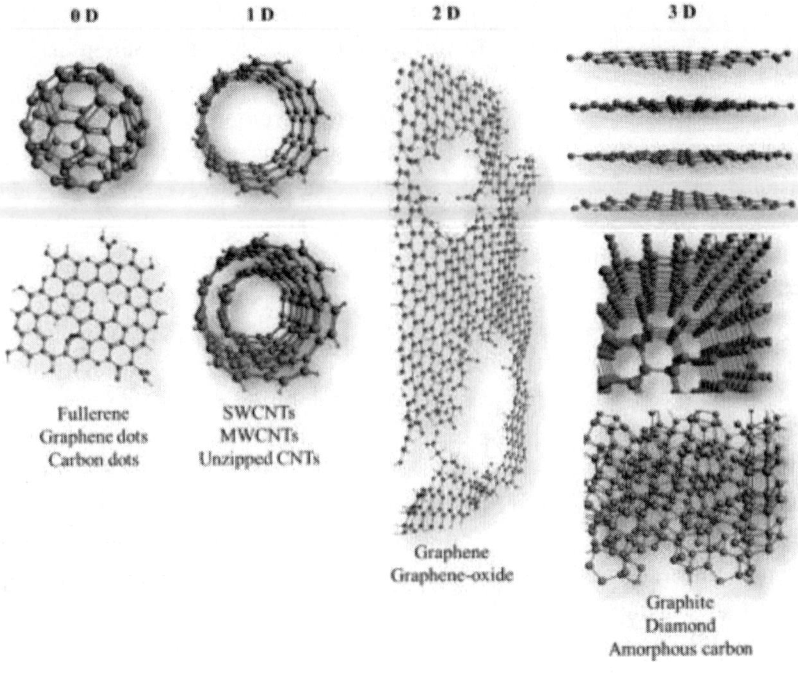

Figure 2. Classification of carbon materials according to their dimensionality.

Typical examples of 1D carbon nanomaterials are single-walled and multi-walled carbon nanotubes (SWCNTs and MWCNTs), with additions to this family like unzipped carbon nanotubes, carbon nanoribbons, and others (Ciric-Marjanovic, Pasti and, Mentus 2015). One dimension of these materials, which we consider their length, is larger than 100 nm. These materials have a well-defined structure, described through the folding scheme of graphene sheets (see further), with a well-defined diameter being constant along the entire nanotube. CNTs are widely commercially available, and reporting their qualities is a rather standardized procedure. According to Lehman et al. (Lehman et al. 2011), regulatory procedures in the US require reporting the following properties:

1. Type of multi-walled carbon nanotube (concentric cylinders, stacked cups or scrolled tubes; the number of walls/tubes).
2. Configuration of nanotube ends (e.g., opened, capped).
3. Description of any branching.
4. Width/diameter of the innermost wall/tube (average and range).
5. Carbon unit cell size and connectivity.
6. Alignment of nanotube along the long axis (straight, bent, buckled).
7. Hexagonal array orientation used in the manufacture of the nanotube.
8. Particle size of catalyst used in the manufacture of the nanotube.
9. Molecular weight (average and range).
10. Particle properties: shape, size (average and distribution), weight (average and distribution), count, surface area (average and distribution), surface-to-volume ratio, aggregation/agglomeration.
11. Structural defect content (non-carbon impurities, lattice defects, adsorbed molecules).

The evaluation of specific surface area (SSA), which is one of the most important properties of carbon materials (see further), for the case of well-defined CNTs renders to a solution of a geometrical problem (Peigney et al. 2001). For SWCNT, the effect of nanotube diameter cancels out, and SSA is constant – 1315 $m^2\ g^{-1}$. For MWCNTs, the addition of extra walls increases the mass more than the exposed surface, and the effect of diameter is present. With every additional wall, SSA decreases; for 20WCNTs, it reaches only ~100 $m^2\ g^{-1}$ and decreases with increasing NT diameter (Peigney et al. 2001). By definition, CNTs have a honeycomb arrangement of carbon atoms (Lehman et al. 2011), i.e., sp^2 hybridization. Thus, the walls of CNTs are generally chemically inert.

In the 2D part of the carbon family, graphene and graphene oxide (GO) stand out as classic examples (Zhu et al. 2010). Graphene is an infinite sheet of carbon arranged into a hexagonal lattice and characterized with some exceptional properties (thermal, mechanical) (Ferrari et al. 2015). Hexagonal structure is governed by sp^2 hybridization of carbon atoms, so the chemical bonds between carbon atoms are directed in-plane. Above and below the graphene plane, π electron cloud forms, making graphene a zero-gap semi-metal (Ferrari et al. 2015). For this reason, pristine graphene is extremely chemically inert (Pasti et al. 2018c), and the main chemical reactivity of graphene arises due to the presence of defects and edges (Pasti et al. 2018b, Yuan et al. 2013), which are always present in real samples. Determination of SSA of graphene is also a geometrical problem, and for pristine graphene, SSA amounts to 2630 m^2g^{-1}. In practice, graphene can be synthesized using bottom-up and top-down approaches (Avouris and Dimitrakopoulos 2012). The best-known and most widely used one is the reduction of GO obtained *via* different modifications of the Hummers method (Chen et al. 2013), but strictly speaking, in this case, one obtains reduced graphene oxide (rGO).

GO is a highly oxidized graphene sheet with O content ranging from 30 to 50 at.% (Dideikin and Vul' 2019). It is a poor conductor or insulator, and its SSA was measured from 2 to 2391 m^2 g^{-1}, depending on the applied approach (Zhang et al. 2020). It is commonly obtained *via* exfoliation of graphite oxide, using the mentioned Hummers method or some other approach. Several models of GO exist in the literature, with the Lerf–Klinowski model being now widely accepted (Lerf et al. 1998). In this model, epoxy- and hydroxyl groups are found on the basal plane, while others, like phenol, ether, and carboxyl, are located at the edges. GO can be controllably reduced to rGO with different amounts of oxygen functional groups using chemical, thermal or electrochemical methods (Shao et al. 2010, Toh et al. 2014). Besides oxygen functional groups, commercial GO could contain small amounts of other heteroatoms (like nitrogen or sulfur). Moreover, the structural disorder in these GO samples is typically very high, and the basal plane can be significantly damaged with a large number of holes reaching several nanometers in diameter (https://www.graphenea.com/collections/graphene-oxide/products/graphene-oxide-4-mg-ml-water-dispersion-1000-ml).

Graphite and diamond represent two extremes of the 3D class of carbon allotropes family. Graphite has a hexagonal lattice, and it is comprised of graphene sheets stacked via weak Van der Walls forces. It is an exceptional conductor, but its conductivity is displayed only in the (0001) plane. In contrast, diamond has a cubic lattice governed by sp^3 hybridization of carbon atoms. It is an insulator and possesses exceptional hardness. Amorphous carbon does not possess long-range order in its lattice but locally contains nanometric domains with graphite- or diamond-like structure. It mostly consists of sp^2 hybridized carbon, and it is usually conductive.

In the preceding discussion, carbon materials have been classified based on their dimensionality, but we considered the forms of carbon with well-defined crystal structures. However, the terms such as carbon nanotubes (nanorods, wires, fibers, platelets, and others) are also used to describe the morphology of the carbon sample (with the addition of "nano" or, more precisely, "nanosized" if dimensions are under 100 nm) irrespective of their crystal structure. A typical example is the conversion of organic materials with different morphologies to carbons, i.e., their carbonization, where the original morphology is preserved in the final carbon material. This scenario is quite usual when polymeric (nano) structures are carbonized. As a rule, the morphology of the polymeric precursor remains preserved (Ciric-Marjanovic, Pasti, and Mentus 2015, Pasti et al. 2018a). The situation is the same when biomaterials with particular morphology, like fibers, are converted to carbons (Breitenbach et al. 2021a, Breitenbach et al. 2021b). All these materials are generally amorphous carbons, but if produced at very high temperatures (> 2500°C), graphitization can take place. Still, even in this case, materials will consist of nanosized graphitic domains, and long-range ordering is not taking place.

So far, several types of nanosized carbon forms have been discussed. However, the material is also considered a nanomaterial if nanosized local structures can describe its internal structure. As most carbon materials have developed pore structures, and the dimensions of these pores are typically in a nanometric range, all carbons can be considered nanomaterials. However, as for 3D carbons, this description comes from their internal pore structure. Therefore, it is more accurate to refer to them as nanostructured materials. The internal pore structure of carbon materials determines most of their SSA, and the SSA of carbon materials is one of the key properties determining their applicability as adsorbents. For achieving the required carbon properties, a large variety of suitable precursor materials is available. In the last two decades, the interest in sustainable carbon precursors exploded, and all sorts of biomass, due to economic and ecological reasons, especially waste streams, have been tested as precursors for activated carbons (Bilal et al. 2013, Danish and Ahmad 2018, Jain, Balasubramanian, and Srinivasan 2016, Mohamad Nor et al. 2013, Tay, Ucar and Karagoz 2009). In addition, a number of chemical and physical routes have been developed for increasing carbon surface areas and developing the pore network in carbon materials (Marsh and Rodriguez-Reinoso 2006, Rodriguez-Reinoso and Molina-Sabio 1992, Rodriguez-

Reinoso, Molina-Sabio and Gonzalez 1995, Wang and Kaskel 2012). This process is called activation, and carbon material that has undergone activation is called activated carbon. All the mentioned classifications and attributes can be combined for naming given carbon material. For example, activated carbon fibers are carbon materials with fibrous morphology that have been activated using some of the possible activation routes (Breitenbach et al. 2020). Furthermore, if we speak about activated carbon nanofibers, this would also mean that diameter of fibers is under 100 nm.

Characterization of Carbon Materials

In order to understand the ways OPs interact with carbon materials and how carbons can be employed to remove OPs from the environment, it is important to address the characterization of carbon materials. The depth of materials characterization is also important to explain the limitations of experimental datasets we use to rationalize the adsorption performance of carbon materials. Considering that adsorption essentially requires the interaction of an OP molecule with the carbon surface, it is important to determine:

1. SSA (as it is linked to the number of sites available for adsorption), porosity, and pore size distribution (as it relates to the availability of the internal carbon surface to an adsorbate)
2. morphology, crystal structure, and the level of disorder (as defects play an important role in surface reactivity),
3. chemical composition (bulk and surface),
4. present surface functional groups (as these are essentially the sites where adsorption takes place).

The main techniques used for these purposes are summarized in Table 3 and depicted in Figure 3.

A direct conclusion of the summary provided in Table 2 is that carbons' physical and chemical properties are practically always determined *ex-situ*. In principle, carbon materials are considered unchanged when in aqueous media in which they interact with adsorbates. While such an approach is not as problematic as in other fields, like electrochemical applications, it is must be considered that carbon surfaces are at least solvated, which could cause the blockage of very narrow pores. The extent of solvation primarily depends on the presence of surface functional groups and their nature, mainly determined using XPS and the solvent used (typically water). Another important point is that many reports present in the literature do not consider all possible techniques (or at least the majority of them) that could provide deep insight into materials' properties. For example, SSA is usually considered the most important quantity for adsorbents, and gas adsorption/desorption measurements are practically always applied. However, the full potential of this technique, besides the determination of pore size distribution and total pore volume, is rarely utilized.

Table 3. The main experimental techniques used for the characterization of carbon materials (adapted from ref. (Pasti et al. 2018a)

Property	Techniques	Remarks
microtexture specific surface and porosity	gas adsorption/desorption measurement, Hg porosimetry	N_2 adsorption/desorption Ar adsorption/desorption CO_2 adsorption/desorption to probe ultramicropores dry material, measurement starts from vacuum conditions
nanotexture morphology	SEM TEM	analysis under vacuum conditions local morphologies and particle sizes
crystal structure	XRD	provides information about bulk properties, and crystallite sizes and
presence of defects	Raman spectroscopy	non-destructive, measurements in air
functional groups	FTIR spectroscopy	measurements in the air; typically, dry materials are measured
chemical composition	elemental analysis EDX	provides information about bulk (C, N, O, H and S), deeper sampling compared to XPS
surface chemical composition	XPS	surface composition and functional groups; measurement is done in a vacuum; cannot determine H

Another standard technique for carbons characterization is Raman spectroscopy. Carbons have rather characteristic Raman spectra, with the so-called G (graphite) and D (disorder) bands appearing in the region 1150 – 1700 cm^{-1} (Ferrari and Robertson 2000). To analyze the spectrum and extract relevant data, like sp^3 content, defect concentration, types of defects (Eckmann et al. 2012, Ferrari and Robertson 2000), this spectral region is typically de-convoluted into five components. This type of Raman spectra processing is a critical point where different approaches are used in the literature, considering the choice of the peak profiles (Gaussian or Lorentian, or some other function), the measure of intensity (peak heights or peak areas), and ultimately positions of these five components. Unfortunately, it is not easy to extract reliable data from Raman spectra de-convolution due to various approaches (different excitation wavelengths, Gaussian or Lorentzian functions used to fit components and other factors).

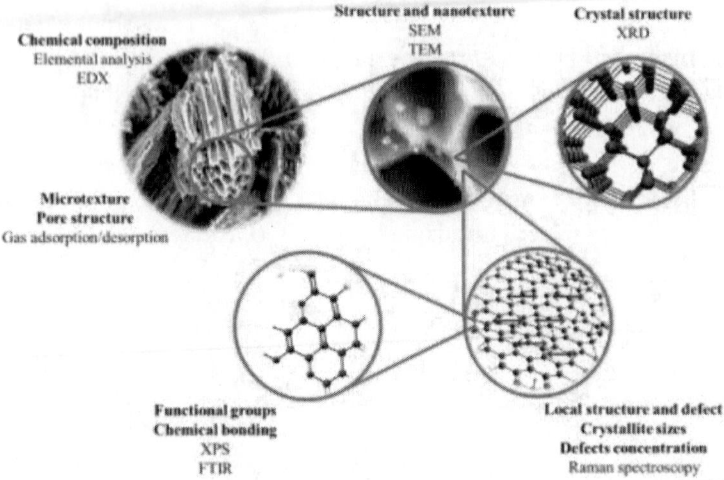

Figure 3. Overview of physical and chemical methods for characterization of carbon materials (Inagaki and Kang 2016).

Another indispensable technique for carbons characterization is XPS, which provides information about the surface composition and the types of functional groups present on the surface (quantitative). Again, such identification is based on the de-convolution of the high-resolution XPS spectra in the spectral region corresponding to each element present on a carbon surface. However, it should be mentioned that using additional chemical derivatization protocols, a highly selective determination of spectroscopically similar functional groups can be achieved (Duchoslav et al. 2018, Everhart and Reilley 1981, Hinterreiter et al. 2019). For example, several functional groups can be identified in the case of oxygen functional groups, which are always present on basically any carbon surface (Figure 4). We have already mentioned that in the case of GO, epoxy, and hydroxyl groups are located at the basal plane. However, this is also the case for other carbons as well. However, local disruption of sp^2 hybridization of carbon atoms in the basal plane also results in the aggregation tendency of these surface groups (Boukhvalov, Katsnelson, and Lichtenstein 2008, Dobrota et al. 2016). Thus, it is expected that carbon planes will be oxidized in patches rather than containing isolated oxygen functional groups.

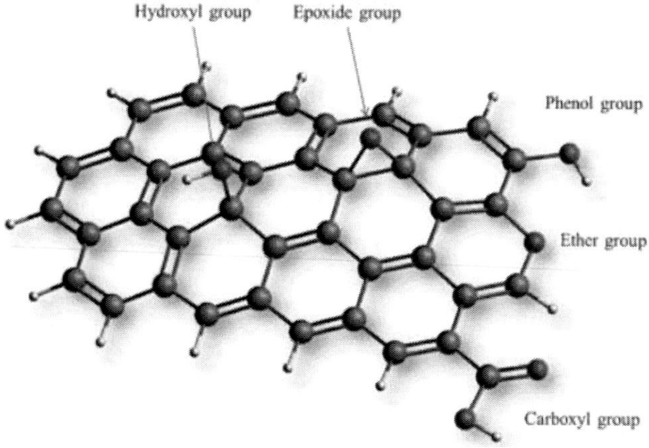

Figure 4. Types of oxygen functional groups present on surfaces of carbon materials (white – H, grey – C, red – O).

Besides oxygen, carbon materials can contain other heteroatoms, most frequently N, B, S, P, and others. These elements are either intentionally added during the synthesis or are naturally present in the carbon precursor (like in the case of biomass-derived carbons or the ones derived from N-containing polymer nanostructures) and remain embedded in carbon structure upon carbonization. The most important heteroatom in the case of carbons is nitrogen which could appear in several forms in carbon materials (Figure 5). These types of nitrogen moieties can be resolved using XPS.

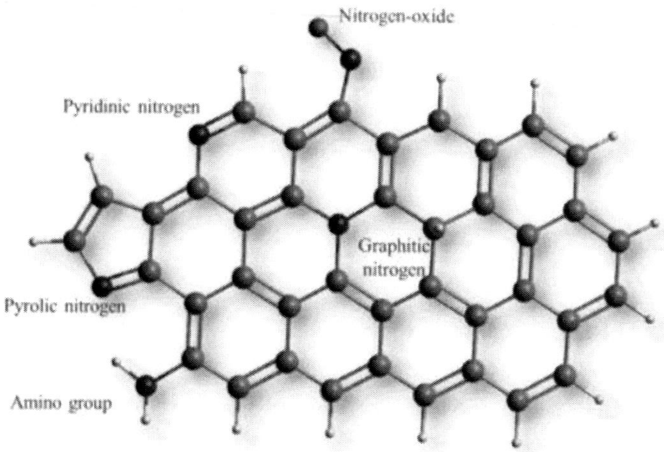

Figure 5. Types of nitrogen moieties present in nitrogen-doped carbons (white – H, grey – C, blue – N).

However, nitrogen functionalities in carbon surfaces should not be considered separately from oxygen functional groups. Theoretical calculations suggest that heteroatoms, being essentially point defects, attract oxygen functional groups (Dobrota et al. 2017) and could also affect the local topology and curvature of carbon planes (Dobrota et al. 2020). These effects are not limited to nitrogen but also appear in the cases of other dopants, like boron, phosphorus, and sulfur. However, it results in the alterations of the local surface hydrophilicity and charge density, which is essential for the interactions of carbons and OP molecules.

Adsorption on Carbons – Thermodynamic and Kinetic Description

Considering the mechanism of adsorption and the forces in action, adsorption is generally divided into chemisorption and physisorption. Strong chemical interactions are in play in the case of chemisorption, and in the case of physisorption, there is no formation of chemical bonds. In both cases, the adsorption can be characterized with an equilibrium constant, and their choice would ultimately lead to different values of thermodynamic functions evaluated for a given adsorption process (Chen, Da, and Ma 2021). We will get to this point later. First, it is important to emphasize that temperature plays a key role in both cases, and adsorption measurements should be done at a constant temperature.

To quantify an OP adsorption process, one can use two approaches. The first one is to define adsorption efficiency or adsorption Uptake. It is evaluated as the relative decrease of OP concentration over time, and in a given moment of time of our experiment (t), it is defined as:

$$\text{Uptake}_t\ (\%) = 100 \times \frac{C_0 - C_t}{C_0} \quad (1)$$

In the equation above, C_0 and C_t are the concentrations of an OP at the beginning of the experiment (initial concentration) and at a given moment of time of the experiment. If an OP is equilibrated with carbon adsorbent, we mark this with "e" in subscript and define equilibrium Uptake as:

$$\text{Uptake}_e\ (\%) = 100 \times \frac{C_0 - C_e}{C_0} \quad (2)$$

Another way to quantify an OP adsorption process is using adsorption capacity (q), which presents the mass of OP (typically in mg) *per* unit mass of adsorbent, i.e., carbon material. At any moment of time during the adsorption of a given OP onto a selected carbon adsorbent, q_t is defined as:

$$q_t = \frac{m_{OP,t}\,(\mathrm{mg})}{m_C\,(\mathrm{g})} \tag{3}$$

In the above equation $m_{OP,t}$ and m_C stand for the mass of adsorbed OP and the mass of carbon in the experiment, respectively. Under equilibrium conditions, maximum (equilibrium) adsorption capacity is reached, and it is given by:

$$q_e = \frac{m_{OP,e}\,(\mathrm{mg})}{m_C\,(\mathrm{g})} \tag{4}$$

In principle, the amount of adsorbed OP is determined indirectly. The quantification is done by determining the concentration of the used OP using, most frequently, high-performance liquid chromatography (HPLC) or its ultra-performance analog (UPLC). Equilibrium is reached during the adsorption process (see further); thus, the concentration of the OP in the solution will never be zero. However, the Uptake can be 100% if the concentration of OP drops under the detection limit of the used analytical method. This fact should be kept in mind when reporting the results and comparing different carbon-based adsorbents in terms of their performance.

Obviously, Uptake and adsorption capacity are interconnected, but moreover, they depend on the concentration of adsorbent used in the experiment. For this reason, the performance of two different adsorbents can only be compared if the experiments are done under identical conditions. This actually means that the same initial concentration of OP and the concentration of carbon adsorbent have to be used, but also the experiments have to be done at the same temperature, as this parameter also determines the distribution of a given OP between the solution and the adsorbent phase. Unfortunately, this is simply not the case by considering the literature survey, and there is a rather wide range of experimental parameters used to evaluate different carbon-based adsorbents. Such a situation in current literature makes the straightforward comparison of different adsorbents practically impossible and limits the identification of key materials properties determining adsorbent performance rather difficult.

Adsorption Thermodynamics

An adsorption process of OP on a given carbon material (Carb) can be described using the following equation:

$$\mathrm{Carb} + \mathrm{OP}_{\mathrm{sol}} \leftrightarrow \mathrm{Carb} - \mathrm{OP}_{\mathrm{ads}} \tag{5}$$

In this process, OP distributes between adsorbed phase (Carb–OP_{ads}) and solution phase (OP_{sol}), and established equilibrium can be described using distribution constant (K_d) defined by the concentration of adsorbed OP and the concentration of OP remaining in the solution. Thus, using the quantities defined in the previous section, distribution constant is defined as:

$$K_d = \frac{q_{OP,e}}{C_e} \tag{6}$$

K_d is a key parameter for the description of adsorption equilibrium, not only in the fundamental experiments investigating adsorption processes but also for describing the distribution of OPs in the environment (Chen, Da and Ma 2021), for example, between soil and aqueous media.

Experimentally, the most important pieces of information are obtained by evaluating equilibrium adsorption capacity using different initial concentrations of OPs at a constant temperature, keeping the concentration of adsorbent also constant. In this way, one obtains the dependence of q_e on C_e, which is called adsorption isotherm. However, as we already mentioned, there are no specific guidelines for choosing the concentration of adsorbent in these experiments and for adsorbents with lower performance, one has to use higher concentration, which further reduces the evaluated adsorption capacity.

In the next step, experimental data are fitted using some adsorption isotherm equation. However, the choice of adsorption isotherm is, again, not unique, and there are many of them available in the literature (Al-Ghouti and Da'ana 2020). For example, a recent paper by Sillanpää group (Mozaffari Majd et al. 2021) assembles 37 adsorption isotherms and over 400 papers published in the period 2010-2020 which use these equations. Interestingly, practically every isotherm reported in this paper is used to describe the adsorption of some compound on a certain carbon material. The question is how to choose a proper isotherm for the description of OP adsorption. First of all, many adsorption isotherms give rather similar q_e on C_e trends, but the overall functional dependence and the shape of the experimental curve (curvature, slope, convexity/concavity, existence of one or more plateaus, and other properties) is the starting point for selection. Then, experimental data are typically fitted using several adsorption isotherms, and the quality of the fit is a decisive parameter. In other words, it is considered that experimental data obey the isotherm with the best fit results. However, this is not the end. Every adsorption isotherm ascribes a certain physical meaning to the equation parameters. Using the obtained parameters, one can judge the nature of the adsorption process (physisorption vs. chemisorption), adsorbate-adsorbate interactions or cooperativity, and the uniformity of adsorption sites on the adsorbent surface, maximum adsorption capacity and other relevant parameters. Several frequently used adsorption isotherms (Table 4) are described further.

The Freundlich model is an empirical model that assumes an energetically heterogeneous surface, where the adsorbed molecules can interact with each other. K_f (expressed in mg g^{-1} (mg $dm^{-3})^{1/n}$) is calledFreundlich constants. For$n = 1$, the adsorption isotherm is a linear function. For $n < 1$, the adsorption is unfavorable; if $n > 1$, the adsorption is favored. However, n is also associated with homogeneity of the adsorbent surface, and for $n = 1$, the surface is considered energetically homogeneous (Mozaffari Majd et al. 2021).

In contrast to the Freundlich model, Langmuir's model assumes adsorption on an energetically homogeneous surface. Once the equilibrium is achieved, a monolayer of adsorbate molecules is formed on the adsorbent surface and there are no interactions between the adsorbed molecules. There are two parameters in the Langmuir model: $q_{L,\max}$ (mg g^{-1}), the theoretical maximum adsorption capacity of the monolayer, and K_L (dm^3 mg^{-1}) called Langmuir constant. In the Langmuir model, this constant is the same as the distribution constant (Eq. (6)).

The Temkin model uses two parameters, K_T, called Temkin constant, and b_T, associated with adsorption heat and adsorbate-adsorbate interactions. The adsorption process is assumed to be multi-layer, and the model works well at intermediate adsorbate concentrations and assumes a linear decrease of differential adsorption heat with increasing adsorbent coverage by adsorbate molecules.

The Dubinin-Radushkevich (DR) model can differentiate between physisorption and chemisorption on the adsorbent surfaces. The parameters of DR isotherm are K_{DR} ($mol^2\ J^{-2}$), DR constant, and q_{DR}, maximum DR capacity. The independent variable is $\varepsilon = RT \times \ln(1 + 1/C_e)$, while adsorption free energy (E_{DR}) in the DR model is related to K_{DR} as $E_{DR} = (K_{DR})^{-1/2}$. If E_{DR} is higher than 8 kJ mol^{-1}, the adsorption process is dominantly chemisorption, while for lower values, physisorption is indicated.

Once the proper model is chosen, different relevant parameters are extracted, and the adsorption process is put in a physical context underlying the applied adsorption isotherm. Among other parameters, standard thermodynamic functions for the adsorption process (ΔG^θ, ΔH^θ, ΔS^θ) can be extracted. In particular, standard Gibbs free energy of adsorption can be evaluated from the standard equilibrium constant (K_d^θ) is linked to ΔG^θ as:

Table 4. Selected equations of adsorption isotherms and their linearized forms. Experimental data can be fitted into the selected adsorption isotherm model using either an explicit form of the adsorption isotherm or the corresponding linearized form

Isotherm	Equation	Linearized form
Freundlich	$q_e = K_f C_e^{1/n}$	$\log q_e = \log K_f + \frac{1}{n} \log C_e$
Langmuir	$q_e = \frac{q_{L,\max} K_L C_e}{1 + K_L C_e}$	$\frac{C_e}{q_e} = \frac{1}{K_L q_{L,\max}} + \frac{C_e}{q_{L,\max}}$
Temkin	$q_e = \frac{RT}{b_T} \ln\left(K_T C_e\right)$	$q_e = \frac{RT}{b_T} \ln\left(K_T\right) + \frac{RT}{b_T} \ln\left(C_e\right)$
Dubinin-Radushkevich	$q_e = q_{DR} \exp(- K_{DR} \varepsilon^2)$	$\ln q_e = \ln q_{DR} - K_{DR} \varepsilon^2$

$$\Delta G^\theta = -RT \times \ln K_d^{\ \theta} \tag{7}$$

Following the detailed thermodynamic analysis of Chen et al. (Chen, Da and Ma 2021), it is clear that K_d^θ has to be dimensionless and cannot be equalized with the distribution constant given by Eq. (6). Therefore, a proper choice of the standard state is needed, and Chen et al. suggest using $C^\theta = 1$ mol $L^{-1,}$ and $q^\theta = 1$ mol kg^{-1} for the standard concentration of adsorbate in the liquid phase and in the adsorbed phase, respectively. In this way, the standard equilibrium constant is linked to the distribution constant as:

$$K_d^{\theta} = K_d \frac{C^{\theta}}{q^{\theta}} = \frac{q_{OP,e}}{C_e} \frac{C^{\theta}}{q^{\theta}} \tag{8}$$

Moreover, various equilibrium constants that appear in different adsorption isotherms models, like K_L, K_T, or K_{DR} listed in Table 4, are not necessarily distribution constants. They have to be converted to it before converting to the standard equilibrium constant and the calculation of ΔG^{θ}. However, we also note that the problem with different dimensions is not only related to the determination of thermodynamic functions, but also different variables used in adsorption isotherms. For example, the Oswin isotherm (Oswin 1946) is given by:

$$q_e = K_{OS} \left(\frac{C_e}{1 - C_e} \right)^{n_{OS}} \tag{9}$$

where K_{OS} and n_{OS}, are the model parameters, with K_{OS} having the meaning analogous to the equilibrium constant. However, it is unclear how to evaluate the right side of Eq. (8), unless the standard state is defined, as "1" and "C_e" have different dimensions.

Adsorption Kinetics

Adsorption thermodynamics is related to equilibrium conditions and, in principle, provides fundamental information about the adsorption process. However, the OP adsorption kinetics analysis is very important from the practical point, specifically relating to adsorption under dynamic conditions, for example, filtration. Clearly, the adsorption kinetics is temperature-dependent, so the experiments must be done at a fixed temperature. Similar to the data fitting into a suitable adsorption isotherm model, experimental data concerning the amounts of adsorbed OP are fitted into the selected reaction kinetics model (Elmi, Mohammadi Damghani and Shokrollahzadeh Taleshi 2020). The best model to which kinetics is ascribed is the one that explains the experimental data the best. Using the quantities defined by Eq. (3) and (4), pseudo-first-order kinetics is described by:

$$\ln(q_e - q_t) = \ln q_e - k_1 t \tag{10}$$

where q_e is experimental equilibrium adsorption capacity. Pseudo-second order model is given by:

$$\frac{t}{q_t} = \frac{1}{k_2 q_e^2} + \frac{1}{q_e} \tag{11}$$

while kinetic equation assuming interparticle diffusion model can be written as:

$$q_t = k_i t^{0.5} + I \tag{12}$$

The constants k_1, k_2, and k_i are pseudo-first, pseudo-second-order, and interparticle diffusion rate constant. Moreover, an analogous set of equations can be set up using the OP concentrations remaining in solution (Ayranci and Hoda 2005), which is particularly suitable for investigating the adsorption of OPs which have very prominent UV-Vis spectra so that the concentration of OP can be monitored in the real-time and UV-Vis signals can be transformed directly to OP concentrations.

Interparticle diffusion mode can be particularly insightful as transformed experimental data can show several linear domains ascribed to different processes. The ratio of the kinetic constants in these linear domains can tell about the relative rates of different processes taking place during OP adsorption. For example (Elmi, Mohammadi Damghani, and Shokrollahzadeh Taleshi 2020), the first step in this model can be related to the diffusion ofOP from the bulk solution to the boundary layer film of the solvent surrounding adsorbent and transport from the solvent film onto the external surface of the adsorbent(film diffusion). The second linear step is ascribed to intraparticle diffusion, corresponding to the transport of OP from the external surface of the adsorbent into the pores. In the final step, OP starts to attach to the internal surface of the adsorbent, finally resulting in adsorption equilibrium. It is important to emphasize that, similarly to the adsorption efficiency, the OP adsorption kinetic parameters also depend on the carbon adsorbent loading or concentration, as clearly shown in the work of Moussavi et al. (Moussavi, Hosseini, and Alahabadi 2013) for the case of diazinon removal from contaminated water by adsorption onto NH_4Cl-induced activated carbon. Interestingly, for the case of pseudo-first-order, the rate constant was found to increase with adsorbent concentration, while for the interparticle diffusion mode, it decayed with increasing concentration of NH_4Cl-induced activated carbon.

OPs–Carbon Interaction Mechanisms

Literature survey shows a very small number of studies addressing the interactions of OPs with carbons from the fundamental point. Nevertheless, the proceeding section will summarize and critically address available works.

First of all, 0D carbons, like fullerenes, GQDs, and CQDs, are not used as adsorbents for OPs. This is probably due to the high costs and complicated synthesis of these materials. Only a few examples of the use of GQDs for pesticide removal, in general, can be found (Agarwal et al. 2016). Higher-dimensional carbons, on the other hand, are frequently used as adsorbents for OPs.

MWCNTs have been employed for the removal of malathion from aqueous solutions (Dehghani et al. 2017). The authors found a strong pH dependence of adsorption efficiency, although it was also noted that malathion decomposes in acidic and alkaline media. At low pH, it was found that adsorption efficiency is low, while it intensifies in alkaline solutions. Such trend was ascribed to the different mechanisms of malathion adsorption at different pH values as the state of the MWCTs surface changes (Figure 6). Authors considered that at low pH, the surface of MWCNTs is positively charged, which can be due to protonation and that the adsorption of malathion takes place *via* electrostatic interactions of the positively charged surface and partially negative atoms in dimethoate molecule, in specific S from P=S moiety and O atoms from methoxy groups attached to the central P atom.

On the other hand, in alkaline media, malathion chemisorption occurs *via* a chemical reaction between surface hydroxyl groups on MWCNTs surface and thiophosphate moiety. According to the reaction scheme provided by Dehghani et al. (Dehghani et al. 2017), two methanol molecules are being released. However, it is not clear whether the authors considered that OH groups are inherently present on the MWCNTs surface or the surface gets hydroxylated at high pH. The latter scenario is quite unlikely as MWCNTs do not engage easily in chemical reactions unless special conditions are reached.

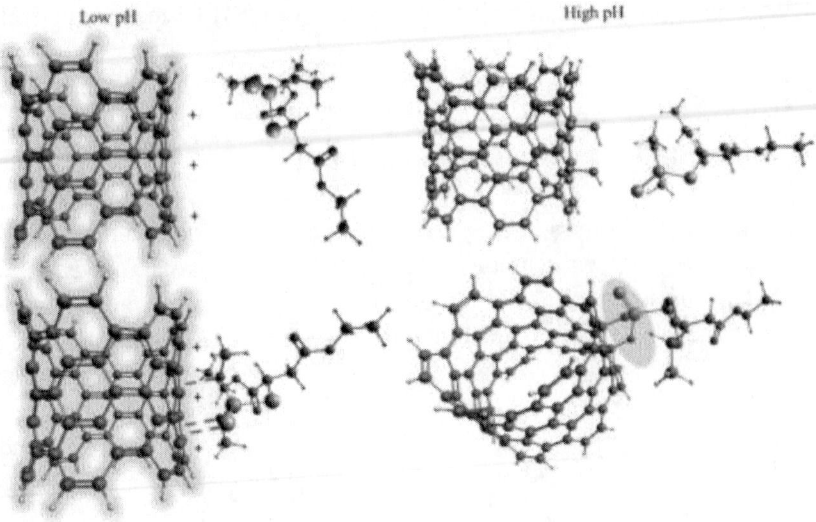

Figure 6. Simplified model of the interaction between MWCNTs and malathion according to ref. (Dehghani et al. 2017). At low pH, the surface of MWCNTs is positively charged, and there are electrostatic interactions with partially negative atoms in the malathion molecule (O and S). At high pH, there is chemisorption due to the chemical reaction of surface OH groups and phosphate moiety of the malathion molecule.

However, the possibility of chemical reactions of OP molecules with carbon materials should not be excluded. Recently, Anićijević et al. (Ancijevic et al. 2021) suggested that dimethoate can react with defect-rich-carbons using two commercially available graphene-oxides as possible substrates. Reaction products do not remain adsorbed on graphene oxide as they were observed using UPLC. While the reaction products were not identified, the oxidation of dimethoate to omethoate was excluded. However, using Density Functional Theory (DFT) calculations, the authors investigated the possible reaction of defect sites with model OP compounds. As models, S=PH_3, S=PF_3 and their oxo analogs were used. In this way, it was possible to investigate the effects of electron density in P=S moiety on the reactivity, as F substituents withdraw more charge from P=S moiety. When these model OPs interacted with pristine graphene, only weak physisorption was observed. Adsorption energies amounted between −0.18 and −0.14 eV. However, S=PH_3 and S=PF_3 reacted with the single vacancy in graphene (Figure 7), while all considered model OPs reacted with the epoxy group (Figure 4) on the graphene basal plane. Finally, the Stone-Wales defect (Figure 7) was inert for chemical reaction with model OPs but provided stronger adsorption than pristine graphene. In this case, also, adsorption was classified as physisorption as adsorption energies did not exceed −0.28 eV.

The first associations to graphene are not defects and oxygen functional groups, although always present on its surface, but a large domain of delocalized π electrons, above and below sp^2 hybridized carbon atoms forming graphene basal plane. An insight into the interaction of pesticides with pristine graphene basal plane can be found in the work of Maliyekkal et al. (Maliyekkal et al. 2013), who investigated graphene as a possible adsorbent for endosulfan (not OP), chlorpyrifos, and malathion. Authors have found astonishing adsorption capacities of graphene for these three pesticides, exceeding 1000 mg g^{-1}, and performed DFT calculations addressing the interactions of studied pesticides with graphene basal plane. The DFT-based analysis of graphene–water–pesticide interactions provided by the authors showed that the direct interactions between graphene and the pesticides are rather weak or unlikely, as the adsorption energies calculated in vacuum were positive. The main conclusion of the mentioned study is that the adsorption is mediated through water molecules. However, it is surprising that the adsorption energies of the pesticide-water complexes on the graphene basal plane amounted to more than −200 kJ mol^{-1}. These values fall in the range of rather strong chemical bonds, and adsorption could be classified as chemisorption, but no formation of new bonds can be seen from the optimized structures provided by the authors, while this work mentions only slight changes in the lengths of polar bonds found in the pesticide molecules upon solvation. Moreover, this value of adsorption energy is close to the strength of the interaction of nascent oxygen with pristine graphene (Pasti et al. 2018c). Therefore, it is unlikely that solvated pesticide molecules can interact so strongly with the pristine graphene basal plane. Nevertheless, we must notice that provided results concerning the physical and chemical characterization of the used adsorbent, which was, actually, a mixture of GO and reduced GO, points to the presence of a large concentration of oxygen functional groups (determined by XPS) and a high concentration of defects (by Raman). Hence, it is questionable whether the model of pristine graphene correctly represents the adsorbent used in the mentioned study.

Concerning OPs adsorption on graphene-based materials, another study has to be mentioned (Lazarevic-Pasti et al. 2018). The study employed three different graphene-based materials – graphene oxide (30 at.% oxygen), near single-layer graphene with approx. 5 at.% of oxygen, and so-called graphene nanoplatelets with only 1.5 at.% of oxygen. While the first two adsorbents had oxygen functional groups distributed at edges and basal planes, the last adsorbent was graphite-like and contained oxygen functional groups only at the edge sites. Moreover, aliphatic dimethoate and aromatic chlorpyrifos were used to investigate the adsorption performance of these graphene-based materials. Ultimately, the adsorption was found to be determined by the structure of the adsorbents and the nature of OP. In particular, aliphatic dimethoate preferred hydrophilic oxidized graphene surfaces. Hence, the best adsorbent was near single-layer graphene, outperforming graphene-oxide due to a much larger specific surface area. On the other hand, aromatic chlorpyrifos adsorbed on the surface of the graphene basal plane with high structural order and preserved the π electron system. Therefore, the best adsorbent was graphene nanoplatelets, while graphene oxide adsorbed chlorpyrifos in minor quantities. Thus, it was concluded that the combination of structural properties of adsorbents and adsorbates defines the adsorption of OP pesticides on graphene-based materials, while the specific surface area of adsorbents is not the crucial factor. However, making the surface of the adsorbent more available for the adsorption of OPs could unquestionably improve adsorption performance, and the ideal combination would be a carbon material with a large specific surface area containing a pre-defined structure for adsorption of a given OP. Even though the mentioned work clearly resolved the impact of graphene-based materials' structure on adsorption performance and links it to the structure of a given OP, there is no definite information about the exact nature of the oxygen functionalities responsible for OPs adsorption. However, if the interactions of OPs with oxygen functionalities were not identified as chemical, all of them likely contributed to a certain extent *via* dispersion and electrostatic interactions.

While not specifically related to OPs adsorption on carbons, the work of Moreno-Castilla (Moreno-Castilla 2004) provides a useful insight into the adsorption of aromatic molecules on carbon surfaces. This work suggests that electrostatic interactions are operative when dissociation can take place under experimental conditions. On the other hand, non-electrostatic interactions arise from dispersive and hydrophobic interactions. Both electrostatic and non-electrostatic interactions are operative during adsorption on carbon surfaces from dilute solutions, while surface functionalities (Figure 4 and Figure 5) could also be suitable for hydrogen bonding. While hydrogen bonds could help adsorption and make it stronger, the primary effect is negative.

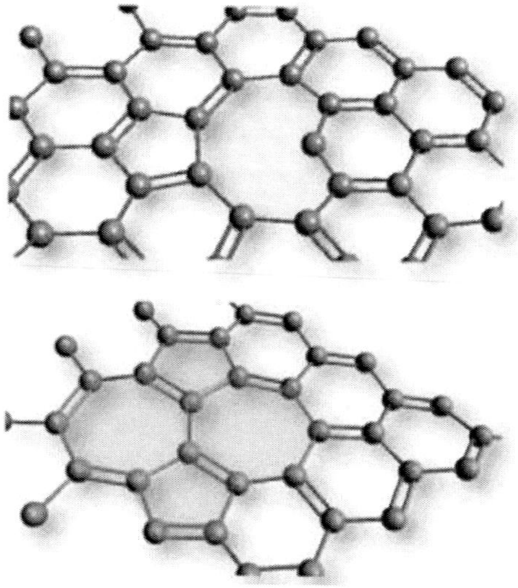

Figure 7. Defects in the graphene basal plane being considered as possible reactive centers for reaction with model OPs (Ancijevic et al. 2021). Upper panel – single vacancy site which is found to be reactive, lower panel – Stone-Wales defect which is chemically inert.

These functionalities form hydrogen bonds with water, and in this way, the surface gets blocked. In principle, hydrogen bonds are stronger than dispersive interactions, so it is energetically unfavorable to break many hydrogen bonds to make adsorption possible. It was concluded that aromatic molecules bind *via* π-π stacking, but some strongly adsorbed aromatic molecules were also identified. Moreno-Castilla suggests that in this case, donor-acceptor or charge-transfer mechanisms are in action. Nevertheless, one of the main conclusions of the mentioned work is that carbons are more effective adsorbents if non-electrostatic interactions are forced by adjusting surface chemistry, pH, and ionic strength of an adsorbent dispersion. In this case, water molecules are pushed away from the surface, maximizing the surface exploitation for adsorption.

Opposed to intuition, it is not a rare conclusion that specific surface area is not the dominant factor for determining adsorbent performance. For example, Vukčević et al. (Vukcevic et al. 2013) produced a series of carbon monoliths with various surface chemistries using chemical treatment of carbon monolith surfaces with HNO_3, H_2O_2 and KOH. Different classes of pesticides were investigated using five selected pesticides (acetamiprid, dimethoate, nicosulfuron, carbofuran and atrazine). While the samples of carbon monoliths differed in terms of specific surface area and pore size distribution, the adsorption performance of the activated carbon monoliths was more affected by the concentration of the surface functional groups. The authors emphasized the role of π-π interactions for the adsorption of pesticides with aromatic structure and the significance of acidic groups in the adsorption of pesticides with aliphatic structure.

The studies mentioned above-investigated adsorption under equilibrium conditions, and the situation might be different under dynamic conditions of OPs adsorption. A clear demonstration of such differences comes from the study of dimethoate and its oxo-analog, omethoate, on ordered mesoporous carbons doped with trace amounts of boron, nitrogen, and phosphorus (Lazarevic-Pasti et al. 2016). While boron-doped ordered mesoporous carbons showed the best adsorption performance, under equilibrium conditions, structural disorder (measured as the ratio of the intensities of D and G band in Raman spectra, I_D/I_G) was found to correlate with adsorption capacity. At the same time, the specific surface area did not have a crucial role in adsorption. However, specific surface areas weighted by I_D/I_G ratios were found to scale with adsorption capacity under dynamic conditions. These results point to a complex picture of OPs interaction with carbon surface, dependent on the experimental conditions.

Finally, it can be quite difficult to comprehend how different properties of carbon materials affect the OP adsorption performance. For example, if carbon material is characterized using gas adsorption/desorption measurements, SEM with EDX, XPS, and Raman spectroscopy, one can easily end up with 25 to 40 different descriptors of a carbon sample, acting, more or less, as independent variables. These variables ultimately determine the performance of adsorbent, evaluated as Uptake or adsorption capacity. It is practically impossible to discriminate between the key parameters determining performance, and a possible solution to this problem with large dimensionality is the use of statistical and chemometric methods or more advanced approaches as artificial intelligence or machine learning. A simple example of this approach was recently demonstrated by Jocić et al. (Jocic et al. 2021), who used multiple linear regression models to reliably predict dimethoate removal from water using a series of viscose-derived activated carbon fibers. In this particular work, dimethoate adsorption, measured as Uptake, was successfully correlated to carbon content (at. %), oxygen content (at. %), and total pore volume ($cm^3\ g^{-1}$). Thus, a data-driven approach could be a fresh strategy for elucidating the role of carbon properties in determining their efficiency as adsorbents for OPs.

Carbons Performance as Adsorbents for OPs Removal - Overview

When comparing the performance of different carbon materials as adsorbents of OPs, Uptake and/or adsorption capacity are typically used. However, we note that both of these quantities are sensitive to the amount of adsorbent used in the experiment. While we have previously described the need and guidelines for establishing standard qualities linked to adsorption thermodynamics, there are no specific guidelines for using a given amount of adsorbent in the experiments used to evaluate Uptake or adsorption capacity. As the adsorption is an equilibrium process, the amount of adsorbed pesticide will depend on the adsorbent concentration. This point is clearly demonstrated in several works. For example, Maliyekkal et al. (Maliyekkal et al. 2013) showed that the measured adsorption capacity of chlorpyrifos is highly sensitive to the adsorbent dose. For graphene (rGO+GO, see the previous section), adsorption capacity increases from < 100 mg g^{-1} for adsorbent concentration 0.25 g dm^{-3}, while it increases over 1000 mg g^{-1} for adsorbent concentration below 3 mg dm^{-3}, if the concentration of chlorpyrifos is maintained constant (2 g dm^{-3}) Similarly, evaluated Uptake and adsorption capacity will depend on the concentration of an OP used in the experiments if the adsorbent concentration is maintained constant. In the work of Lazarević-Pašti et al. (Lazarevic-Pasti et al. 2016), it was shown that dimethoate Uptake decreases with increasing concentration of dimethoate, while the trend in evaluated adsorption capacity is the opposite. As the initial concentration of dimethoate increases, the adsorption capacity also increases, reaching almost 500 mg g^{-1} for a dimethoate concentration of 2.5×10^{-2} mol dm^{-3}.

An additional factor contributing to the complex problem of comparison of different adsorbents is the pH. As explained, depending on the nature of OP, their stability is different under different pH conditions of adsorption experiments. On the other hand, surface functional groups, primarily oxygen ones (Figure 4), can get protonated or deprotonated, changing their charge at various pH values. Figure 6 shows how such varying conditions could affect malathion adsorption on oxygen-functionalized MWCNTs. Thus, the presence of oxygen functional groups will inevitably affect measured adsorption capacities.

For all these reasons, comparing different carbon materials as adsorbents for OPs is not straightforward, and it is, actually, quite difficult. Therefore, our intention is not to provide an exhaustive overview of available literature reports but to show great variability of carbon performance in OPs removal from aqueous media. The data are summarized in Table 5. In addition, some recent reviews can be used for a more detailed overview, including other types of adsorbents as well (Karthik et al. 2021, Kassiri et al. 2021, Mojiri et al. 2020).

As shown in Table 5, there is a significant scattering in evaluated adsorption capacities of different carbon materials towards different OPs, ranging within five orders of magnitude. As explained at the beginning of this section, this is not solely the consequence of different properties of carbon-based adsorbents but also different experimental conditions used by different research groups. While carbon materials can undoubtedly provide high adsorption capacities for OPs, it is necessary to standardize experimental procedures for the evaluation of adsorption performance in order to make the comparison of different carbons possible.

Table 5. Summary of literature reports regarding carbon adsorbents performance for OPs removal from aqueous solutions

Pesticide/adsorbent	**Adsorption capacity (mg g^{-1}) or Uptake (%)**	**Notes and reference**
Chlorfenvinphos/Graphene-coated silica	4.9 mg g^{-1}	Phenol ring and N and S atoms contribute bonding (Liu et al. 2013)
Chlorpyrifos/ Cellulose/Graphene oxide	150 mg g^{-1}	Electron donating from S and P determine adsorption (Suo et al. 2018)
Chlorpyrifos/Graphene	Up to 1200 mg g^{-1}	Water-mediated adsorption (Maliyekkal et al. 2013)
Chlorpyrifos/biochars	4.3 to 14.8 mg g^{-1}	Higher concentration of aromatic moieties and de-ashing improve adsorption performance (Zheng et al. 2019)
chlorpyrifos-methyl/ biochars	15.0 to 50.5 mg g^{-1}	
Chlorpyrifos/graphene nanoplatelets	140 mg g^{-1}	Adsorption depends on a specific combination of adsorbent/adsorbate systems and their corresponding structures; specific surface area determines the number of sites available for adsorption (Lazarevic-Pasti et al. 2018)
Chlorpyrifos/near-single-layer graphene	79 mg g^{-1}	
Dimethoate/graphene oxide	5.2 mg g^{-1}	
Dimethoate/near-single-layer graphene	37 mg g^{-1}	
Dimethoate and omethoate/doped ordered mesoporous carbons	Up to 100% removal of omethoate and 80% for dimethoate	Defects in carbon material and specific surface area have the highest impact on adsorption, dimethoate concentration as high as 5×10^{-3}mol dm^{-3} (Lazarevic-Pasti et al. 2016)
Dimethoate/viscose-derived activated carbon fibers	Up to 400 mg g^{-1}	Carbon and oxygen content and total pore volume can be used to predict Uptake, physisorption (Jocic et al. 2021)

Pesticide/adsorbent	Adsorption capacity (mg g^{-1}) or Uptake (%)	Notes and reference
Dimethoate/activated carbon monoliths	0 – 46 μg g^{-1}	Specific surface area is not the key factor for adsorption; surface chemistry determines the adsorption performance (Vukcevic et al. 2013)
Diazinon/Walnut shell-modified activated carbon	4.9 to 156 mg g^{-1}	Physisorption (Bayat, Alighardashi and Sadeghasadi 2018)
Diazinon/NH_4Cl induced activated carbon	Up to 250 mg g^{-1}	Langmuir adsorption isotherm, interparticle diffusion kinetics (Moussavi, Hosseini and Alahabadi 2013)
Malathion/granular activated carbon	Up to 900 mg g^{-1}	Investigated the effects of particle size, the dosage of carbons, and the initial concentration of malathion, fitted into Langmuir and Freundlich isotherms (Jusoh et al. 2011)

While the discussion of the use of carbons in OPs detection is beyond the scope of this work, we have to mention this possible application, as it is closely related to the environmental issues related to OPs accumulation while relying on high adsorption capacity of carbons, either individually or as a part of a composite material (Aghaie et al. 2019). In particular, carbon-based electrochemical sensors for OPs detection and monitoring are frequently employed, relying on mentioned advantages of carbon-based materials (Kanoun et al. 2021). However, in this case, one more property of carbons is of crucial importance – conductivity. In general, carbon materials produced at high temperatures have high conductivities, which is associated with removing oxygen functional groups and improved graphitization and ordering. However, in this case, adsorption properties have to be properly tuned, especially if carbon material is intended to provide pre-concentration of an OP analyte on the electrode surface using a suitably chosen electrochemical technique.

Conclusion

Very rich surface chemistry of carbon materials, being available in different forms and dimensionalities, combined with desirable pore structure and large specific surface area, set them as good candidates for adsorbents for OPs removal from aqueous media. As OPs can be easily oxidized to more toxic forms, oxons, it is important to note that this process was not observed in the case of OPs removal using carbon materials, although this issue is not systematically addressed in the literature. OPs usually interact with carbon materials *via* electrostatic, hydrophobic, dispersion, and π-π stacking interactions. The type of pre-dominant interactions depends on the structure of a given OP, the structure of carbon adsorbent, and pH. For this reason, it is of utmost importance to characterize carbon adsorbents in detail, starting from morphology, microtexture, and specific surface area, to chemical composition and local chemical bonding on carbon surfaces.

It seems that a critical mass of literature reports pointing that surface chemistry rather than the specific surface area of carbon adsorbents determines adsorption performance. However, it is more clear that a large specific surface area is necessary to provide a high number of adsorption sites. However, in most available reports, the formation of strong chemical bonds between OPs and carbon surfaces is not indicated. The dominant mechanism is physisorption, but the accurate determination of thermodynamic quantities associated with adsorption of OPs is problematic, as there is no rigorously defined standard state for adsorption studies. Moreover, the literature survey indicates that different research groups use various adsorbent doses and OPs concentrations for adsorption measurements and different techniques to determine equilibrium OPs concentrations, resulting in a wide range of reported adsorption capacities. Such practices also make the comparison of adsorption capacities reported for different carbon materials by different research groups practically impossible.

For the reasons above, we suggest that it is necessary to standardize procedures for OPs adsorption analysis on carbons (but also other adsorbents) in a way that:

(i) The concentration of adsorbent is defined to a given value, typically in the order of several g dm^{-3}.
(ii) The range of OPs concentrations is defined for adsorption isotherm construction, while temperature (range) for adsorption isotherms is also fixed.
(iii) The detection limit for the method used to evaluate the equilibrium OP concentration is explicitly stated.
(iv) If thermodynamic data for adsorption are reported, the standard state must be clearly defined. We suggest that the work of Chen et al. (Chen, Da and Ma 2021) should be followed.
(v) If the kinetics of OPs adsorption is investigated, the points (i) – (iii) should also apply.

Besides standardizing adsorption measurements, adsorbent characterization should also follow certain recommendations to understand OP-adsorbent interactions and correlate them with adsorption measurements. In the case of carbon adsorbents, characterization should report at least:

(i) Bulk chemical composition (using elemental analysis or EDX) and surface chemical composition (using XPS)
(ii) Specific surface area and relevant microtexture properties like pore size distribution, micropore surface, micropore volume, mesopore surface and the corresponding volume, and the total pore volume
(iii) Level of the structural disorder (typically assessed with Raman spectroscopy)

All the suggested points should provide reliable quantitative data which can be used for data-driven analysis of OPs interactions with carbon surface. Under these conditions, such methodology would enable the development of models capable of explaining the mechanisms of OPs interactions with carbon surface and designing novel advanced carbon-based adsorbents for OPs removal from aqueous media.

Acknowledgments

I.A.P. acknowledges the support provided by the Serbian Ministry of Education, Science and Technological Development (Contract number: 451-03-68/2020-14/200146). S.B., C.U. and C.F. gratefully acknowledge funding from the European Regional Development Fund (EFRE) and the province of Upper Austria through the program IWB 2014-2020 (project BioCarb-K).

References

Agarwal, S., Sadeghi, N., Tyagi, I., Gupta, V. K. and Fakhri, A. 2016. Adsorption of toxic carbamate pesticide oxamyl from liquid phase by newly synthesized and characterized graphene quantum dots nanomaterials. *Journal of Colloid and Interface Science*, 478: 430-438.

Aghaie, A., Khanmohammadi, A., Hajian, A., Schmid, U. and Bagheri, H. 2019. Nonenzymatic Electrochemical Determination of Paraoxon Ethyl in Water and Fruits by Graphene-Based NiFe Bimetallic Phosphosulfide Nanocomposite as a Superior Sensing Layer. *Food Analytical Methods*, 12(7): 1545-1555.

Al-Ghouti, M. A. and Da'ana, D. A. 2020. Guidelines for the use and interpretation of adsorption isotherm models: A review. *Journal of Hazardous Materials*, 393: 122383.

Ancijevic, V., Jelic, M., Jovanovic, A. Z., Potkonjak, N., Pasti, I. A. and Lazarevic-Pasti, T. D. 2021. Organophosphorous pesticide removal from water by graphene-based materials - Only adsorption or something else as well? *Journal of the Serbian Chemical Society*, 86(7-8): 699-710.

Anićijević, V. J., Petković, M., Pašti, I. A. and Lazarević-Pašti, T. 2021. *Decomposition of Dimethoate and Omethoate in Aqueous Solutions – Half-Life, Neurotoxicity and Mechanism of Hydrolysis*. Preprints 2021, 2021080515: This content is a preprint and has not been peer-reviewed.

Avouris, P. and Dimitrakopoulos, C. 2012. Graphene: synthesis and applications. *Materials Today*, 15(3): 86-97.

Ayranci, E. and Hoda, N. 2005. Adsorption kinetics and isotherms of pesticides onto activated carbon-cloth. *Chemosphere*, 60(11): 1600-1607.

Ayyalusamy, S., Mishra, S. and Suryanarayanan, V. 2018. Promising post-consumer PET-derived activated carbon electrode material for non-enzymatic electrochemical determination of carbofuran hydrolysate. *Scientific Reports*, 8(1): 13151.

Barberis, A., Spissu, Y., Fadda, A., Azara, E., Bazzu, G., Marceddu, S., Angioni, A., Sanna, D., Schirra, M. and Serra, P. A. 2015. Simultaneous amperometric detection of ascorbic acid and antioxidant capacity in orange, blueberry and kiwi juice, by a telemetric system coupled with a fullerene- or nanotubes-modified ascorbate subtractive biosensor. *Biosens Bioelectron*, 67: 214-223.

Bayat, M., Alighardashi, A. and Sadeghasadi, A. 2018. Fixed-bed column and batch reactors performance in removal of diazinon pesticide from aqueous solutions by using walnut shell-modified activated carbon. *Environmental Technology & Innovation*, 12: 148-159.

Belenkov, E. A. and Greshnyakov, V. A. 2013. Classification schemes for carbon phases and nanostructures. *New Carbon Materials*, 28(4): 273-282.

Bilal, M., Shah, J. A., Ashfaq, T., Gardazi, S. M. H., Tahir, A. A., Pervez, A., Haroon, H. and Mahmood, Q. 2013. Waste biomass adsorbents for copper removal from industrial wastewater-A review. *Journal of Hazardous Materials*, 263: 322-333.

Boukhvalov, D. W., Katsnelson, M. I. and Lichtenstein, A. I. 2008. Hydrogen on graphene: Electronic structure, total energy, structural distortions and magnetism from first-principles calculations. *Physical Review B*, 77(3): 035427.

Breitenbach, S., Gavrilov, N., Pasti, I., Unterweger, C., Duchoslav, J., Stifter, D., Hassel, A. W. and Fürst, C. 2021a. *Biomass-Derived Carbons as Versatile Materials for Energy-Related Applications: Capacitive Properties vs. Oxygen Reduction Reaction Catalysis*. *C*, 7(3): 55.

Breitenbach, S., Lumetzberger, A., Hobisch, M. A., Unterweger, C., Spirk, S., Stifter, D., Fürst, C. and Hassel, A. W. 2020. *Supercapacitor Electrodes from Viscose-Based Activated Carbon Fibers: Significant Yield and Performance Improvement Using Diammonium Hydrogen Phosphate as Impregnating Agent*. *C*, 6(2): 17.

Breitenbach, S., Unterweger, C., Lumetzberger, A., Duchoslav, J., Stifter, D., Hassel, A. W. and Fürst, C. 2021b. Viscose-based porous carbon fibers: improving yield and porosity through optimization of the carbonization process by design of experiment. *Journal of Porous Materials*, 28(3): 727-739.

Broomfield, C. A., Lenz, D. E. and MacIver, B. 1986. The stability of soman and its stereoisomers in aqueous solution: Toxicological considerations. *Archives of Toxicology*, 59(4): 261-265.

Chen, J., Yao, B., Li, C. and Shi, G. 2013. An improved Hummers method for eco-friendly synthesis of graphene oxide. *Carbon*, 64: 225-229.

Chen, T., Da, T. and Ma, Y. 2021. Reasonable calculation of the thermodynamic parameters from adsorption equilibrium constant. *Journal of Molecular Liquids*, 322: 114980.

Ciric-Marjanovic, G., Pasti, I. and Mentus, S. 2015. One-dimensional nitrogen-containing carbon nanostructures. *Progress in Materials Science*, 69: 61-182.

Colovic, M. B., Krstic, D. Z., Lazarevic-Pasti, T. D., Bondzic, A. M. and Vasic, V. M. 2013. Acetylcholinesterase inhibitors: pharmacology and toxicology. *Current neuropharmacology*, 11(3): 315-335.

Danish, M. and Ahmad, T. 2018. A review on utilization of wood biomass as a sustainable precursor for activated carbon production and application. *Renewable and Sustainable Energy Reviews*, 87: 1-21.

Dehghani, M. H., Niasar, Z. S., Mehrnia, M. R., Shayeghi, M., Al-Ghouti, M. A., Heibati, B., McKay, G. and Yetilmezsoy, K. 2017. Optimizing the removal of organophosphorus pesticide malathion from water using multi-walled carbon nanotubes. *Chemical Engineering Journal*, 310: 22-32.

Dideikin, A. T. and Vul', A. Y. 2019. Graphene Oxide and Derivatives: The Place in Graphene Family. *Frontiers in Physics*, 6(149).

Dobrota, A. S., Pasti, I. A., Mentus, S. V. and Skorodumova, N. V. 2016. A general view on the reactivity of the oxygen-functionalized graphene basal plane. *Physical Chemistry Chemical Physics*, 18(9): 6580-6586.

Dobrota, A. S., Pasti, I. A., Mentus, S. V. and Skorodumova, N. V. 2017. A DFT study of the interplay between dopants and oxygen functional groups over the graphene basal plane - implications in energy-related applications. *Physical Chemistry Chemical Physics*, 19(12): 8530-8540.

Dobrota, A. S., Pasti, I. A., Mentus, S. V., Johansson, B. and Skorodumova, N. V. 2020. Altering the reactivity of pristine, N- and P-doped graphene by strain engineering: A DFT view on energy related aspects. *Applied Surface Science*, 514: 145937.

Duchoslav, J., Kehrer, M., Hinterreiter, A., Duchoslav, V., Unterweger, C., Fürst, C., Steinberger, R. and Stifter, D. 2018. Novel protocol for highly efficient gas-phase chemical derivatization of surface amine groups using trifluoroacetic anhydride. *Applied Surface Science*, 443: 244-254.

Eckmann, A., Felten, A., Mishchenko, A., Britnell, L., Krupke, R., Novoselov, K. S. and Casiraghi, C. 2012. Probing the Nature of Defects in Graphene by Raman Spectroscopy. *Nano Letters*, 12(8): 3925-3930.

Elmi, F., Mohammadi Damghani, F. and Shokrollahzadeh Taleshi, M. 2020. Kinetic and Isotherm Studies of Adsorption of the Metribuzin Herbicide on an Fe_3O_4/CNT@PDA

Hybrid Magnetic Nanocomposite in Wastewater. *Industrial & Engineering Chemistry Research*, 59(20): 9604-9610.

Everhart, D. S. and Reilley, C. N. 1981. Chemical derivatization in electron spectroscopy for chemical analysis of surface functional groups introduced on low-density polyethylene film. *Analytical Chemistry*, 53(4): 665-676.

Ferrari, A. C. and Robertson, J. 2000. Interpretation of Raman spectra of disordered and amorphous carbon. *Physical Review B*, 61(20): 14095-14107.

Ferrari, A. C., Bonaccorso, F., Fal'ko, V., Novoselov, K. S., Roche, S., Bøggild, P., Borini, S., Koppens, F. H. L., Palermo, V., Pugno, N., Garrido, J. A., Sordan, R., Bianco, A., Ballerini, L., Prato, M., Lidorikis, E., Kivioja, J., Marinelli, C., Ryhänen, T., Morpurgo, A., Coleman, J. N., Nicolosi, V., Colombo, L., Fert, A., Garcia-Hernandez, M., Bachtold, A., Schneider, G. g. F., Guinea, F., Dekker, C., Barbone, M., Sun, Z., Galiotis, C., Grigorenko, A. N., Konstantatos, G., Kis, A., Katsnelson, M., Vandersypen, L., Loiseau, A., Morandi, V., Neumaier, D., Treossi, E., Pellegrini, V., Polini, M., Tredicucci, A., Williams, G. M., Hee Hong, B., Ahn, J.-H., Min Kim, J., Zirath, H., van Wees, B. J., van der Zant, H., Occhipinti, L., Di Matteo, A., Kinloch, I. A., Seyller, T., Quesnel, E., Feng, X., Teo, K., Rupesinghe, N., Hakonen, P., Neil, S. R. T., Tannock, Q., Löfwander, T. and Kinaret, J. 2015. Science and technology roadmap for graphene, related two-dimensional crystals, and hybrid systems. *Nanoscale*, 7(11): 4598-4810.

Gomaa, H. M., Suffet, I. H. and Faust, S. D. 1969. Kinetics of hydrolysis of diazinon and diazoxon. *Residue Rev*, 29: 171-190.

Gruden, R., Buchholz, A. and Kanoun, O. 2014. Electrochemical analysis of water and suds by impedance spectroscopy and cyclic voltammetry. *J. Sens. Sens. Syst.*, 3(2): 133-140.

Hanrahan, G., Patil, D. G. and Wang, J. 2004. Electrochemical sensors for environmental monitoring: design, development and applications. *Journal of Environmental Monitoring*, 6(8): 657-664.

Hinterreiter, A. P., Duchoslav, J., Kehrer, M., Truglas, T., Lumetzberger, A., Unterweger, C., Fürst, C. and Stifter, D. 2019. Determination of the surface chemistry of ozone-treated carbon fibers by highly consistent evaluation of X-ray photoelectron spectra. *Carbon*, 146: 97-105.

Hossain, S. M. Z. and Mansour, N. 2019. Biosensors for on-line water quality monitoring - a review. *Arab Journal of Basic and Applied Sciences*, 26(1): 502-518.

https://inchem.org/documents/pims/chemical/pimg001.htm#SectionTitle:1.1%20%20Substance.

https://www.graphenea.com/collections/graphene-oxide/products/graphene-oxide-4-mg-ml-water-dispersion-1000-ml.

Inagaki, M. and Kang, F. 2016. *Materials Science and Engineering of Carbon*, pp. 1-6, Butterworth-Heinemann.

Jain, A., Balasubramanian, R. and Srinivasan, M. P. 2016. Hydrothermal conversion of biomass waste to activated carbon with high porosity: A review. *Chemical Engineering Journal*, 283: 789-805.

Jocic, A., Breitenbach, S., Bajuk-Bogdanović, D., Pašti, I., Unterweger, C., Fürst, C. and Lazarevic Pasti, T. 2021. Viscose-derived activated carbons fibers as highly efficient adsorbents for dimethoate removal from water. *ChemRxiv. Cambridge: Cambridge Open Engage*: This content is a preprint and has not been peer-reviewed.

Jusoh, A., Hartini, W. J., Ali, N. and Endut, A. 2011. Study on the removal of pesticide in agricultural run off by granular activated carbon. *Bioresour Technol*, 102(9): 5312-5318.

Kanoun, O., Lazarevic-Pasti, T., Pasti, I., Nasraoui, S., Talbi, M., Brahem, A., Adiraju, A., Sheremet, E., Rodriguez, R. D., Ben Ali, M. and Al-Hamry, A. 2021. A Review of Nanocomposite-Modified Electrochemical Sensors for Water Quality Monitoring. *Sensors*, 21(12): 4131.

Karthik, V., Selvakumar, P., Senthil Kumar, P., Vo, D.-V. N., Gokulakrishnan, M., Keerthana, P., Tamil Elakkiya, V. and Rajeswari, R. 2021. Graphene-based materials for environmental applications: a review. *Environmental Chemistry Letters*, 19(5): 3631-3644.

Kassiri, H., Rabbani, D., Mohebi, F., Dehghani, R. and Takhtfiroozeh, S. 2021. A review on the removal methods of organophosphate insecticide malathion from the environment. *Journal of entomological research*, 2021 v.45 no.1(no. 1): pp. 145-152.

Lazarevic-Pasti, T., Anicijevic, V., Baljozovic, M., Anicijevic Vasic, D., Gutic, S., Vasic, V., Skorodumova, N. V. and Pasti, I. A. 2018. The impact of the structure of graphene-based materials on the removal of organophosphorus pesticides from water. *Environmental Science: Nano*, 5(6): 1482-1494.

Lazarevic-Pasti, T. D., Bondzic, A. M., Pasti, I. A., Mentus, S. V. and Vasic, V. M. 2013. Electrochemical oxidation of diazinon in aqueous solutions via electrogenerated halogens - Diazinon fate and implications for its detection. *Journal of Electroanalytical Chemistry*, 692: 40-45.

Lazarevic-Pasti, T. D., Pasti, I. A., Jokic, B., Babic, B. M. and Vasic, V. M. 2016. Heteroatom-doped mesoporous carbons as efficient adsorbents for removal of dimethoate and omethoate from water. *RSC Advances*, 6(67): 62128-62139.

Lehman, J. H., Terrones, M., Mansfield, E., Hurst, K. E. and Meunier, V. 2011. Evaluating the characteristics of multiwall carbon nanotubes. *Carbon*, 49(8): 2581-2602.

Lerf, A., He, H., Forster, M. and Klinowski, J. 1998. Structure of Graphite Oxide Revisited. *The Journal of Physical Chemistry B*, 102(23): 4477-4482.

Li, H., Yan, X., Qiao, S., Lu, G. and Su, X. 2018. Yellow-Emissive Carbon Dot-Based Optical Sensing Platforms: Cell Imaging and Analytical Applications for Biocatalytic Reactions. *ACS Appl Mater Interfaces*, 10(9): 7737-7744.

Liu, N., Xu, Z., Morrin, A. and Luo, X. 2019. Low fouling strategies for electrochemical biosensors targeting disease biomarkers. *Analytical Methods*, 11(6): 702-711.

Liu, X., Zhang, H., Ma, Y., Wu, X., Meng, L., Guo, Y., Yu, G. and Liu, Y. 2013. Graphene-coated silica as a highly efficient sorbent for residual organophosphorus pesticides in water. *Journal of Materials Chemistry A*, 1(5): 1875-1884.

Lockridge, O., Verdier, L. and Schopfer, L. M. 2019. Half-life of chlorpyrifos oxon and other organophosphorus esters in aqueous solution. *Chemico-Biological Interactions*, 311: 108788.

Lu, L., Zhou, L., Chen, J., Yan, F., Liu, J., Dong, X., Xi, F. and Chen, P. 2018. Nanochannel-Confined Graphene Quantum Dots for Ultrasensitive Electrochemical Analysis of Complex Samples. *ACS Nano*, 12(12): 12673-12681.

Macalady, D. L. and Wolfe, N. L. 1983. New perspectives on the hydrolytic degradation of the organo phosphorothioate insecticide chlorpyrifos. *Journal of Agricultural and Food Chemistry*, 31(6): 1139-1147.

Maliyekkal, S. M., Sreeprasad, T. S., Krishnan, D., Kouser, S., Mishra, A. K., Waghmare, U. V. and Pradeep, T. 2013. Graphene: A Reusable Substrate for Unprecedented Adsorption of Pesticides. *Small*, 9(2): 273-283.

Marsh, H. and Rodriguez-Reinoso, F. 2006. *Activated Carbon*, Elsevier Science Ltd, Oxford.

Mohamad Nor, N., Lau, L. C., Lee, K. T. and Mohamed, A. R. 2013. Synthesis of activated carbon from lignocellulosic biomass and its applications in air pollution control-a review. *Journal of Environmental Chemical Engineering*, 1(4): 658-666.

Mojiri, A., Zhou, J. L., Robinson, B., Ohashi, A., Ozaki, N., Kindaichi, T., Farraji, H. and Vakili, M. 2020. Pesticides in aquatic environments and their removal by adsorption methods. *Chemosphere*, 253: 126646.

Moreno-Castilla, C. 2004. Adsorption of organic molecules from aqueous solutions on carbon materials. *Carbon*, 42(1): 83-94.

Moussavi, G., Hosseini, H. and Alahabadi, A. 2013. The investigation of diazinon pesticide removal from contaminated water by adsorption onto NH_4Cl-induced activated carbon. *Chemical Engineering Journal*, 214: 172-179.

Mozaffari Majd, M., Kordzadeh-Kermani, V., Ghalandari, V., Askari, A. and Sillanpaa, M. 2021. Adsorption isotherm models: A comprehensive and systematic review (2010â"2020). *Science of The Total Environment*, 151334.

Oswin, C. R. 1946. The kinetics of package life. III. The isotherm. *Journal of the Society of Chemical Industry*, 65(12): 419-421.

Panjan, P., Ohtonen, E., Tervo, P., Virtanen, V. and Sesay, A. M. 2017. Shelf Life of Enzymatic Electrochemical Sensors. *Procedia Technology*, 27: 306-308.

Panjan, P., Virtanen, V. and Sesay, A. M. 2017. Determination of stability characteristics for electrochemical biosensors via thermally accelerated ageing. *Talanta*, 170: 331-336.

Pasti, I. A., Janosevic Lezaic, A., Gavrilov, N. M., Ciric-Marjanovic, G. and Mentus, S. V. 2018a. Nanocarbons derived from polymers for electrochemical energy conversion and storage - A review. *Synthetic Metals*, 246: 267-281.

Pasti, I. A., Jovanovic, A., Dobrota, A. S., Mentus, S. V., Johansson, B. and Skorodumova, N. V. 2018b. Atomic adsorption on graphene with a single vacancy: systematic DFT study through the periodic table of elements. *Physical Chemistry Chemical Physics*, 20(2): 858-865.

Pasti, I. A., Jovanovic, A., Dobrota, A. S., Mentus, S. V., Johansson, B. and Skorodumova, N. V. 2018c. Atomic adsorption on pristine graphene along the Periodic Table of Elements - From PBE to non-local functionals. *Applied Surface Science*, 436: 433-440.

Peigney, A., Laurent, C., Flahaut, E., Bacsa, R. R. and Rousset, A. 2001. Specific surface area of carbon nanotubes and bundles of carbon nanotubes. *Carbon*, 39(4): 507-514.

Qian, Z. S., Shan, X. Y., Chai, L. J., Ma, J. J., Chen, J. R. and Feng, H. 2014. DNA nanosensor based on biocompatible graphene quantum dots and carbon nanotubes. *Biosens Bioelectron*, 60: 64-70.

Rodriguez-Reinoso, F. and Molina-Sabio, M. 1992. Activated carbons from lignocellulosic materials by chemical and/or physical activation: an overview. *Carbon*, 30(7): 1111-1118.

Rodriguez-Reinoso, F., Molina-Sabio, M. and Gonzalez, M. T. 1995. The use of steam and CO2 as activating agents in the preparation of activated carbons. *Carbon*, 33(1): 15-23.

Shao, Y., Wang, J., Engelhard, M., Wang, C. and Lin, Y. 2010. Facile and controllable electrochemical reduction of graphene oxide and its applications. *Journal of Materials Chemistry*, 20(4): 743-748.

Suo, F., Xie, G., Zhang, J., Li, J., Li, C., Liu, X., Zhang, Y., Ma, Y. and Ji, M. 2018. A carbonised sieve-like corn straw cellulose - graphene oxide composite for organophosphorus pesticide removal. *RSC Advances*, 8(14): 7735-7743.

Tay, T., Ucar, S. and Karagoz, S. 2009. Preparation and characterization of activated carbon from waste biomass. *Journal of Hazardous Materials*, 165(1): 481-485.

Toh, S. Y., Loh, K. S., Kamarudin, S. K. and Daud, W. R. W. 2014. Graphene production via electrochemical reduction of graphene oxide: Synthesis and characterisation. *Chemical Engineering Journal*, 251: 422-434.

Van Scoy, A., Pennell, A. and Zhang, X. 2016. Environmental Fate and Toxicology of Dimethoate. *Rev Environ Contam Toxicol*, 237: 53-70.

Vukcevic, M., Kalijadis, A., Babic, B., Lausevic, Z. and Lausevic, M. 2013. Influence of different carbon monolith preparation parameters on pesticide adsorption. *Journal of Serbian Chemical Society*, 78(10): 1617-1632.

Wang, J. and Kaskel, S. 2012. KOH activation of carbon-based materials for energy storage. *Journal of Materials Chemistry*, 22(45): 23710-23725.

Wang, Z., Hu, T., Liang, R. and Wei, M. 2020. Application of Zero-Dimensional Nanomaterials in Biosensing. *Frontiers in Chemistry*, 8(320).

Yuan, W., Zhou, Y., Li, Y., Li, C., Peng, H., Zhang, J., Liu, Z., Dai, L. and Shi, G. 2013. The edge- and basal-plane-specific electrochemistry of a single-layer graphene sheet. *Scientific Reports*, 3(1): 2248.

Zhang, C., He, J., Zhang, Y., Chen, J., Zhao, Y., Niu, Y. and Yu, C. 2018. Cerium dioxide-doped carboxyl fullerene as novel nanoprobe and catalyst in electrochemical biosensor for amperometric detection of the CYP2C19*2 allele in human serum. *Biosens Bioelectron*, 102: 94-100.

Zhang, S., Wang, H., Liu, J. and Bao, C. 2020. Measuring the specific surface area of monolayer graphene oxide in water. *Materials Letters*, 261: 127098.

Zheng, H., Zhang, Q., Liu, G., Luo, X., Li, F., Zhang, Y. and Wang, Z. 2019. Characteristics and mechanisms of chlorpyrifos and chlorpyrifos-methyl adsorption onto biochars: Influence of deashing and low molecular weight organic acid (LMWOA) aging and co-existence. *Sci Total Environ*, 657: 953-962.

Zheng, M., Ruan, S., Liu, S., Sun, T., Qu, D., Zhao, H., Xie, Z., Gao, H., Jing, X. and Sun, Z. 2015. Self-Targeting Fluorescent Carbon Dots for Diagnosis of Brain Cancer Cells. *ACS Nano*, 9(11): 11455-11461.

Zhu, Y., Murali, S., Cai, W., Li, X., Suk, J. W., Potts, J. R. and Ruoff, R. S. 2010. Graphene and Graphene Oxide: Synthesis, Properties, and Applications. *Advanced Materials*, 22(35): 3906-3924.

Chapter 10

MOFs Compounds as Effective Tools for Detection of Organophosphate Compounds

Aleksandra Bondžić*
Department of Physical Chemistry,
"Vinča" Institute of Nuclear Sciences- National Institute of the Republic of Serbia,
University of Belgrade, Belgrade, Serbia

Abstract

Organophosphate compounds (OPs) are highly acute toxic compounds used in agriculture production as pesticides. They bind to acetylcholinesterase leading to nerve paralysis and organ failure in humans. Hence, it has become mandatory on a global level to detect and remove organophosphate compounds from the environment. There are many conventional methods (nanoscale materials or sophisticated instruments) available to detect these compounds. However, these methods are time-consuming, expensive, and require an analytical laboratory with skilled staff. These constraints were the driving force for developing sensing platforms capable of detecting pesticides with excellent sensitivity. Metal-organic frameworks (MOFs) belong to porous coordination polymers fabricated from metal clusters/ions and multifunctional organic ligands. The tunable size and shape of pores, stable framework and the opportunity for desirable post-synthetic modifications classify them as promising candidates for potential use in detecting OPs at low nanomolar levels. This chapter will discuss the application of MOFs for the detection of selected organophosphate compounds. The special emphasis will be on MOF-based sensors, their sensitivity, selectivity, advantages and disadvantages in the OPs detection process.

* Corresponding Author's Email: alekmile@yahoo.com.

In: Organophosphates
Editor: Tamara Lazarević-Pašti
ISBN: 978-1-68507-652-8

Keywords: organophosphate pesticides, MOFs, sensors, AChE

Introduction

The noticeable growth of the human population in the world imposes the need for an increasing amount of food that can only be achieved by improving agricultural productivity. Consequently, the utilization of organophosphate compounds (OPs) has to have a pivotal role in agriculture because approximately one-third of global production depends on them (Lykogianni et al. 2021). However, OPs are highly acute toxic compounds that are able to bind to acetylcholinesterase (AChE), leading to nerve paralysis and organ failure, while chronic exposure to OPs often leads to cancer, asthma, diabetes and neuroinflammation (Čolović, Krstić, Vasić, et al. 2013; Čolović, Krstić, Lazarević-Pašti, et al. 2013; Kim, Kabir, and Jahan 2017; Evenset et al. 2016). Hence, it is necessary to detect and quantify OPs in different environmental matrices such as surface, ground and wastewater as well as in the soil. The conventional methods for OPs detection include analytical techniques such as gas chromatography (GC) and high-performance liquid chromatography (HPLC) (Farina et al. 2017; Harshit, Charmy, and Nrupesh 2017). However, the above-mentioned techniques are often time-consuming, requiring sample pretreatment and preparation, an analytical laboratory with skilled staff and expensive equipment. The chromogenic and luminescent chemosensors could partially overcome the aforementioned drawbacks, but they are usually limited by complicated syntheses, substandard photostability and interferences with the other analytes. The enzyme-linked immunosorbent assays (ELISA), as well as AChE and organophosphorus hydrolase (OPH) enzyme sensors, have also been recognized methods for pesticides detection (Pundir and Chauhan 2012; Zhang et al. 2008; Mulyasuryani and Prasetyawan 2015). However, ELISA methods are limited by the availability of specific antibodies, while the major drawback of the AChE biosensor is the lack of sensitivity and selectivity. Additionally, AChE and OPH biosensors require enzymes that are often commercially expensive or unavailable, and their purification procedures are complex and time-consuming.

In the light of these limitations, there is the need for the development of rapid and efficient methods/technologies for the OPs detections. The development of sophisticated materials such as carbon materials, quantum dots, molecularly imprinted polymers and metal-organic frameworks (MOFs) compounds enabled improvement of sensing methods for OPs detection.

Among these listed materials, MOFs are highlighted as advanced materials for analyte sensing. Metal-organic frameworks (MOFs) represent crystalline porous materials with highly ordered structures, where metal-containing units (metal ion or metal cluster) are surrounded by organic “linker” molecules forming repeating cage-like structures. MOF’s geometric structures depend on the coordination number of metal ions included and could be linear, square planar, cubic, pyramidal, trigonal bi-pyramidal, T-/Y-shaped, tetrahedral and octahedral (Burnett, Barron, and Choe 2012). The metal ions commonly involved in MOFs structure are zinc, cobalt, iron, aluminum, titanium and zirconium, while commonly involved organic linkers could be carboxylates, amines, nitrites, phosphates and sulfonates such as the glutaric acid; 1,4-benzene dicarboxylic acid (BDC); 1,3,5-benzene tricarboxylic acid (BTC); 4,4’-bipyridine; pyrazine; imidazole; 4,5-imidazole dicarboxylic acid (H_3Imdc); triazole; 4’-biphenyl dicarboxylic acid (BPDC) and 1,4-bis(4’-pyrazolyl) benzene (Xu et al. 2021) (Figure 1). MOFs synthesis is based on the self-assembly of metal ions and organic ligands during the hydrothermal, solvothermal, microwave, electrochemical, mechanochemical, or ultrasonic polymerization processes. The MOFs’ synthesis could be divided into a direct synthesis in aqueous or non-aqueous media and the combined use of aqueous/non-aqueous media in particular ratios depending on the type of medium in which synthesis occurs. Water, methanol, ethanol and N, N-dimethylformamide (DMF) are the most commonly used solvents. Currently, according to the Cambridge Structural Database, more than 6000 MOFs structures are being synthesized and divided into several groups, including MIL MOF (Material of Institute of Lavoisier) such as MIL-53, MIL-100, MIL-101; zeolitic imidazolate framework, ZIF MOFs that include ZIF-10, ZIF-68, and ZIF-69 and UiO MOF (University of Oslo) based on Zr metal ions such as UiO-66; UiO-67 for instance (Xu et al. 2021). The MOFs possess high thermal and chemical stability, which are correlated with the metal ion oxidation state. However, due to the MOF’s tendency to hydrolysis, some of them are more or less sensitive to water which limits their practical application. Owing to MOFs’ prominent surface properties such as tunable pore size and shape, the stable framework, and the possibility for post-synthetic modifications, MOFs can be used for drug delivery (Lawson, Walton, and Chan 2021; Abánades Lázaro and Forgan 2019; Sun et al. 2020; Markopoulou et al. 2020), chemical separation (Shi et al. 2020; Mukherjee et al. 2020; Gu et al. 2021)gas storage(Connolly et al. 2020; Suresh et al. 2021), detoxification of chemical warfare agents (Zhou et al. 2021; Chen and Snurr 2020; Xiong, Wang, et al. 2021), water remediation (Xiong, Li et al. 2021) and in sensing application (Sacchetti, Rossetti, and Martí-Rujas 2021; Yan 2017).

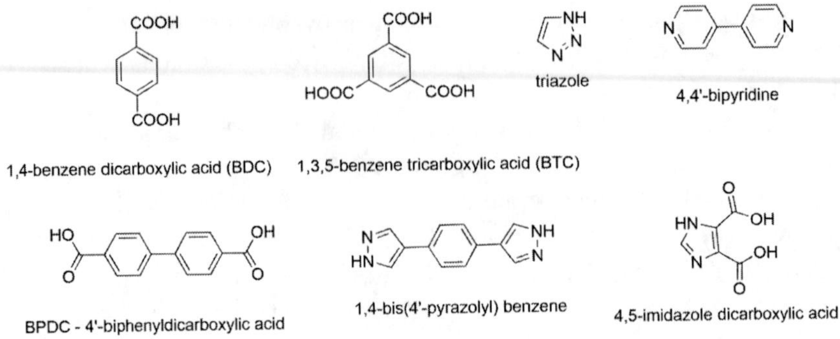

Figure 1. Different types of commonly used organic linkers.

MOFs as Sensors for Organophosphate Compound Detection

MOFs-based sensors for OPs detection could be divided into three main groups: luminescent, electrochemical and biosensors. The classification into the luminescent and electrochemical sensors was performed depending on the type of signal which will be detected, while biosensors represent a special class of sensors. In order to improve the sensor performance, electrochemical MOFs-based sensors can be conjugated with different materials, including nanoparticles, carbon materials, polymers, leading to the production of numerous MOF sensor subgroups (Figure 2).

Herein, different MOFs-based sensors for pesticides detection are presented and compared with each other with the aim to highlight their advantage/disadvantages in the OPs detection.

MOFs as Luminescent Sensors

Luminescent metal-organic framework (LMOF) is a class of MOFs that possesses intrinsic fluorescence and represents one of the most attractive groups of MOF compounds because of numerous applications such as sensing (Yan 2017; Yang et al. 2020; Xu et al. 2018), photocatalysis (Wu et al. 2016; Wang et al. 2015), non-linear optics (Yu et al. 2012), etc. Their fluorescence emission could originate from metal ion, organic ligand or metal-ligand charge transfer, whereas the changes in fluorescence (quenching or enhancing) after interaction with analyte could be exploited for its detection in different media. The pesticides detection in water is based mainly on fluorescence quenching, as a consequence of the electron or energy transfer process, while the sensitivity of fluorescence sensing depends on the ability of LMOFs to donate electrons to the electron-deficient analyte. This ability could be increased by introducing electron-rich π conjugated moieties to the organic linkers of MOFs such as carbazole (Xu, Zhang, and Zhang 2019), different aromatic derivatives of benzene (tetrakis phenyl benzene (Xu et al. 2018); 1,4-benzene dicarboxylate (Kumar, Paul, and Deep 2014b), derivatives of naphthalene, etc. On the other side, the presence of the electron-withdrawing group in the OPs structures contributes to the selectivity of the detection. In contrast to traditional instrumental detection methods, fluorescence sensing is characterized by high sensitivity, fast response, low cost and simple operation.

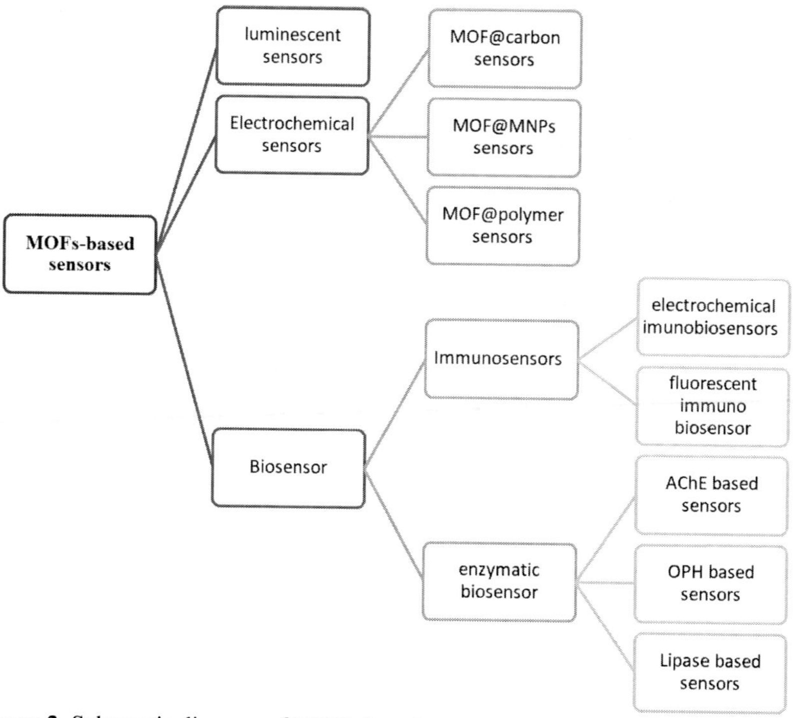

Figure 2. Schematic diagram of MOFs-based sensors.

LMOFs for Detection of a Nitro Group Containing Organophosphate Pesticides

The nitro group (-NO_2) is a strongly electron-withdrawing group, whose presence in the aromatic structure of OPs reduces electronic density and consequently enables photo-induced electron transfer from electron-donating LMOF to electron-deficient OPs. This transfer is responsible for the quenching phenomenon observed after LMOF's interaction with selected OPs. Among the various OPs, parathion, parathion-methyl, paraoxon and fenitrothion belong to the nitro group containing organophosphate pesticides. In this section, the recently synthesized LMOFs are presented, and their advantages/disadvantages, the limit of detection (LOD) and selectivity in the OPs detection are highlighted.

Kumar et al. synthesized luminescent metal-organic framework, MOF-5, namely $Zn_4O(BDC)_3$, where BDC is 1,4-benzene dicarboxylate for direct chemosensing of the nitro group containing OPs including parathion, methyl-parathion, paraoxon and fenitrothion(Kumar, Paul, and Deep 2014b). The desired MOF-5 was synthesized by stirring under nitrogen gas an appropriate amount of $Zn(NO_3)_2$ and BDC in the presence of diethyl formamide (DEF) solvent. This synthetic procedure gave MOFs' lower surface area (895 m^2/g) and total pore volume (0.986 cm^3/g) compared to the other method due to the interpenetrated nature of the synthesized particles (Liu 2011; Li et al. 2004). MOF-5 dispersion in ethanol shows strong emission at 493 nm after excitation at 330 nm, attributed to the presence of 1,4-benzene dicarboxylate linker in the MOF-5 structure. It was shown that the nitro group containing OPs effectively quenched MOF-5 emission at 493 nm in contrast to the other investigated pesticides, malathion, dichlorvos and monocrotophos. Hence, it was proposed that the selectivity in quenching was associated with the presence of the electron-withdrawing group. Moreover, in the same work, the bimolecular quenching constants (k_q) values were in good correlation with the oxidation potential of the investigated pesticides, indicating that the selectivity of quenching depends upon the redox potential of OPs. Using this MOF-5, it is possible to detect -NO_2 containing OPs in the concentration range of 5 - 600 ppb while the detection limit was 5 ppb. Although this MOF has great potential in the selective detection of NO_2 containing pesticides, recycled MOF shows a 20% decreased initial fluorescence intensity with each regeneration cycle. Hence, the regeneration and reuse of MOF-5 were limited only to 3 - 4 cycles. Same year these authors published synthesis of $[Cd(atc)(H_2O)_2]_n$ (Cd-NMOF1) possessing electron donor -NH_2 group in the structure of linker used in the previous report, terephthalic acid, and tested it for sensing of same pesticides (Kumar, Paul, and Deep 2014a). As a previous report, strong NMOF1 emission was associated with the linker, 2-amino terephthalic acid, as well as with the presence of surface defects in the MOF structure, such as ionizable oxygen vacancies or Cd(II) ion vacancies on the surface. The fluorescence quenching was the consequence of electrostatic interactions between the photoexcited Cd-MOF and electron-withdrawing $-NO_2$ moiety, while high selectivity was achieved toward pesticides without $-NO_2$ group in the structure. Compared to previously synthesized MOF-5 with a similar linker, the introduction of electron donor group in the structure of MOF's linker led to an increased detection limit up to 1 ppb. The linear detection range from 1 to 500 ppb was also observed.

Better sensing capabilities were achieved using 1,4-naphthalene dicarboxylic acid in the structure of Cd-MOF. Yang et al. (Yang et al. 2020) developed Cd-MOF, namely [Cd(1,4-ndc)(DMA)] (1,4-H_2ndc =1,4-naphthalene dicarboxylic, DMA = N, N'-dimethylacetamide) luminescent sensor for detection of structurally different OPs and tested its efficiency with spiked river water samples. It was found that 2×10^{-4} M parathion-methyl quenched fluorescence of this MOF with 95.4% efficiency. Additionally, the calculated low limit of detection (LOD) value for detecting this pesticide was 3 nM (0.879 ppb), indicating extreme sensitivity. Superior sensitivity observed during fluorescence detection was explained by the synergetic effect of photo-induced electron transfer and resonance energy transfer. The presence of electron-withdrawing group ($-NO_2$) in the structure of pesticide contributed to photo-induced electron transfer from electron-donating Cd-MOF to electron-deficient parathion methyl adsorbed on the surface of Cd-MOF. Moreover, the high spectral overlapping between the fluorescent emission spectra of Cd-MOF and the absorption spectra of parathion-methyl promote a large extent to quenching phenomenon via FRET process (Förster Resonance Energy Transfer). Notably, the authors found that after ten cycles, the initial fluorescence was unchanged, pointing out superior properties of this Cd-MOF as an agent for parathion-methyl detection.

The great work was published by Xu et al. where superior sensitivity for parathion methyl was achieved (Xu et al. 2018). These authors utilized ZnPO-MOF with 1,2,4,5-tetrakis(4-carboxyphenyl) benzene (H_4TCPB) linker as a highly sensitive and selective sensor for rapid detection of parathion-methyl. ZnPO-MOF has strong blue emission after excitation at 365 nm, a large BET surface area (4073.9 m^2/g) and strong thermal stability. Pesticide detection was based on monitoring the changes in ZnPO-MOF fluorescence intensity in the presence of pesticides. The authors investigated 18 different pesticides, but the best sensitivity was achieved for nitro-containing pesticides (parathion and parathion-methyl). It was found that the optimal ZnPO-MOF concentration for satisfactory photoluminescence response was 10 μg/ml. Additionally, LOD for parathion-methyl with ZnPO-MOF was 0.456 nM (0.134 ppb), much lower than the maximum residue limits (MRL) in the European Union pesticides databank. As in the previous reports, the pesticide detection mechanism was related to the presence of the nitroaromatic group in the structure of the pesticide, which led to photo-induced electron transfer. ZnPO-MOF was also tested in the real matrix using local lake water, and excellent recovery was obtained, indicating its potential application with real water samples.

MOFs as Electrochemical Sensors

Electrochemical sensors represent devices in which the electrical signals originating from the electrochemical reaction are detected by amperometry, potentiometry or conductometry. These sensors are applied in environmental monitoring, food quality control, medical diagnostics, and chemical treatment detection (Akyüz and Koca 2019; Sakdarat et al. 2021). As mentioned in the previous section, the simple syntheses and the MOFs' physicochemical properties make them suitable materials for sensor construction. However, despite the tremendous potential for sensor fabrication, the weak electrical conductivity of MOFs is a major drawback for application as electrochemical sensors. In order to overcome this deficiency, during the MOFs synthesis, organic linkers with conductor's properties can be used. In addition, post-synthetic modifications on the MOFs' surface using conducting materials such as metal nanoparticles (MNPs) (Xia et al. 2019), polymers (Shanahan, Kissel, and Sullivan 2020) and carbon materials can be performed (Zhou et al. 2015).

MOF@carbon Materials as an Electrochemical Sensor

The low price, high conductivity, tensile strength and durability put carbon materials in the group of endowed materials for MOFs' conductivity enhancement. Besides this, MOFs/carbon materials integration brings about enhancement of MOFs' stability. In the study of Chang et al., the fabrication of an electrochemical MOFs/carbon-based sensor for detection of methyl partition and the other nitro compounds by differential pulse voltammetry (DPV) was shown (Liu, Bo, and Guo 2019). The novel composite JUC-62@pOMC contained Cu-based MOF, named JUC-62 with 3’,3’,5’,5’-azobenzene tetracarboxylic acid (H_4ABTC) as linker ligand and platelet ordered mesoporous carbon (pOMC). This carbon material possesses a large amount of short and well-defined mesopores, which are more exposed than in the case of conventional carbon materials. Investigation of optimal conditions for biosensor operations revealed improved conductivity efficiency in the presence of the moderate amount of pOMC (pOMC-3) in phosphate buffer at pH 7. Additionally, 160 s as accumulation time, and 0.0 V as accumulation potential were selected. As a working electrode, a glassy carbon electrode (GCE) was used. Fabricated JUC-62@pOMC-3/GCE showed improved electrochemical properties toward the reduction of nitroaromatic compounds compared to individual components. Also, a good linear range (0.15 – 250 μM) and 12 nM LOD were achieved.

For methyl parathion detection, the MOF@carbon material sensor with better performance was constructed by Gao et al. (Gao et al. 2021). These authors utilized Zr-based MOF with terephthalic acid as a ligand (BDC) which contains the Zr-O group. This group has a high affinity for phosphate group binding, enabling OPs effective removal from the environment. In order to improve electrochemical properties, reduced graphene oxide (rGO) was used as carbon materials. Although rGO shows significant OPs removal efficiency, its conjugation with Zr-based MOFs enables excellent selectivity in methyl parathion detection. The electrochemical response was investigated by cyclic voltammetry (CV) and square wave voltammetry (SWV) measurements. In both cases, a good electrochemical response for methyl parathion is obtained. On the CV curve, reversible (at 0.0 V) and irreversible peaks (at -0. 53V) were observed, proving methyl parathion adsorption on Zr-BDC-rGO. Additionally, a better electrochemical response for methyl parathion was obtained for constructed electrochemical sensors compared to individual electrodes. The optimization of Zr-BDC-rGO/GCE sensor was done by investigating the influence of the Zr-BDC concentration, rGO concentration, pH, and contact time on the sensor performance. The best electrochemical response was achieved for the sensor constructed using 1.5mg/ml Zr-BDC and 0.1 mg/ml rGO. The higher Zr-BDC concentration increases resistance and consequently decreases current. On the other side, the higher rGO concentration leads to reducing of sensitivity due to the background current. The best sensor operation was achieved at pH 5 and for 12 min of contact time. This sensor showed a wider linear range of methyl parathion detection, 0.001

– 3 μg/ml (0.0038 – 11.4 μM) and a lower limit of detection, 0.5 ng/ml (1.9 nM) compared to the aforementioned electrochemical MOF@carbon material based sensor.

MOF@MNPs as an Electrochemical Sensor

MOF-magnetic nanoparticles conjugate, MOF@MNPs, can also be utilized for electrochemical methyl parathion detection by DPV. In their report, Hu et al. incorporated Fe_3O_4 nanoparticles with $Co_2(BTC)_2$ MOF (BTC = 1,3,5-benzene tricarboxylic) for preparation of magnetic Fe_3O_4/MOF conjugate for sensitive and selective methyl parathion detection(Hu et al. 2014). The electrochemical behavior of methyl parathion with the fabricated sensor was investigated by CV and DPV using glassy carbon and magnetic glassy carbon electrode (MGCE). The authors showed that adding the Fe_3O_4/MOF conjugate to the solution led to amplifying of electrical signal due to the methyl parathion adsorption on the Fe_3O_4/MOF conjugate. The authors obtained a better electrochemical response using MGCE, which was explained by the inability of a glassy carbon electrode to collect methyl parathion around itself, in contrast to magnetic GCE.

MOF@molecularly Imprinted Polymer as an Electrochemical Sensor

One of the interesting approaches for the construction of MOF film-based selective electrochemical sensor was published in the study of Do et al. (Do et al. 2015). The authors developed a selective and sensitive electrochemical sensor for glyphosate (Gly) detection by fabrication of molecularly imprinted polymer gold-MOF (MIP-MOF) on the surface of the gold electrode. The molecularly imprinted gold MOF was obtained by electropolymerization of a previously formed p-amino thiophenol monolayer on the gold electrode, and p-amino thiophenol functionalized gold nanoparticles (AgNPs) in the solution that contained Gly as a template molecule. By extraction of Gly molecules from MIP-MOF by incubation in PBS, pH 7.2, the formation of cavities with size complementary to Gly molecules was achieved. In this way, high selectivity and sensitivity of the sensor were attained. The Gly electrochemical detection was achieved using the linear sweep voltammetry (LSW) in the presence of $(Fe(CN)_6)^{3-/4-}$ redox probe. Such fabricated electrochemical MOF sensor showed the wide linear range of Gly detection from 1 pg/L to 1 μg/L and a LOD of 0.8 pg/L. The investigation of MIL-MOF sensor potential to detect Gly metabolite, aminomethylphosphonic acid, revealed significant selectivity properties of this sensor. Additionally, the authors showed successful application and excellent recovery of the developed sensor in two tap water samples.

MOF Thin Film for Improving Stripping Voltammetry Detection of OPs

Owing to MOFs' potential to sorb OPs on their surface, MOFs can be additionally applied to improve the OPs detection by stripping voltammetry (SV). Since SV requires analyte preconcentration on the electrode surface, the MOFs represent the sorbent matrix that is conjugated with the electrode for solid-phase extraction of OPs. The formation of thin MOFs' film on the glassy carbon electrode surface led to improved detection reflected in the enhancement of SV signal. In the study of Wen et al., SV detection of methyl parathion with LOD of 0.006 μg/ml was attained by formation of thin $(Cd(2,2',4,4'\text{-bptcH2}))_n$ MOF (where bptc = 2,2',4,4'-biphenyl tetracarboxylic acid) film on the glassy carbon electrode (Wen et al. 2010). The methyl parathion was sorbed onto the MOF's thin film by π-π stacking interactions and hydrogen bonding to MOF. Such prepared sensor, Cd-MOF/GCE, showed the best performance with 12 min of the immersion time.

MOFs Based Biosensors

The OPs biosensors represent the class of sensors in which biological components interacting with the pesticides lead to the detectible physicochemical event, which is proportional or inversely proportional to OPs concentration in the investigated samples (Karadurmus, Kaya, and Ozkan 2021). The biological component has to be specific toward the analyte, enabling high sensitivity of the constructed biosensor. Immobilization of the biological components is performed on different supports such as semi-conductor nanomaterials, screen-printed electrodes, quantum dots (Pundir and Chauhan 2012). The process of immobilization could be achieved by physical adsorption, physical entrapment, covalent coupling, self-assembly monolayer, oriented immobilization or electropolymerization.

In the MOF-based biosensors, the appropriate structures of MOFs compounds enable attaching/incorporation of the biological component on/into the surface/interior of MOF mainly by physical adsorption or covalent binding. By the type of biological recognition component attached to the MOF, these biosensors could be divided into two classes: immune biosensor, where antibody has a role of biological component and enzyme biosensors, where it belongs to enzymes. On the other side, each of these classes could be divided into optical and electrochemical biosensors according to the used detection methods.

MOF Based Immunosensors

MOF-based immunosensors are a class of biosensors based on the use of antibodies or antigens as the specific sensing element that provides concentration-dependent signals. Based on the principles of the solid-phase immunoassays, antibodies or antigens are immobilized onto the MOF as a suitable carrier, while the pesticides binding on the antibody can be converted into a relevant detectable signal by a transducer. The advantages of these sensors are reflected in the low volume of reagents and samples, portability, simplicity, high-specificity and -sensitivity, as well as small-time consumption.

As aforementioned, MOF-based immunosensors can be divided into luminescent and electrochemical sensors depending on the pesticides detection method.

MOF Based Fluorescent Immuno-Biosensor

Recent studies demonstrated the practical utilization of MOFs as immune-fluorescent sensors in pesticides detection. This type of sensor is based on the formation of fluorescent MOF/antibody complex where interaction between pesticide and antibody leads to the decrease of its fluorescence. These sensors possess higher sensitivity compared to simple fluorescence sensors.

As mentioned in the previous section, NMOF1 possesses significant photoluminescence properties that were employed for the fast and selective detection of nitro-containing pesticides with a detection limit up to 1 ppb (Kumar, Paul, and Deep 2014a). Meanwhile, in order to obtain lower LOD for parathion-methyl, Kumar et al. developed an immunosensor based on the same

MOF, NMOF1, and anti-parathion antibody (Kumar et al. 2016). Biosensing substrate was formed by activation of –COOH group of the NMOF1 organic linker (2-amino terephthalic acid) and its amide linkage with anti-parathion antibody. Conjugates formation was confirmed by UV/Vis and IR spectroscopy. The bathochromic shift from 305 nm - 350 nm in absorption spectra was associated with a reduction in the surface charge of the NMOF1 after attaching charged anti-parathion antibody molecules. Additionally, the bathochromic shift of bands at 1610, 2342, 2360 and 2382 cm^{-1} was observed with the synchronous blue shift of the band at 3451 cm^{-1}. These results pointed out the strong binding of anti-parathion molecules on the outside of the NMOF1. Formed conjugate possessed strong fluorescence emission at 436 nm, which is reduced up to 14% compared to the pure metal-organic framework due to the binding of the antibody. As in the case of a simple NMOF1 fluorescence sensor, the addition of methyl-parathion leads to a decrease of immunosensor's fluorescence but with a wider linear detection range (1-1000 ppb) and LOD down to 1ppb, indicating better sensing properties of such formed NMOF1.

MOF Based Electrochemical Immuno-Biosensor

The biosensor based on electrochemical transducing of the biochemical events to electrical signals belongs to the class of electrochemical biosensors. In this type of sensor, the biomolecule (antibody or enzyme) is immobilized on the electrode through physical adsorption, covalent attachment or physical entrapment/encapsulation. Due to this close coating of the electrode, the high performance of the sensor in terms of analytical sensitivity is achieved. Since the sensitivity and reusability of enzyme electrochemical biosensors are mainly dependent on successful enzyme immobilization on the electrode, MOFs with their large surface area and tunable pore size might be applied as efficient enzyme immobilization matrices. In this way, by encapsulation of the enzyme in the MOFs' mesopores, the problem of enzyme aggregation and leaching from electrodes could be avoided.

In the recently published work of Chansi et al., MOF-5 was embedded into the immune-nano-hybrid Chi-AuNP-rIgG-BSA, composed from chitosan capped gold nanoparticles (Chi-AuNPs) and anti-rabbit IgG antibody to fabricate multi-layered pesticide sensor (Chansi et al. 2020). Bovine serum albumin (BSA) was used to block nonspecific binding sites to antibodies, while chitosan's role was avoiding of AuNPs agglomeration. The porous

MOF-5 provided a large surface area for immobilization of nano immune hybrid. Before MOF-5 incorporation into nano-immuno hybrid, MOF/ITO electrode was constructed, where ITO is indium tin oxide electrode. ITO was used as a working electrode, Ag/AgCl as the referent electrode in PBS containing 5mM $(Fe(CN)6)^{3-}/^{4-}$ as a redox species. The developed electrochemical nano-immuno biosensor was employed to detect different pesticides, including chlorpyrifos, dimethoate, imidacloprid, bifenthrin. Its linear detection range is from 4 to 100 ng/L with a sensitivity of 25.4 $\mu Ang^{-1}Lcm^{-2}$.

MOF Based Enzymatic Biosensor

The enzyme-based biosensors are one of the most widely adopted sensors for pesticide detection. Most of these biosensors are based on acetylcholine esterase inhibition (vide infra), but in the recent literature data, it is possible to find reports about biosensors developed on enzymes such as butyrylcholinesterase, organophosphorous hydrolase, organophosphorous acid anhydrolase, and lipase (Lazarević-Pašti et al. 2012; Lazarević-Pašti et al. 2013; Ma et al. 2018; Mulyasuryani and Prasetyawan 2015)

MOF@enzyme Fluorescence Biosensor

In contrast to above mentioned, simple MOF fluorescence sensors, where observed fluorescence changes originated from the interaction of MOFs with pesticide, in fluorescence biosensors, they are the consequence of the interaction of the product of the enzymatic reaction with external fluorescence compound. Accordingly, there is no need for utilization of luminescent MOFs, while enzyme immobilization prorogates the biosensor stability.

The organophosphorus hydrolase (OPH) is the enzyme responsible for the catalysis of hydrolysis of P-O, P-F and P-S bonds in the organophosphorus pesticides. However, OPH exhibited the highest specificity toward the P-O bond with k_{cat} values a few orders of magnitude higher (3170 s^{-1} for paraoxon and k_{cat} 630 s^{-1} for methyl parathion) compared to the pesticides with P-F and P-S bonds (Lyagin and Efremenko 2021; Efremenko and Sergeeva 2001). Wang et al. exploited this fact for developing an organophosphorus-hydrolase biosensor for amperometric detection of OP nerve agents (Wang et al. 2003). A few years later, Deo et al. study developed a carbon nanotube/

organophosphorus hydrolase electrochemical biosensor in order to determine parathion by amperometric detection of p-nitrophenol as an oxidation product (Deo et al. 2005). Pedrosa et al. furthered the same biosensor by covalent immobilization strategy achieving much higher OPH activity and stability over 7 months (Pedrosa et al. 2010). In the recently published report, Mehta et al. exploited the same strategy to develop the MOF-based fluorescence biosensor using the hexahistidine-tagged organophosphorus hydrolase (OPH^{6His}) as a biological component of the biosensor (Mehta et al. 2019). They used Zr-MOF, namely UiO-66-NH_2, onto which surface OPH^{6His} had been immobilized. This was the first report on covalent immobilization of OPH^{6His} onto Zr-MOF. In contrast to the previously mentioned biosensor, the catalytically converted methyl parathion was detected by the addition of an external fluorescent compound, cumarin1, while the concentration of methyl parathion is correlated with the degree of cumarin1's fluorescence quenching. By covalent immobilization, 40% more enzymatic activity was achieved compared to the free form of the enzyme. The methyl parathion was detectable down to 0.19 nM, while constructed biosensor was easily regenerated and could be used for multiple cycles.

MOF@enzyme Electrochemical Biosensor

In the electrochemical biosensors, the MOFs have the role of the linker between the enzyme (biological recognition element) and the working electrode. Due to MOFs' structural properties, using them as linkers decreases the possibility of enzyme leaching from the electrode surface while the operational enzyme stability is increased. In addition, immobilization on the MOFs structure also enables long-term the use of such biosensors.

MOF Biosensor Based on Inhibition Mechanism of Acetylcholinesterase

The acetylcholinesterase, AChE, belongs to cholinergic enzymes, which are responsible for the maintaining of the acetylcholine (ACh) level at postsynaptic neuromuscular junctions hydrolyzing it into acetic acid and choline. The reversible inhibitors of AChE represent promising potential drugs in neurodegenerative diseases (Bondžić, Lazarević-Pašti, et al. 2020; Bondžić, Senćanski, et al. 2020; Bondžić et al. 2021; Čolović, Krstić, Lazarević-Pašti, et al. 2013) while AChE irreversible inhibitors such as organophosphate pesticides are highly neurotoxic compounds (Čolović, Krstić, Lazarević-Pašti, et al. 2013; Čolović, Krstić, Vasić, et al. 2013). OPs biosensors based on AChE work based on measuring the percent inhibition of AChE activity due to exposure to pesticides (Dhull et al. 2013; Ivanov et al. 2000; Lazarević-Pašti et al. 2012; Lazarević-Pašti et al. 2013). Depending on the used transducer, the obtained signal is directly or indirectly (electrochemical sensor) proportional to the percent of AChE inhibition. In the electrochemical AChE sensor, the signal originates from the anodic oxidation of thiocholine, which is obtained by AChE catalyzed hydrolysis of acethylthiocholine. The obtained signal is in indirect proportion to the concentration of organophosphate pesticides.

In recently published reports, Nagabooshanam et al. fabricated two similar MOF@AChE biosensors for rapid and sensitive amperometric detection of chlorpyrifos (Nagabooshanam et al. 2021; Nagabooshanam et al. 2019). In both biosensors, the authors have used bioconjugate of Basolite Z1200 MOF and acetylcholinesterase, which was applied to the surface of the working electrodes. In one of these biosensors, bioconjugate was grafted on the gold microelectrode, previously assembled on the microfluidic platform. The other biosensor contained a carbon electrode as a working electrode, fabricated in-house by screen printing technology. To ensure a portable tool for on-site analysis, these platforms were also integrated with portable electronics. The linear range of chlorpyrifos detection was 10 to 100 ng/L with a low limit of detection of 6 ng/L, while the sensitivity was 0.598 $\mu A/ngL^{-1}$ for biosensor which utilized the gold electrode. On the other side, a lower detection limit (3 ng/L) and sensitivity for chlorpyrifos were obtained using carbon electrodes. Both biosensors required a very low sample volume, 2μL, while their stability was at least one month.

In a study by Mahmoudi et al., a hybrid AChE amperometric biosensor for OPs detection was developed using Ce doped UiO66 and multi-walled-carbon nanotube (MWCNTs) (Mahmoudi et al. 2019). In this study, the authors examined paraoxon detection using different types of carbon-based materials, such as graphene oxide, carbon black, and MWCNTs. The biosensor with 7% Ce and 30% MWCNTs had the best performance, which was attributed to the good AChE's loading in vacant sites, increased electron transfer as well decreased thiocholine's diffusion path to the electrode. The glassy carbon electrode was used as a working electrode, while platinum and Ag/AgCl as the auxiliary electrode and the reference electrode, respectively.

The paraoxon's linear detection range was from 0.01 to 150 nM, while the low detection limit was 0.004 nM.

MOF Biosensor Based on Lipase-Catalyzed Hydrolysis

Unlike above mentioned, often used enzymes in biosensor fabrication (AChE and OPH), the commercially available lipase is cheaper (Do et al. 2015). The additional advantage is its presence in easily available form and simple purification procedures, which consequently influences the cost-effectiveness of biosensor fabrication. Cheong et al. found that Burkholderiacepacia lipase, BCL, can be successfully immobilized onto zeolitic imidazolate frameworks, ZIF-8 (BCL@ZIF-8) and amine-functionalized ZIF-8 (BCL@An-ZIF-8) nanoparticles with retaining the catalytic activity of the free enzyme (Cheong et al. 2017). The authors showed that fabricated conjugates were stable for up to 20 days and could be reused in at least 10 consecutive cycles. In the later report (Ma et al. 2018), the same group of authors employed these lipase@MOF conjugates for the fabrication of electrochemical biosensor for the detection of methyl parathion. The lipase@MOF conjugates are adhered to a glassy carbon electrode by chitosan. Methyl parathion detection is based on enzyme-catalyzed hydrolysis of pesticide and direct measuring of redox current of the released p-nitrophenol by differential pulse voltammetry. These biosensors detected methyl parathion with a wide linear range, good, recovery and repeatability while LOD was as low as 0.28 μM. Using the same enzyme, Wang et al. (Wang et al. 2019) developed a MOF nanofiber biosensor with lower LOD. In this case, the authors used Asp-Cu MOFs, which contains Cu ion as metal ion and aspartic acid as linker (Zhao, Shen, and Han 2015). As in the previous cases (BCL)@MOF biosensor was made from three electrodes system, glassy carbon electrode as working electrode, calomel electrode as referent electrode and platinum wire as the auxiliary electrode. This biosensor possesses high selectivity for methyl-parathion, a wide linear detection ranging from 0.1-38 μM as well as a low limit of detection, 0.067μM. Additionally, stability up to 21 days was obtained if the biosensor was stored at 4°C.

Conclusion

In this chapter, various MOF-based sensors for OPs detection are outlined. Among them, luminescent MOF compounds have been highlighted as excellent candidates in the application for the detection of pesticides with the electron-withdrawing group, such as $-NO_2$ without additional MOF modification. Higher sensitivity, as well as satisfactory reusability of such "raw" MOF, can be achieved by utilization of larger electron-rich π conjugated system as organic linkers in MOFs structures. MOFs can also be applied as electrochemical sensors. The weak electrical conductivity of MOFs could be overcome by post-synthetic modifications on their surface and by the formation of different conjugates with conducting materials such as carbon materials, metal nanoparticles and some polymers. MOFs can also be conjugated with biological materials such as antibodies and enzymes, forming MOF biosensors. By immobilization of enzymes or antibodies on the MOF's surface, increased operational enzyme stability and long-term use of biosensors are achieved.

Taking into account all the above MOF-based sensor examples, it can be concluded that MOF compounds represent advanced materials with promising potential for use in pesticide detection. Their potential is built on their unique structure, e.g., large surface area, adjustable porosity and functional areas, as well as adjustable luminescence.

Acknowledgment

This research was funded by the Ministry of Education, Science and Technological Development of the Republic of Serbia, grant number 451-03-9/2022-14/200017.

References

Abánades Lázaro, I., and R. S. Forgan. 2019. "Application of zirconium MOFs in drug delivery and biomedicine." *Coordination Chemistry Reviews*, 380: 230-259. doi: https://doi.org/10.1016/j.ccr.2018.09.009.

Akyüz, D., and A. Koca. 2019. "An electrochemical sensor for the detection of pesticides based on the hybrid of manganese phthalocyanine and polyaniline." *Sensors and Actuators B: Chemical*, 283: 848-856. doi: https://doi.org/10.1016/j.snb.2018.11.155.

Bondžić, A. M., T. D. Lazarević-Pašti, A. R. Leskovac, S. Ž. Petrović, M. B. Čolović, T. N. Parac-Vogt, and G. V. Janjić. 2020. "A new acetylcholinesterase allosteric site responsible for binding voluminous negatively charged molecules – the role in the mechanism of AChE inhibition." *European Journal of Pharmaceutical Sciences*, 151: 105376. doi: https://doi.org/10.1016/j.ejps.2020.105376.

Bondžić, A. M., M. V. Senćanski, A. V. Vujačić Nikezić, M. V. Kirillova, V. André, A. M. Kirillov, and B. P. Bondžić. 2020. "Aminoalcoholate-driven tetracopper(II) cores as dual acetyl and butyrylcholinesterase inhibitors: Experimental and theoretical elucidation of mechanism of action." *Journal of Inorganic Biochemistry*, 205: 110990. doi: https://doi.org/10.1016/j.jinorgbio.2019.110990.

Bondžić, A. M., J. J. Žakula, L. B. Korićanac, O. D. Keta, G. V. Janjić, I. S. Đorđević, and S. U. Rajković. 2021. "Cytotoxic activity and influence on acetylcholinesterase of series dinuclear platinum(II) complexes with aromatic nitrogen-containing heterocyclic bridging ligands: Insights in the mechanisms of action." *Chemico-Biological Interactions*, 109708. doi: https://doi.org/10.1016/j.cbi.2021.109708.

Burnett, B. J., P. M. Barron, and W. Choe. 2012. "Recent advances in porphyrinic metal–organic frameworks: materials design, synthetic strategies, and emerging applications." *Cryst Eng Comm*, 14 (11): 3839-3846. doi: 10.1039/c2ce06692k.

Chansi, R. Bhardwaj, R. P. Rao, I. Mukherjee, P. K. Agrawal, T. Basu, and L. M. Bharadwaj. 2020. "Layered construction of nano immuno-hybrid embedded MOF as an electrochemical sensor for rapid quantification of total pesticides load in vegetable extract." *Journal of Electroanalytical Chemistry*, 873: 114386. doi: https://doi.org/ 10.1016/j.jelechem.2020.114386.

Chen, H., and R. Q. Snurr. 2020. "Insights into Catalytic Gas-Phase Hydrolysis of Organophosphate Chemical Warfare Agents by MOF-Supported Bimetallic Metal-Oxo Clusters." *ACS Applied Materials & Interfaces*, 12 (13): 14631-14640. doi: 10.1021/acsami.9b19484.

Cheong, L.-Z., Y. Wei, H. Wang, Z. Wang, X. Su, and C. Shen. 2017. "Facile fabrication of a stable and recyclable lipase@amine-functionalized ZIF-8 nanoparticles for esters hydrolysis and transesterification." *Journal of Nanoparticle Research*, 19. doi: 10.1007/s11051-017-3979-3.

Connolly, B. M., D. G. Madden, A. E. H. Wheatley, and D. Fairen-Jimenez. 2020. "Shaping the Future of Fuel: Monolithic Metal–Organic Frameworks for High-Density Gas Storage." *Journal of the American Chemical Society*, 142 (19): 8541-8549. doi: 10.1021/jacs.0c00270.

Čolović, M. B., D. Z. Krstić, T. D. Lazarević-Pašti, A. M. Bondžić, and V. M. Vasic. 2013. "Acetylcholinesterase Inhibitors: Pharmacology and Toxicology." *Current Neuropharmacology*, 11 (3): 315-335. doi: http://dx.doi.org/10.2174/1570159X11311 030006.

Čolović, M. B., D. Z. Krstić, V. M. Vasić, A. M. Bondžić, G. S. Uščumlić, and S. D. Petrović. 2013. "Organophosphorus insecticides: Toxic effects and bioanalytical tests for evaluating toxicity during degradation processes." *Hemijska industrija*, 67 (2): 217-230. doi: https://doi.org/10.2298/HEMIND120323060C.

Deo, R. P., J. Wang, I. Block, A. Mulchandani, K. A. Joshi, M. Trojanowicz, F. Scholz, W. Chen, and Y. Lin. 2005. "Determination of organophosphate pesticides at a carbon nanotube/organophosphorus hydrolase electrochemical biosensor." *Analytica Chimica Acta*, 530 (2): 185-189. doi: https://doi.org/10.1016/j.aca.2004.09.072.

Dhull, V., A. Gahlaut, N. Dilbaghi, and V. Hooda. 2013. "Acetylcholinesterase Biosensors for Electrochemical Detection of Organophosphorus Compounds: A Review." *Biochemistry Research International*, 2013: 731501. doi: 10.1155/2013/731501.

Do, M. H., A. Florea, C. Farre, A. Bonhomme, F. Bessueille, F. Vocanson, N.-T. Tran-Thi, and N. Jaffrezic-Renault. 2015. "Molecularly imprinted polymer-based electrochemical sensor for the sensitive detection of glyphosate herbicide." *International Journal of Environmental Analytical Chemistry*, 95 (15): 1489-1501. doi: 10.1080/03067319.2015.1114109.

Efremenko, E. N., and V. S. Sergeeva. 2001. "Organophosphate hydrolase — an enzyme catalyzing degradation of phosphorus-containing toxins and pesticides." *Russian Chemical Bulletin*, 50 (10): 1826-1832. doi: 10.1023/a:1014377912147.

Evenset, A., I. G. Hallanger, M. Tessmann, N. Warner, A. Ruus, K. Borgå, G. W. Gabrielsen, G. Christensen, and P. E. Renaud. 2016. "Seasonal variation in accumulation of persistent organic pollutants in an Arctic marine benthic food web." *Science of The Total Environment*, 542: 108-120. doi: https://doi.org/10.1016/j.scitotenv.2015.10.092.

Farina, Y., M. P. Abdullah, N. Bibi, and W. M. A. W. M. Khalik. 2017. "Determination of pesticide residues in leafy vegetables at parts per billion levels by a chemometric study using GC-ECD in Cameron Highlands, Malaysia." *Food Chemistry*, 224: 55-61. doi: https://doi.org/10.1016/j.foodchem.2016.11.113.

Gao, N., R. Tan, Z. Cai, H. Zhao, G. Chang, and Y. He. 2021. "A novel electrochemical sensor via Zr-based metal organic framework–graphene for pesticide detection." *Journal of Materials Science*. doi: 10.1007/s10853-021-06436-6.

Gu, C., Z. Yu, J. Liu, and D. S. Sholl. 2021. "Construction of an Anion-Pillared MOF Database and the Screening of MOFs Suitable for Xe/Kr Separation." *ACS Appl Mater Interfaces*, 13 (9): 11039-11049.

Harshit, D., K. Charmy, and P. Nrupesh. 2017. "Organophosphorus pesticides determination by novel HPLC and spectrophotometric method." *Food Chemistry*, 230: 448-453. doi: https://doi.org/10.1016/ j.foodchem.2017.03.083.

Hu, L., N. Wu, J. Zheng, J. Xu, M. Zhang, and P. He. 2014. "Preparation of a magnetic metal organic framework composite and its application for the detection of methyl parathion." *Anal Sci*, 30 (6): 663-8.

Ivanov, A. N., G. A. Evtugyn, R. E. Gyurcsányi, K. Tóth, and H. C. Budnikov. 2000. "Comparative investigation of electrochemical cholinesterase biosensors for pesticide determination." *Analytica Chimica Acta*, 404 (1): 55-65. doi: https://doi.org/10.1016/S0003-2670(99)00683-2.

Karadurmus, L., S. I. Kaya, and S. A. Ozkan. 2021. "Recent advances of enzyme biosensors for pesticide detection in foods." *Journal of Food Measurement and Characterization*, 15 (5): 4582-4595. doi: 10.1007/s11694-021-01032-3.

Kim, K. H., E. Kabir, and S. A. Jahan. 2017. "Exposure to pesticides and the associated human health effects." *The Science of the total environment*, 575: 525-535.

Kumar, P., K.-H. Kim, V. Bansal, A. K. Paul, and A. Deep. 2016. "Practical utilization of nanocrystal metal organic framework biosensor for parathion specific recognition."

Microchemical Journal, 128: 102-107. doi: https://doi.org/10.1016/j.microc.2016.04.008.

Kumar, P., A. K. Paul, and A. Deep. 2014a. "A luminescent nanocrystal metal organic framework for chemosensing of nitro group containing organophosphate pesticides." *Analytical Methods*, 6 (12): 4095-4101. doi: 10.1039/c3ay42189a.

Kumar, P., A. K. Paul, and A. Deep. 2014b. "Sensitive chemosensing of nitro group containing organophosphate pesticides with MOF-5." *Microporous and Mesoporous Materials*, 195: 60-66. doi: https://doi.org/10.1016/j.micromeso.2014.04.017.

Lawson, H. D., S. P. Walton, and C. Chan. 2021. "Metal–Organic Frameworks for Drug Delivery: A Design Perspective." *ACS Applied Materials & Interfaces*, 13 (6): 7004-7020. doi: 10.1021/acsami.1c01089.

Lazarević-Pašti, T. D., A. M. Bondžić, I. A. Pašti, S. V. Mentus, and V. M. Vasić. 2013. "Electrochemical oxidation of diazinon in aqueous solutions via electrogenerated halogens – Diazinon fate and implications for its detection." *Journal of Electroanalytical Chemistry*, 692: 40-45. doi: https://doi.org/10.1016/j.jelechem.2013.01.005.

Lazarević-Pašti, T. D., A. M. Bondžić, I. A. Pašti, and V. M. Vasić. 2012. "Indirect electrochemical oxidation of organophosphorous pesticides for efficient detection via acetylcholinesterase test." *Pesticide Biochemistry and Physiology*, 104 (3): 236-242. doi: https://doi.org/10.1016/j.pestbp.2012.09.004.

Li, C., C. Wang, Y. Ma, and S. Hu. 2004. "Voltammetric Determination of Trace Amounts of Fenitrothion on a Novel Nano-TiO2 Polymer Film Electrode." *Microchimica Acta*, 148 (1): 27-33. doi: 10.1007/s00604-004-0257-8.

Liu, C., X. Bo, and L. Guo. 2019. "A novel electrochemical sensing platform of JUC-62 metal-organic framework/platelet ordered mesoporous carbon for high selective detection of nitro-aromatic compounds." *Sensors and Actuators B: Chemical*, 297: 126741. doi: https://doi.org/10.1016/j.snb.2019.126741.

Liu, X. 2011. "Electrochemical Sensor for Determination of Parathion Based on Electropolymerization Poly(Safranine) Film Electrode." *International Journal of Electrochemistry*, 2011: 986494. doi: 10.4061/2011/986494.

Lyagin, I., and E. Efremenko. 2021. "Enzymes, Reacting with Organophosphorus Compounds as Detoxifiers: Diversity and Functions." *Int J Mol Sci*, 22 (4).

Lykogianni, M., E. Bempelou, F. Karamaouna, and K. A. Aliferis. 2021. "Do pesticides promote or hinder sustainability in agriculture? The challenge of sustainable use of pesticides in modern agriculture." *Science of The Total Environment*, 795: 148625. doi: https://doi.org/10.1016/ j.scitotenv.2021.148625.

Ma, B., L.-Z. Cheong, X. Weng, C.-P. Tan, and C. Shen. 2018. "Lipase@ZIF-8 nanoparticles-based biosensor for direct and sensitive detection of methyl parathion." *Electrochimica Acta*, 283: 509-516. doi: https://doi.org/10.1016/j.electacta.2018.06.176.

Mahmoudi, E., H. Fakhri, A. Hajian, A. Afkhami, and H. Bagheri. 2019. "High-performance electrochemical enzyme sensor for organophosphate pesticide detection using modified metal-organic framework sensing platforms." *Bioelectrochemistry*, 130: 107348. doi: https://doi.org/10.1016/j.bioelechem.2019.107348.

Markopoulou, P., N. Panagiotou, A. Li, R. Bueno-Perez, D. Madden, S. Buchanan, D. Fairen-Jimenez, P. G. Shiels, and R. S. Forgan. 2020. "Identifying Differing Intracellular Cargo Release Mechanisms by Monitoring *In Vitro* Drug Delivery from MOFs in Real Time." *Cell Reports Physical Science*, 1 (11): 100254. doi: https://doi.org/10.1016/j.xcrp.2020.100254.

Mehta, J., S. Dhaka, A. K. Paul, S. Dayananda, and A. Deep. 2019. "Organophosphate hydrolase conjugated UiO-66-NH2 MOF based highly sensitive optical detection of methyl parathion." *Environmental Research*, 174: 46-53. doi: https://doi.org/10.1016/j.envres.2019.04.018.

Mukherjee, S., Y. He, D. Franz, S.-Q. Wang, W.-R. Xian, A. A. Bezrukov, B. Space, Z. Xu, J. He, and M. J. Zaworotko. 2020. "Halogen–C2H2 Binding in Ultramicroporous Metal–Organic Frameworks (MOFs) for Benchmark C2H2/CO2 Separation Selectivity." *Chemistry – A European Journal*, 26 (22): 4923-4929. doi: https://doi.org/10.1002/chem.202000008.

Mulyasuryani, A., and S. Prasetyawan. 2015. "Organophosphate Hydrolase in Conductometric Biosensor for the Detection of Organophosphate Pesticides." *Analytical Chemistry Insights*, 10: ACI.S30656. doi: 10.4137/aci.s30656.

Nagabooshanam, S., S. Roy, A. Mathur, I. Mukherjee, S. Krishnamurthy, and L. M. Bharadwaj. 2019. "Electrochemical micro analytical device interfaced with portable potentiostat for rapid detection of chlorpyrifos using acetylcholinesterase conjugated metal organic framework using Internet of things." *Scientific Reports*, 9 (1): 19862. doi: 10.1038/s41598-019-56510-y.

Nagabooshanam, S., S. Sharma, S. Roy, A. Mathur, S. Krishnamurthy, and L. M. Bharadwaj. 2021. "Development of Field Deployable Sensor for Detection of Pesticide From Food Chain." *IEEE Sensors Journal*, 21 (4): 4129-4134. doi: 10.1109/jsen.2020.3030034.

Pedrosa, V. A., S. Paliwal, S. Balasubramanian, D. Nepal, V. Davis, J. Wild, E. Ramanculov, and A. Simonian. 2010. "Enhanced stability of enzyme organophosphate hydrolase interfaced on the carbon nanotubes." *Colloids and Surfaces B: Biointerfaces*, 77 (1): 69-74. doi: https://doi.org/10.1016/j.colsurfb.2010.01.009.

Pundir, C. S., and N. Chauhan. 2012. "Acetylcholinesterase inhibition-based biosensors for pesticide determination: A review." *Analytical Biochemistry*, 429 (1): 19-31. doi: https://doi.org/10.1016/j.ab.2012.06.025.

Sacchetti, A., A. Rossetti, and J. Martí-Rujas. 2021. "Chapter Three - Metal-organic frameworks (MOFs) for sensing." In *Advances in Chemical Engineering*, edited by Luca Magagnin and Filippo Rossi, 91-122. Academic Press.

Sakdarat, P., J. Chongsuebsirikul, C. Thanachayanont, S. Prichanont, and P. Pungetmongkol. 2021. "Development of a Nonenzymatic Electrochemical Sensor for Organophosphate Pesticide Detection Using Copper (II) Oxide Nanorod Electrodes." *Journal of Nanomaterials*, 2021: 6623668. doi: 10.1155/2021/6623668.

Shanahan, J., D. S. Kissel, and E. Sullivan. 2020. "PANI@UiO-66 and PANI@UiO-66-NH2 Polymer-MOF Hybrid Composites as Tunable Semiconducting Materials." *ACS Omega*, 5 (12): 6395-6404. doi: 10.1021/acsomega.9b03834.

Shi, M., R. Huang, W. Qi, R. Su, and Z. He. 2020. "Synthesis of superhydrophobic and high stable Zr-MOFs for oil-water separation." *Colloids and Surfaces A:*

Physicochemical and Engineering Aspects, 602: 125102. doi: 10.1016/j.colsurfa.2020.125102.

Sun, Y., L. Zheng, Y. Yang, X. Qian, T. Fu, X. Li, Z. Yang, H. Yan, C. Cui, and W. Tan. 2020. "Metal–Organic Framework Nanocarriers for Drug Delivery in Biomedical Applications." *Nano-Micro Letters*, 12 (1): 103. doi: 10.1007/s40820-020-00423-3.

Suresh, K., D. Aulakh, J. Purewal, D. J. Siegel, M. Veenstra, and A. J. Matzger. 2021. "Optimizing Hydrogen Storage in MOFs through Engineering of Crystal Morphology and Control of Crystal Size." *Journal of the American Chemical Society*, 143 (28): 10727-10734. doi: 10.1021/jacs.1c04926.

Wang, G., Q. Sun, Y. Liu, B. Huang, Y. Dai, X. Zhang, and X. Qin. 2015. "A Bismuth-Based Metal–Organic Framework as an Efficient Visible-Light-Driven Photocatalyst." *Chemistry – A European Journal*, 21 (6): 2364-2367. doi: https://doi.org/10.1002/chem.201405047.

Wang, J., R. Krause, K. Block, M. Musameh, A. Mulchandani, and M. J. Schöning. 2003. "Flow injection amperometric detection of OP nerve agents based on an organophosphorus–hydrolase biosensor detector." *Biosensors and Bioelectronics*, 18 (2): 255-260. doi: https://doi.org/10.1016/S0956-5663(02)00178-1.

Wang, Z., B. Ma, C. Shen, and L. Z. Cheong. 2019. "Direct, selective and ultrasensitive electrochemical biosensing of methyl parathion in vegetables using Burkholderia cepacia lipase@MOF nanofibers-based biosensor." *Talanta*, 197: 356-362.

Wen, L.-L., F. Wang, X.-K. Leng, C.-G. Wang, L.-Y. Wang, J.-M. Gong, and D.-F. Li. 2010. "Efficient Detection of Organophosphate Pesticide Based on a Metal−Organic Framework Derived from Biphenyltetracarboxylic Acid." *Crystal Growth & Design*, 10 (7): 2835-2838. doi: 10.1021/cg1004352.

Wu, Y.-P., X.-Q. Wu, J.-F. Wang, J. Zhao, W.-W. Dong, D.-S. Li, and Q.-C. Zhang. 2016. "Assembly of Two Novel Cd3/(Cd3 + Cd5)-Cluster-Based Metal–Organic Frameworks: Structures, Luminescence, and Photocatalytic Degradation of Organic Dyes." *Crystal Growth & Design*, 16 (4): 2309-2316. doi: 10.1021/acs.cgd.6b00093.

Xia, Z., J. Fang, X. Zhang, L. Fan, A. J. Barlow, T. Lin, S. Wang, G. G. Wallace, G. Sun, and X. Wang. 2019. "Pt nanoparticles embedded metal-organic framework nanosheets: A synergistic strategy towards bifunctional oxygen electrocatalysis." *Applied Catalysis B: Environmental*, 245: 389-398. doi: https://doi.org/10.1016/j.apcatb.2018.12.073.

Xiong, J., A. Li, Y. Liu, L. Wang, X. Qin, and J. Yu. 2021. "Scalable and hierarchically designed MOF fabrics by netting MOFs into nanofiber networks for high-performance solar-driven water purification." *Journal of Materials Chemistry A*, 9 (37): 21005-21012. doi: 10.1039/d1ta05589e.

Xiong, J., L. Wang, X. Qin, and J. Yu. 2021. "Acid-promoted synthesis of defected UiO-66-NH2 for rapid detoxification of chemical warfare agent simulant." *Materials Letters*, 302: 130427. doi: https://doi.org/10.1016/ j.matlet.2021.130427.

Xu, N., Q. Zhang, and G. Zhang. 2019. "A carbazole-functionalized metal–organic framework for efficient detection of antibiotics, pesticides and nitroaromatic compounds." *Dalton Transactions*, 48 (8): 2683-2691. doi: 10.1039/c8dt04558e.

Xu, X., Y. Guo, X. Wang, W. Li, P. Qi, Z. Wang, X. Wang, S. Gunasekaran, and Q. Wang. 2018. "Sensitive detection of pesticides by a highly luminescent metal-organic framework." *Sensors and Actuators B: Chemical*, 260: 339-345. doi: https://doi.org/10.1016/j.snb.2018.01.075.

Xu, Y., H. Wang, X. Li, X. Zeng, Z. Du, J. Cao, and W. Jiang. 2021. "Metal–organic framework for the extraction and detection of pesticides from food commodities." *Comprehensive Reviews in Food Science and Food Safety*, 20 (1): 1009-1035. doi: https://doi.org/10.1111/1541-4337.12675.

Yan, B. 2017. "Lanthanide-Functionalized Metal–Organic Framework Hybrid Systems To Create Multiple Luminescent Centers for Chemical Sensing." *Accounts of Chemical Research*, 50 (11): 2789-2798. doi: 10.1021/acs.accounts.7b00387.

Yang, L., Y.-L. Liu, C.-G. Liu, F. Ye, and Y. Fu. 2020. "A luminescent sensor based on a new Cd-MOF for nitro explosives and organophosphorus pesticides detection." *Inorganic Chemistry Communications*, 122: 108272. doi: https://doi.org/10.1016/j.inoche.2020.108272.

Yu, J., Y. Cui, C. Wu, Y. Yang, Z. Wang, M. O'Keeffe, B. Chen, and G. Qian. 2012. "Second-Order Nonlinear Optical Activity Induced by Ordered Dipolar Chromophores Confined in the Pores of an Anionic Metal–Organic Framework." *Angewandte Chemie International Edition*, 51 (42): 10542-10545. doi: https://doi.org/10.1002/anie.201204160.

Zhang, Q., Q. Sun, B. Hu, Q. Shen, G. Yang, X. Liang, X. Sun, and F. Liu. 2008. "Development of a sensitive ELISA for the analysis of the organophosphorous insecticide fenthion in fruit samples." *Food Chemistry*, 106 (3): 1278-1284. doi: https://doi.org/10.1016/j.foodchem.2007.07.049.

Zhao, C., C. Shen, and W. Han. 2015. "Metal–organic nanofibers as anodes for lithium-ion batteries." *RSC Advances*, 5 (26): 20386-20389. doi: 10.1039/c4ra16416d.

Zhou, C., S. Zhang, H. Pan, G. Yang, L. Wang, C.-a. Tao, and H. Li. 2021. "Synthesis of macroscopic monolithic metal–organic gels for ultra-fast destruction of chemical warfare agents." *RSC Advances*, 11 (36): 22125-22130. doi: 10.1039/d1ra01703a.

Zhou, J., X. Li, L. Yang, S. Yan, M. Wang, D. Cheng, Q. Chen, Y. Dong, P. Liu, W. Cai, and C. Zhang. 2015. "The Cu-MOF-199/single-walled carbon nanotubes modified electrode for simultaneous determination of hydroquinone and catechol with extended linear ranges and lower detection limits." *Analytica Chimica Acta*, 899: 57-65. doi: https://doi.org/10.1016/ j.aca.2015.09.054.

Chapter 11

Spiders and Organophosphate Pesticides

Boris D. Dudić*
Faculty of Biology, University of Belgrade, Belgrade, Serbia

Abstract

Organophosphate insecticides are a group of chemicals widely used in agriculture, horticulture, and veterinary medicine. They act in the same way by blocking acetylcholinesterases and rapidly undergo hydrolytic degradation in sunlight, which makes them a plausible alternative to other groups of pesticides, especially organochlorine and carbamate ones. Today, organophosphates are extensively used throughout the world due to their efficiency. However, they can also affect beneficial soil invertebrates that have an important role to play in the food web, decomposition of organic substances, and maintenance of soil structures. Among such soil invertebrates are spiders. Comprising one of the most abundant and diverse groups of natural enemies, spiders inhabit various types of agroecosystems and take part to a significant extent in pest control. Unfortunately, extensive use of pesticides can disturb the role of spiders, so the purpose of this chapter would be to provide an overview of data published about the effect exerted on these soil invertebrates by organophosphates, which—like their overall ecotoxicology—is a subject that has received rather limited attention to date.

Keywords: organophosphates, pesticides, spiders, agroecosystems, toxicity

* Corresponding Author's Email: boris.dudic@bio.bg.ac.rs.

In: Organophosphates
Editor: Tamara Lazarević-Pašti
ISBN: 978-1-68507-652-8

Introduction

Pesticides represent a wide range of chemical substances that are utilized to eliminate, counteract or control various populations of pests. Depending on the types of pests they target, pesticides are categorized as (Aktar et al., 2009):

- rodenticides (against rodents),
- insecticides (against insects),
- fungicides (against fungi),
- molluscicides (against mollusks),
- nematocides (against nematodes), and
- herbicides (against weeds).

A separate group of chemicals is marked as plant growth regulators, affecting developmental or metabolic processes in higher plants, mostly at low dosages (Rademacher, 2015). Pesticides are widely applied throughout the world, protecting a market worth billions of dollars, playing an essential role in suppressing pest activities, which would otherwise significantly reduce productivity, and play an important part in fulfilling the demands of a growing world population (Wilson & Tisdell, 2001). Pesticides have been extensively used to advance agricultural efficiency by controlling the number of pests and pathogens. Based on their chemical composition, pesticides can be classified into four main categories: organochlorines, carbamates, pyrethroids, and organophosphates (Dar et al., 2020).

Their highly degradable nature and strong effectiveness against pests have made organophosphorus pesticides (OPs) an important substitute to more persistent agrochemicals (such as organochlorines) and rendered them a group of compounds widely used for pest control. Today, OPs represent around 38% of the total quantity of pesticides consumed globally, and they are extensively applied in agriculture, veterinary medicine, and horticulture, as well as for domestic purposes (Dar et al., 2020; Singh et al., 2006). However, extensive and indiscriminate use of pesticides in agriculture has often led to harmful effects due to their high biological activity and toxicity, especially in the case of OPs, which affect the parasympathetic, sympathetic, and central nervous systems, particularly on the synapses between neurons, by reducing the activity of serine esterases (Ecobichon & Joy, 1993; Paudyal, 2008).

In that regard, pesticides and their residues (including OPs) can have a significant impact on the biodiversity of primary and secondary decomposers (collembolans, mites, nematodes, mollusks, insects, isopods) and predators (spiders, beetles) in the agroecosystems (Dinter & Poehling, 1992; Desneux et al., 2007; Giglio et al., 2017). For example, spiders represent the most abundant and significant group of invertebrate predators in terrestrial agroecosystems, and they have a major beneficial role in all temperate agricultural habitats as regulators of insect pests and as bioindicators (Ossamy et al., 2016; Perkins et al., 2018; Ikonov et al., 2019). However, their ability to reduce the number of a variety of pests can be disturbed if agrochemicals with an adverse effect are intensively applied (Pekár, 2012).

Spiders in Agroecosystems

Spiders are a group of predators widely distributed in most terrestrial habitats, and several hundred species inhabit different kinds of agroecosystems around the world (Ikonov et al., 2019). Consequently, they can dominate cultivated environments where other groups of predatory arthropods (like ants) are usually rare or absent. Around 40 000 spider species are currently known, the majority of them feeding on insects. Still, they exhibit great behavioral diversity concerning their hunting strategies, dispersal modes, and ability to overcome adverse environmental conditions, which explains their high species diversity in both cultivated and natural areas (Peck et al., 1998; Marc et al., 1999).

The group of small, ground-dwelling, web-building spiders from the family Linyphiidae prevails among the arachnofauna in Western European crop fields, with several species such as *Oedothorax apicatus* (Blackwall, 1850), *Erigone atra* Blackwall, 1833, *Erigone dentipalpis* (Wider, 1834), and *Tenuiphantes tenuis* (Blackwall, 1852) usually comprising 93-99% of all spider individuals (Nyffeler & Sunderland, 2003) (Figure 1).

Linyphiid spiders catch prey by making webs over small depressions in the soil and are usually evenly distributed on arable land. At the same time, a tendency towards aggregation is evident in parts of the agroecosystem where the prey is present in large numbers (Nyffeler & Sunderland, 2003). The taxonomic composition somewhat changes in Central Europe, where lycosids and other medium-sized hunting foragers have a greater share (around 76%), with the species *Pardosa agrestis* (Westring, 1861) increasing in numbers along a northwest-to-southeast gradient (Samu et al., 2011; Dudić et al., 2013; Dudić, 2015) (Figure 2).

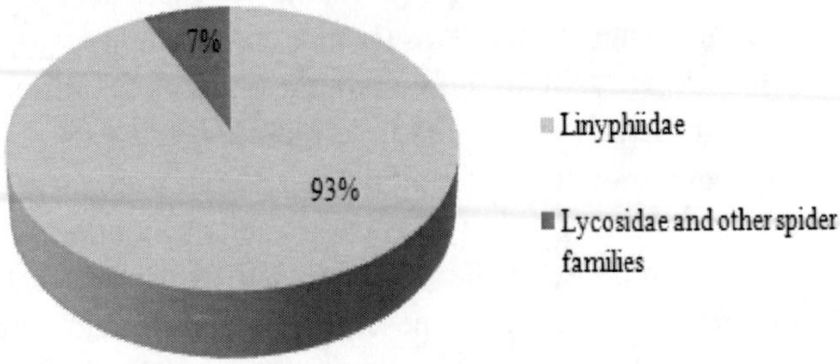

Figure 1. General taxonomic structure of the spider fauna on crops in Western Europe.

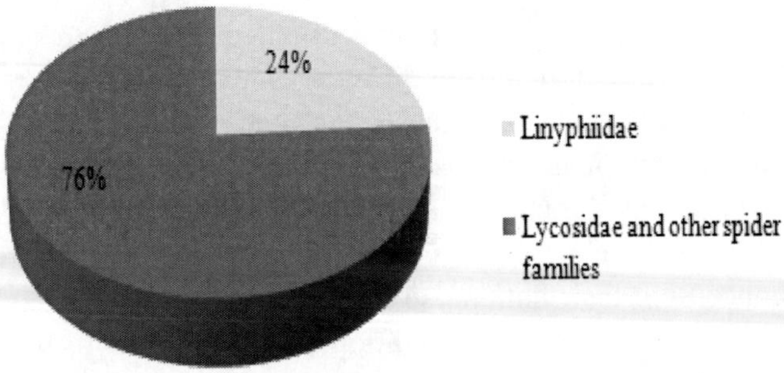

Figure 2. General taxonomic structure of the spider fauna on crops in Central Europe.

Compared to lyniphiids, whose length does not exceed 3 mm, lycosids are medium-sized (4-6 mm) spiders that actively hunt on the ground. Lycosid populations are predominantly present in peripheral zones of the agroecosystems (Samu et al., 2011).

In the North American agroecosystems, active hunters from the family Oxyopidae dominate the spider agrobiont fauna. At the same time, the share of linyphiids is significantly lower (below 25% of the total number of recorded individuals) (Nyffeler & Sunderland, 2003; Uetz et al., 1999).

The composition of the spider fauna in orchards is different compared to the araneofauna of field crops and with clear stratification. The upper branches of trees are inhabited by spiders from the families Oxyoipidae, Thomisidae, Araneidae, Theridiidae, Clubionidae, and others, while the ground layer is dominated by species of the families Lycosidae and Ghanphosidae (Mansour & Whitecomb, 1986; Marc et al., 1999; Dudić, 2015).

The type of agricultural land, its microstructure, and proximity to refugial habitats can determine the mode of colonization by agrobiont spiders. In agroecosystems exposed to frequent disturbance, spiders will often inhabit the air due to the aeronautical behavior of many juvenile specimens from the families Linyphiidae, Lycosidae, and Araneidae (Marc et al., 1999). Cereal crops are subjected to intensive cultivation, exposed to frequent use of pesticides, and with rare, mostly remote, uncultivated refuges in the neighborhood, so they are not a suitable environment for the survival of species with low mobility. Research has shown that only Linyphiidae, which adapt well to the space-time pattern of agroecosystem disturbances and have a high dispersion potential, can successfully inhabit such terrains (Dudić, 2015). In orchards, where conditions are more stable, the composition of the spider fauna is different (Mansour & Whitecomb, 1986).

Spiders exhibit a significant degree of diversity in different types of agroecosystems, adapting their survival strategy to the drastic changes in conditions that occur at regular intervals in these habitat types. They achieve that goal by synchronizing their life cycle with periodic changes in the state of arable surfaces (primarily by adjusting the time of their reproduction, wintering, and colonization) (Samu & Szinetár, 2002). Numerous experiments have pointed to the fact that spiders have a great influence on the size of insect populations in the agroecosystems, with the effect on harmful species being of particular importance. Various studies have shown that spiders on arable land feed mostly on aphids (*Metopolophium dirhodum* (Walker, 1849), *Rhopalosiphum padi* (Linnaeus, 1758), *Sitobion avenae* (Fabricius, 1775)) and Diptera (*Oscinella frit* (Linnaeus, 1758), *Mayetiola destructor* (Say, 1817), *Contarinia tritici* (Kirby, 1798), *Sitodiplosis mosellana* (Gehin, 1857)), many of which are known as pests. Several authors have noted that spiders will consume Collembola rather than aphids, if their numbers in the field are high (Marc et al., 1999; Harwood et al., 2002). A similar effect was observed in the case of Lepidoptera populations in orchards, where spiders feed on their larvae (Mansour & Whitecomb, 1986). Studies have shown that in an agroecosystem, spiders can act regulatorily through the pressure of one or two species towards a specific type of prey or through the action of the whole group on the biocenosis of insects, including pests. However, it should be noted that not all spider species are effective against all types of prey. There are also opposite opinions about the importance of spiders in pest biocontrol. Nyffeler and Benz (1979b) concluded that the population density of spiders in wheat and rapeseed fields is too low (a total of 0.1—1 individuals / m^2) to have any great impact on insect numbers (Nyffeler & Benz, 1978). Riechert and Lockley

(1984) considered that their solitary behavior and the nature of their life cycle make spiders unsuitable for pest biocontrol. Competition with other predators as well as among spiders themselves can significantly reduce their effect on insects (Riechert & Lockley, 1984).

Bioindicators are one of the main parameters used for determining the ecological value of a biotope and assessing the impact of human factors on ecosystem functioning and faunal diversity. Spiders have a high abundance in terrestrial habitats, and their potential significance as bioindicators has been confirmed in many studies (Marc et al., 1999). The main advantages of using spiders as a bioindicator can be summarized as follows (Dudić, 2015):

- they are found in large numbers from the ground levels in the habitat, all the way up to the highest layers of vegetation,
- they possess specific ecological requirements in relation to the biotope where they occur,
- changes in the composition of the arneofauna can be observed in a small area within a particular biotope,
- spiders occupy a special place within the food chain because they can be predators, but at the same time, also serve as prey.

In order to evaluate the level of human disturbance or the quality of natural habitat, spider community composition can be a good indicator because it quickly and strongly reacts to sudden and acute changes in environmental conditions such as prey structure and the presence of soil pollutants or pesticides (Marc et al., 1999).

Organophosphate Pesticides

Organophosphates are composed of phosphorus, oxygen, and carbon (P-O-C) bonds and usually represent esters and thiol or amide derivatives of phosphoric acid where the terminal oxygen is connected to phosphorus by a double bond (Yang et al., 2005). The general chemical structure of organophosphorus compounds is illustrated in Figure 3.

X represents any leaving group, commonly aliphatic, homocyclic, or heterocyclic, bonded through -O- or -S- to phosphorus, while R_1 and R_2 are usually methyl or ethyl groups singly bonded to phosphorus, which makes the formation of different compounds possible through various atoms (Vale & Lotti, 2015). Some types of OPs are presented in Figure 4.

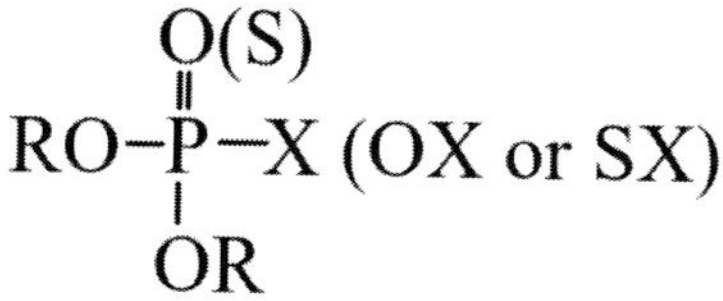

Figure 3. The general structure of organophosphates.

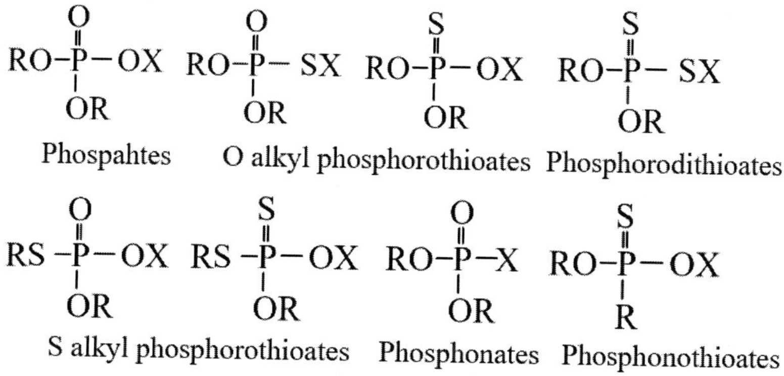

Figure 4. Some types of organophosphorus pesticides.

Synthetic and biogenic OPs with C-P linkage are thermally and chemically inert and more resistant to thermal hydrolysis, photolytic degradation, and chemical decomposition than are similar compounds which contain more reactive S-P, O-P, or N-P linkages (Greaves & Letcher, 2017; O'Brien, 2016). They are the main ingredient of herbicides, pesticides, and insecticides and are widely used for pest control, horticulture, agriculture, industrial products, vector control, plastic making, and domestic purposes (Singh & Prasad, 2018; Yadav et al., 2017; Sidhu et al., 2019). Organophosphate compounds are often used as chemical warfare agents and represent the main component of nerve gas (Adeyinka & Pierre, 2018). They are water-soluble and can rapidly degrade in natural conditions by abrasion, dissolution, or volatilization (Wang et al., 2014; Balderacchi et al., 2013). For that reason, OPs represent a plausible alternative to other types of pesticides, especially organochlorine ones, and today they are widely used worldwide. Between 2011 and 2015, the use of OPs grew to an amount of almost 7 million kilograms in Asia and the Pacific region, North America and Europe (Wang et al., 2015).

Due to their ability to inhibit the activity of acetylcholinesterase, OPs can produce a significant level of neurotoxicity among terrestrial and aquatic animals (Muhammad, 2017). In humans, they can increase the risk of different endocrine, neurological, genotoxic, metabolic, teratogenic (bladder cancer and leukemia), and psychiatric disorders (Sidhu et al., 2019; Kumar et al., 2016). In plants, OPs have adverse effects on photosynthesis, oxidative stress resistance, and various metabolic and biosynthetic processes (Serra et al., 2013; Zobiole et al., 2016; Vagi et al., 2018). Besides neuroteratogenic and genotoxic effects on plants, humans, and animals, organophosphate pesticides also have a wider negative ecological and environmental impact (Sun et al., 2016).

Sublethal Effects of Organophosphate Pesticides on the Spider Fauna

Different types of OPs can produce various effects, usually on life-history traits, especially on the behavioral and psychological levels (sublethal effects). Neurotoxic influence disrupts activities that are controlled by complex neural interactions (prey capture, movement, reproductive biology, dispersal, defense), while behavioral changes are usually manifested on the psychological level (Pekár, 2012). Spiders can survive the exposure, but their functional response and numbers would be significantly reduced, which can result in decreased pest control ability. For spiders in fields, the possibility of encountering sublethal doses of pesticides is assumed to be higher than the likelihood of encountering a lethal one. Dilution of pesticides in water, the filter effect of plants, and spray drift can lead to sparse contamination in crop fields (Pekár, 2013). In some cases, spiders can encounter sublethal doses several days after pesticide application. Spider webs can also reduce contact between pesticide droplets and individuals (Pekár, 1999). Recovery from sublethal effects usually happens after few days (Pekár, 2012).

So far, research on the sublethal effects of OPs has been conducted on 22 spider species (Table 1).

Table 1. List of studies treating sublethal effects of organophosphorus pesticides on spider species

Family/species	Insecticide/acaricide	Herbicide	Reference
Agelenidae			
Agelena labyrinthica	Dimethoate		(Wilczek, 2005)
Araneidae			
Alpaida venillae	-	Glyphosate	(Benamú et al., 2010)
Larionioides scolpetarius	Diazinon	-	(Langwiler & Benz, 1994)
Linyphiidae			
Frontinella pyramitela	Malathion	-	(Tietjen & Cady, 2007)
Hylyphantes graminicola	Dimethoate, Methamidophos	-	(Deng et al., 2006,2003; Peng et al., 2010)
Lycosidae			
Anoteropsis hilaris	Diazinon	-	(Van Erp et al., 2002)
Pardosa amentata	Dimethoate	-	(Toft & Jansen, 1998)
Pardosa milvina	-	Glyphosate	(Evans et al., 2010,Wrinn et al., 2012)
Pardosa palustris	-	Glyphosate	(Michalková, & Pekár, 2009)
Pardosa agricola		Glyphosate	(Michalková, & Pekár, 2009)
Pardosa pseudoannulata	Methamidophos Paraoxon Omethoate	-	(Zhi et al., 2006; Wang et al., 2019, Meng et al., 2015; Zhang et al., 2014)
Pardosa prativaga	Dimethoate	-	(Pedersen et al., 2002)
Rabidosa rabida	Malathion	-	(Tietjen, 2006; Tietjen & Cady, 2007)
Schizocosa ocreata	Malathion	-	(Tietjen & Cady, 2007)
Xerolycosa nemoralis	Dimethoate	-	(Stalmach et al., 2015; Wilczek et al., 2016; Wilczek et al., 2013)
Pirata subpiraticus	Methamidophos	-	(Li eta al., 2015)
Pardosa agrestis	-	Glyphosate	(Leccia ate al., 2016; Korenko et al., 2016)
Pardosa lugubris	Dimethoate	Glyphosate	(Wilczek, 2005; Niedobová et al., 2019)
Hogna bivittata	-	Glyphosate	(Lacava et al., 2021)
Cheiracanthiidae			
Cheiracanthium mildei	Malathion	-	(Mansour, 1984)
Salticidae			
Salticus scenicus	Malathion	-	(Tietjen & Cady, 2007)
Sparassidae			
Polybetes pythagoricus	Fenitrothion	-	(Tietjen & Cady, 2007)
Oxyopidae			
Oxyopes salticus	Malathion	-	(Brown et al., 2015)

Effects on Enzymatic Activity

Organophosphorus pesticides have a significant effect on physiological processes, especially enzymatic activity. Studies conducted on the linyphiid spider *Hylyphantes graminicola* (Sundevall, 1830) (Peng et al., 2010) and lycosids *Anoteropsis hilaris* (L. Koch, 1877) (Van Erp et al., 2002) and *Pardosa pseudoannulata* (Bösenberg & Strand, 1906) (Zhang et al., 2014; Meng et al., 2015; Wang et al., 2019) showed that OP application led to reduced activity of acetylcholinesterase (AChE), an enzyme responsible for terminating transmission at cholinergic synapses by hydrolyzing acetylcholine. Cholinesterases are key enzymes in the nervous system, and their inhibition or reduced activity can cause physiological perturbations, leading to hyperactivity and eventually result in the death of the individual. Due to a similar chemical structure, OP's bind readily with AChE, which inhibits its activity and prevents acetylcholine hydrolysis. Neurotransmitter accumulation occurs in the nervous system, causing overstimulation of nerves and stopping of neurotransmission to organs (Paudyal, 2008). The reaction of OPs with AChE is schematically presented in Figure 5. In studies conducted on *Pardosa pseudoannulata*, high sensitivity of the enzyme AChE1 to omethoate exposure was observed, indicating its role as a major synaptic enzyme in that spider species (Zhang et al., 2014).

Degradation of the AChE-OP complex and reactivation of the enzyme depends primarily on the type of OP compound. The regenerative process can be a spontaneous, slow reactivation or so-called "aging" when the AChE-OP complex with the passage of time becomes unresponsive towards reactivating agents. A second reaction usually depends on other factors, too, like temperature and pH (Dar et al., 2020). The regenerative ability of AChE can also be related to the nutritional state of spiders. Starvation and nutrient imbalance reduce the rate at which AChE is re-synthesized by causing a deficiency of the nutrients necessary for AChE biosynthesis in the lycosid *Pardosa prativaga* (L. Koch, 1870) (Pedersen et al., 2002). The tolerance of *Xerolycosa nemoralis* (Westring, 1861) to stressing factors is also determined by nutritional conditions. Females have a greater amount of deposited storage substances and better cope with stressors (Wilczek et al., 2013).

In order to combat exposure to OP intoxication, the spider body produces detoxification enzymes, viz., glutathione S transferase (GST) and glutathione peroxidase, which can be present throughout the year, as in the case of *Pardosa amentata* (Clerck, 1757) (Evans et al., 2010). A low level of detoxifying enzymes in the spider body or their complete absence also results in greater exposure to toxins.

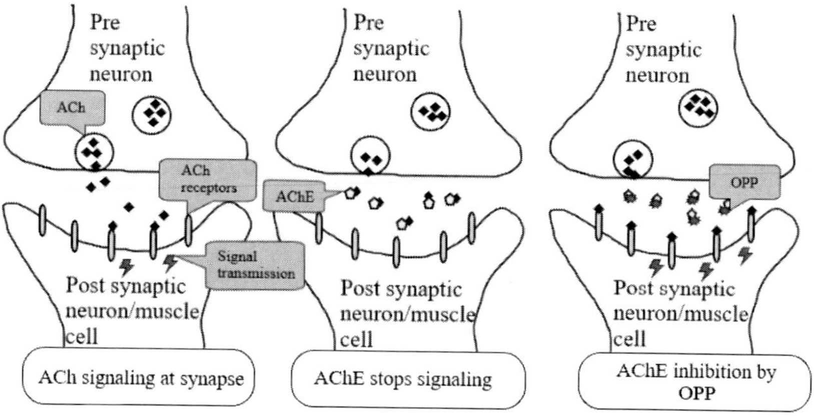

Figure 5. Scheme of OP reaction with AChE.

Research conducted on *Anoteropsis hilaris* showed that glutathione peroxidase could be a more important compound in combating exposure to toxins. Diazinon application did not induce the activity of glutathione S transferase but did induce that of glutathione peroxidase (Van Erp et al., 2002). Higher concentrations of glutathione peroxidase during exposure to dimethoate were also recorded in *Xerolycosa nemoralis.* Male individuals displayed an increased level of parameters, which could indicate that gender-related glutathione metabolism has an important role in the detoxifying process (Stalmach et al., 2015; Wilczek et al., 2016). This can be linked to behavioral and physiological differences between female and male spiders that are mainly connected with reproduction.

Males stop feeding after reaching the adult stage and concentrate mainly on mating, which depletes their energy reserves needed for defensive mechanisms, while in females, protection against toxins is crucial for keeping genetic material safe (Li et al., 2015). Increased levels of metallothionein, superoxide dismutase, and catalase were noticed in the lycosid spiders *Pardosa lugubris* (Walckenaer, 1802), *Xerolycosa nemoralis,* and the agelenid *Agelena labyrinthica* (Clerck, 1757) when exposed to dimethoate, which indicates the role played by those compounds in defense against toxins (Wilczek, 2005; Stalmach et al., 2015). Another investigation, conducted on *Pardosa pseudoannulata* specimens collected from biologically and chemically controlled rice fields, demonstrated higher activity of carboxylesterase in individuals exposed to methamidophos treatment, which indicated some level of resistance to OPs developed by spiders (Wang et al., 2006). It has also been shown that populations of the miturgid *Chiracanthium mildei* L. Koch, 1864 from a citrus grove treated frequently with malathion were more tolerant to the pesticide than specimens collected from a cotton field (Mansour, 1984).

Other enzymes apart from those present in the nervous system can also be affected by OPs. Application of a high dosage of methamidophos inhibited protease activity in the midgut of the lycosid *Pardosa pseudoannulata* (Wang et al., 2019). Methamidophos also decreased the activity of mitochondrial Ca^{2+}-Mg^{2+}-ATPase in the wolf spider *Pirata subpiraticus*, (Bösenberg & Strand, 1906), but no significant effect was observed on Na^{+}-K^{+}-ATPase (Li et al., 2015).

Genotoxic, Immunological and Physiological Effects

Compounds of organophosphorus pesticides can cause molecular damage to DNA by alkylation of nitrogenous bases (mainly thymidine and guanine), affecting the activities of antioxidative enzymes and the cell redox system and generating reactive oxidative species (Lukaszewicz-Hussain, 2010; Dogan et al., 2011; Ali et al., 2014). Double and single DNA helix breaks were produced by the deoxyribose reaction with hydroxyl radicals (Cooke et al., 2003). In the case of *Xerolycosa nemoralis*, exposure to dimethoate caused DNA damage in cells of the midgut gland, an important detoxifying and digestive organ, and in hemocytes, a key element of the humoral and cellular immune response of spiders (Foelix, 2011). Hemolymph cells proved to be more sensitive to genotoxic effects of OPs, with males affected in a higher percentage than females. A decrease in hemocytes levels can diminish the immunological response of spiders and thereby expose them to microbial invasion (Wilczek et al., 2016). Dimethoate caused apoptotic changes in midgut gland cells of *Pardosa lugubris* and *Agelena labyrinthica*. In comparison with web-building agelenids, in wolf spiders, which inhabit open and sunny areas, apoptosis of midgut gland cells was less intense, possibly due to more effective regenerative mechanisms (Wilczek, 2005).

It has been reported that OPs tend to modify cell permeability by accumulating in membranes, which leads to misfunction in physiological processes. In the sparassid *Polybetes pythagoricus* (Holmberg, 1875), fenitrothion altered lipid dynamics and decreased the capacity for oxygen binding and uptake of palmitic acid (Tietjen & Cady, 2007).

Effects on Spider Locomotion, Prey Capture, and Defense

The most profound effect of contact with sublethal OP doses are changes in the locomotor activity of spiders, a fundamental behavioral characteristic essential for retention of diverse lifestyles and for capturing prey. Cursorial species are especially sensitive because by walking, they are immediately exposed to sublethal effects (Pekar, 2013). The lycosid *Pardosa palustris* (Linnaeus, 1758) showed an uncoordinated walking pattern and moved less actively on a glyphosate-treated surface (Michalková & Pekár, 2009). The salticid *Salticus scenicus* (Clerck, 1757) and lycosid *Rabidosa rabida* (Walckenaer, 1837) became active earlier than normal on a malathion-treated surface. Their altered circadian rhythm was possibly caused by changed interaction between their circadian clock and light receptors, as was suggested by the authors (Tietjen, 2006). In some cases, contact with an OP-treated surface elevated the activity of spiders. Glyphosate- and malathion-treated surfaces increased activity in the lycosids *Pardosa milvina* (Hentz, 1844), *Schizocosa ocreata* (Hentz, 1844) and *Rabidosa rabida*, the lyniphiid *Frontinella pyramitela* (Walckenaer, 1841), and the salticid *Salticus scenicus* (Evans et al., 2010). Spiders recognized pesticide residues and actively avoided contact with them. Repulsion-induced locomotion, but whether OPs can induce dispersal is yet to be studied (Pekár, 2012). Juveniles of *Pardosa* sp. did not avoid a glyphosate-treated surface (Michalková & Pekár, 2009). Increased locomotion can also be a result of abnormal neurotransmitter activity in the protocerebral ganglion, which can cause impaired visual processing (Pekár, 2013).

Application of OPs caused reduced predation efficiency in several spider species, probably as a result of a decreased locomotor activity or possible neurotoxic and anti-nutritional effects of the applied pesticides (Pekár, 2012). The web-building araneid *Alpaida veniliae* refused to feed on prey affected by glyphosate application, while oral uptake of OPs in juveniles of the same species changed their web construction pattern in terms of a reduced number of spirals and a smaller radius (Benamú et al., 2010). Female specimens of the lyniphiid *Hylyphantes graminicola* showed drastically decreased searching efficiency and handling time after topical application of methamidophos, which led to a smaller body size among the offspring (Deng et al., 2007).

Glyphosate-based herbicides in a mixture with surfactants reduced predatory activity of the cursorial lycosid spider *Pardosa lugubris*. The authors assumed that surfactants disrupted surface tensions of the protective hairs on the spider´s body, thereby allowing pesticides to penetrate it (Niedobová et al., 2019). Negative effects of glyphosate-based herbicides on total and cumulative predatory activity were exhibited in the lycosid *Pardosa agrestis* (Westring, 1861), especially in fresh wet residues (Korenko et al., 2016). Alternatively, some studies found limited or no effect of OP application on spider prey efficiency or web architecture. On the lycosid *Hogna bivittata* (Mello-Leitão, 1939), glyphosate had a short-term, pest-specific effect on the spider´s functional response (Lacava et al., 2021), while diazinon treatment only slightly changed the web architecture of the araneid *Larinioides sclopetarius* (Clerck, 1757) (Lengwiler & Benz, 1994). Malathion did not reduce the foraging success of the oxyopid *Oxyopes salticus* Hentz, 1845, although 31% of individuals died after five days of exposure (Brown et al., 2015). Also, the lycosid *Pardosa agricola* (Thorell, 1856) did not show any change in the rate of capture and consumption of flies after exposure to glyphosate residues (Michalková & Pekár, 2009). The differences observed in OP effects can be the result of product composition, method of application, dosage, and the spider species (Niedobová et al., 2019).

The neurotoxic effect of pesticides can make spiders susceptible to predation by decreasing their movement and consequently reducing their ability to escape (Pekár, 2013). It has been found that OPs interrupt chemical communication, which represents the oldest way of communicating in spiders and is a key part of secondary defense strategies (Tristram, 2003). The response of *Pardosa milvina* (Hentz, 1844) to kairomones of the intraguild predator *Tigrosa helluo* (Walckenaer, 1837) was not drastically affected by the herbicide glyphosate, but it altered the spider´s response to cues of the carabid *Scarites quadriceps* Chaudoir, 1843 (Wrinn et al., 2012).

Effects on Reproductive Biology and Development

All stages of the reproductive biology of spiders can be affected by pesticides, from courtship behavior and mate finding to gametogenesis and brood care. Methamidophos and dimethoate applications reduced egg sac production in the linyphiid *Hylyphantes graminicola* (Deng et al., 2006), while malathion decreased the mating rate in *Rabidosa rabida*, although it had no effect on courtship behavior (Tietjen & Cady, 2007). In the araneid *Alpaida veniliae*, treatment with glyphosate lowered fecundity and fertility because the spiders refused to feed on herbicide-contaminated prey. Reduced feeding, especially prior to oviposition, can also lead to changes in the reproductive biology of spiders (Benamú et al., 2010). The glyphosate-based herbicide Roundup reduced the ability to find a mate in males of the lycosid *Pardosa agrestis* by making it hard for them to detect pheromone, airborne and dragline cues of females (Leccia et al., 2016).

Organophosphorous compounds were also found to prolong the duration of the early embryonic stages in *Hylyphantes graminicola* and *Alpaida veniliae*. Juveniles of *H. graminicola* had a smaller body mass, possibly due to reduced food intake (Peng at al., 2010). After being exposed to a glyphosate herbicide through intoxicated prey, females of *A. veniliae* produced an increased number of abnormal eggs (Benamú et al., 2010).

Lethal Effects

The lethality of pesticides depends not only on biotic and abiotic factors but also on the pesticide dosage, duration of exposure, bioavailability, and the route of uptake. Standard doses of pesticides introduced orally, topically or through residues usually cause the highest mortality rate, especially in the case of insecticides and acaricides (Vagi et al., 2018). A broad spectrum of organophosphates are known to be very toxic to spiders, and the lethal effects of 12 OPs on 25 spider species have been studied to date (Table 2).

High mortality caused by OPs can be attributed to the following things:

- Reduced cell permeability as a consequence of OP accumulation in membranes and changes in their lipid structure;
- Diminished capacity for oxygen biding; and
- Effects on neurotransmitters (Tietjen & Cady, 2007).

The level of mortality is also influenced by the biology of spider species, their way of life, sex, and the developmental stage (Pekár, 2012). Differences among species are noticeable. The tetragnathid *Tetragnatha maxillosa* Thorell, 1895 was more susceptible to diazinon than were studied species from the families Lycosidae and Linyphiidae, while the lycosid *Pardosa pesudoannulata* was more strongly affected by phenthoate than was *T. maxillosa* (Tanaka et al., 2000). The phosphorothioate pirimiphos-methyl had a very toxic effect on the theridiid *Phyloneta impressa*. L. Koch, 1881 Combinations of organophosphate and pyrethroid compounds also caused high mortality among theridiid and clubionid spiders (Wise, 1995; Mansour, 1988). Nurelle D, which consists of the organophosphate chlorpyrifos and the pyrethroid cypermethrin, caused 100% mortality in all of the studied spider species (Pekár & Beneš, 2008). The nutritional condition is also of great importance. Starved individuals of *Pardosa prativaga* showed a higher mortality rate after the application of dimethoate than in those that were well-nourished (Pedersen et al., 2002). This finding can be important because a spider's mortality is higher in fields, where they are often malnourished and starved (Wise, 1995).

Table 2. List of studies treating lethal effects of organophosphorus pesticides on spider species

Family/species	**Insecticide/acaricide**	**Herbicide**	**Reference**
Agelenidae			
Coelotes terrestris	Fenitrothion, Vamidothion	-	(Albert & Bogenschütz, 1984)
Anyphaenidae			
Hibana futilis	Malathion,	-	(Elzen & Pfannenstiel, 2009)
Hibana velox	Ethion	-	(Amalin et al., 2000)
Araneidae			
Araneus ventricosus	Malathion	-	(Hukusima & Kondo, 1962)
Argiope argentata	Methamidophos	-	(Mansour & Nentwig, 1988)
Clubionidae			
Clubiona japonica	Malathion	-	(Hukusima & Kondo, 1962)
Linyphiidae			
Agyneta lila	Malathion	-	(Hukusima & Kondo, 1962)
Bathyphantes gracilis	Dimethoate, Parathion	-	(De Clercq et al., 1991)
Erigone atra	Dimethoate, Parathion	-	(De Clercq et al., 1991;Dinter, 1991; Dinter et al., 1998)
Gnathonarium exsiccatum	Diazinon, Isoxathion, Phenthoate	-	(Tanaka et al., 2000)
Hylyphantes graminicola	Dimethoate	-	(Peng et al., 2010)
Linyphia triangularis	Methamidophos	-	(Mansour & Nentwig, 1988)

Family/species	Insecticide/acaricide	Herbicide	Reference
Oedothorax apicatus	Dimethoate, Fenitrothion, Parathion	-	(De Clercq et al., 1991;Dinter, 1997; Dinter et al., 1998)
Tenuiphantes tenuis	Dimethoate, Parathion	Glyphosate	(Haughton et al., 2001)
Ummeliata insecticeps	Diazinon, Phenthoate	-	(Tanaka et al., 2000)
Lycosidae			
Anoteropsis hilaris	Diazinon	-	(Van Erp et al., 2002)
Pardosa milvina		Glyphosate	(Evans et al., 2010)
Lycosidae			
Pardosa pseudoannulata	Diazinon, Isoxathion, Phenthoate	-	(Tanaka et al., 2000)
Cheiracanthiidae			
Cheiracanthium mildei	Methamidophos, Methidathion	Glyphosate	(Mansour & Nentwig, 1988;Mansour, 1988; Mansour et al., 1991)
Philodromidae			
Philodromus aureolus	Methamidophos	-	(Mansour & Nentwig, 1988)
Salticidae			
Carrhotus xanthogramma	Malathion	-	(Hukusima & Kondo, 1962)
Tetragnathidae			
Tetragnatha maxillosa	Diazinon, Isoxathion, Phenthoate	-	(Tanaka et al., 2000)
Theridiidae			
Chrysso octomaculata	Malathion	-	(Hukusima & Kondo, 1962)
Phyloneta impressa	Thiometon, Phosalone	-	(, 1999; Michalková & Pekár, 2009; Pekár & Haddad, 2005; Pekár, 2002)
Thomisidae			
Ebrechtella tricuspidata	Malathion	-	(Hukusima & Kondo, 1962)

Herbicides are generally less lethal. Even when applied in the highest recommended field concentrations, glyphosate was almost harmless to the lyiniphiid *Tenuiphantes tenuis* (Blackwall, 1852) and the cheiracanthiid *Cheiracanthium mildei,* which implies that acute toxicity of this herbicide for field populations of spiders is low (Haughton et al., 2001).

Hormesis

Hormesis represents a relatively favorable biological response to low exposure to toxins or other types of stressors (Mattson, 2008). In several studies, a low dosage of organophosphates improved the performance of spider species in different ways. The lyniphiid *Hylyphantes graminicola* (Deng et al., 2007) and lycosids *Pardosa pseudoannulata* and *Pardosa amentata* captured more prey due to increased searching efficiency. It should be noted that the captured prey wasn´t consumed (Wang eta al., 2006). OPs also have been found to induce physiological processes. Protease activity increased in *P. pseudoannulata* after exposure to low levels of methamidophos, and a slightly higher hatching rate was observed in *H. graminicola* (Deng eta al., 2006; Zhi et al., 2006).

The phenomenon of hormesis has still been insufficiently investigated in spiders since available data are scarce and additional research is required.

Conclusion

Due to the increasing world population and expansion of urban areas, the production and consumption of pesticides are growing tremendously. Because of their efficiency and degradability, organophosphates are a widely used class of pesticides today, but their unregulated and inappropriate utilization can pose a threat to beneficial non-target organisms and lead to loss of biodiversity.

In that regard, investigation of the effects of OPs on the spider fauna is very important. Available studies have shown that those effects are wide and impact almost every aspect of the spider's life, from physiological processes and reproduction biology to movement and prey capture. Pesticides can have lethal or sublethal effects. So far, ecotoxicological studies have been mostly focused on agrobiont species (which is understandable in view of their importance for the management of insect pests in the agroecosystems). Other spider guilds have been less thoroughly investigated, although it is reasonable to assume that the effects of pesticides on them could be the same or at least similar in nature.

Changes in the spider phenotype resulting from repeated OP applications are unknown. Also, the possibility of a mutagenic effect remains uninvestigated. There is an assumption that growing pesticide use can change the habitat preference of spiders and cause a lower abundance of agrobiont species. Studies comparing lethal and sublethal effects of pesticides on the population dynamics of spiders are also scarce. Further research is therefore required to interpret the long-term influence of OPs on the spider fauna as an important part of biological control in ecosystems.

Acknowledgment

This work was supported by the Serbian Ministry of Education, Science and Technological Development (Grant No. 451-03-9/2021-14/ 200178).

References

Adeyinka, A., and Pierre, L., 2018. *Organophosphates*. Ino [Internet]. StatPearls Publishing. PMID: 29763035.

Aktar, W., Sengupta, D. and Chowdhuri A., 2009. Impact of pesticides use in agriculture: their benefits and hazards. *Interdisciplinary Toxicology* 2 (1), 1-12

Albert, R. V. and, H., 1984. Prüfung der Wirkung von Pflanzenbehandlungsmitteln auf die Nutzarthropode *Coelotes terrestris* (Wider) (Araneida, Agelenidae) mit Hilfe eines Glasplattentests [Testing the effect of plant treatment products on the beneficial arthropod Coelotes terrestris (Wider) (Araneida, Agelenidae) using a glass plate test]. *Anzeiger für Schädlingskunde, Pflanzenschutz, Umweltschutz*, 57(6), 111-117.

Ali, D., Gend Kumar, P., Kumar, S. and Ahmed, M., 2014. Evaluation of genotoxic and oxidative stress response to dimethoate in freshwater fish *Channa punctatus* (Bloch). *Chemical Speciation & Bioavailability*, 26(2), 111-118.

Amalin, D. M., Pena, J. E., Yu, S. J. and McSorley, R., 2000. Selective toxicity of some pesticides to *Hibana velox* (Araneae: Anyphaenidae), a predator of citrus leafminer. *Florida Entomologist*, 254-262.

Balderacchi, M., Benoit, P., Cambier, P., Eklo, O., M., Gargini, A., Gemitzi, A., Trevisan, M., 2013. Groundwater pollution and quality monitoring approaches at the European level. *Critical Reviews in Environmental Science and Technology*, 43, 323–408.

Benamú, M. A., Schneider, M. I. and Sanchez, N. E., 2010. Effects of the herbicide glyphosate on biological attributes of *Alpaida veniliae* (Araneae, Araneidae), in laboratory. *Chemosphere*, 78 (7), 871-876.

Brown, C., Hanna, C. J. and Hanna, C. J., 2015. The importance of pesticide exposure duration and mode on the foraging of an agricultural pest predator. *Bulletin of nvironmental contamination and toxicology*, 94(2), 178-182.

Cooke, M. S., Evans, M. D., Dizdaroglu, M. and Lunec, J., 2003. Oxidative DNA damage: mechanisms, mutation, and disease. *The FASEB Journal*, 17(10), 1195-1214.

Dar, M. A., Kaushik, G. and Chiu, J. F. V., 2020. Pollution status and biodegradation of organophosphate pesticides in the environment. In *Abatement of Environmental Pollutants* (pp. 25-66). Elsevier.

De Clercq, R., Casteels, H. and Janssens, J., 1991. Influence of pesticides on the epigeal arthropod fauna in laboratory tests. *Bulletin OILB SROP*, 14(4), 110-114.

Deng, L., Dai, J., Cao, H. and Xu, M., 2006. Effects of an organophosphorous insecticide on survival, fecundity, and development of *Hylyphantes graminicola* (Sundevall) (Araneae: Linyphiidae). *Environmental Toxicology and Chemistry: An International Journal*, 25(11), 3073-3077.

Deng, L., Dai, J., Cao, H. and Xu, M., 2007. Effects of methamidophos on the predating behavior of *Hylyphantes graminicola* (Sundevall) (Araneae: Linyphiidae). *Environmental Toxicology and Chemistry: An International Journal*, 26(3), 478-482.

Desneux, N., Decourtye, A. and Delpuech, J., M., 2007. The sublethal effects of pesticides on beneficial arthropods. *Annual Review of Entomology* 52: 81–106.

Dinter, A. and Poehling, H. M., 1992. Spider populations in winter wheat fields and the side-effects of insecticides. *Aspects of Applied Biology* 31(1):77–85.

Dinter, A., 1997. Laboratory testing of the effects of pesticides on two dominant spider species (*Erigone atra* (Blackwall) and *Oedothorax apicatus* (Blackwall)) of the European agricultural ecosystem, and predictions from laboratory testing to the field situation. In *New studies in ecotoxicology*. 19-22.

Dinter, A., Mausch, H. and Brauckhoff, U., 1998. Risk assessments for the linyphiid spiders *Erigone atra* and *Oedothorax apicatus* and the pyrethroid Sumicidin 10 resulting from laboratory dose-response studies. *IOBC WPRS Bulletin*, 21, 77-88.

Dogan, D., Can, C., Kocyigit, A., Dikilitas, M., Taskin, A. and Bilinc, H., 2011. Dimethoate-induced oxidative stress and DNA damage in *Oncorhynchus mykiss*. *Chemosphere*, 84(1), 39-46.

Dudić, B., D., 2015. Life Cycle characteristics of some European agrobiont spider species (Aranea: Arachnida). Doctoral dissertation, Unversity of Belgarde, Belgrade. 209 pp.

Dudić, B., Tomić, V., Sivčev, I., Büchs, W., Sivčev, L., Graora, D. and Gotlin-Čuljak, T., 2013. New data on spider fauna from Northern Serbia. *Archives of Biological Sciences*, 65(4), 1669-1673.

Ecobichon, D. J., Joy, R. M., 1993. *Pesticides and Neurological Diseases*. CRC Press.

Elzen, G. W. and Pfannenstiel, R. S., 2009. Extreme Susceptibility of *Hibana futillis* Spiderlings to Selected nsecticides in a laboratory Bioassay. *Southwestern Entomologist*, 34(1), 103-106.

EPA, January 2017. *What Is a Pesticide?* http://www.epa.gov.

Evans, S. C., Shaw, E. M. and Rypstra, A. L., 2010. Exposure to a glyphosate-based herbicide affects agrobiont predatory arthropod behaviour and long-term survival. *Ecotoxicology*, 19(7), 1249-1257.

Foelix, R., 2011. *Biology of spiders*. OUP USA.

Giglio, A., Cavaliere, F., Giulianini, P., G., Mazzei, A., Talarico, F., Vommaro, M., A. and Brandmayr, P., 2017. Impact of agrochemicals on non-target species: *Calathus fuscipes* Goeze 1777 (Coleoptera: Carabidae) as model. *Ecotoxicology and Environmental Safety* 142: 522–529.

Greaves, A. K., and Letcher, R. J., 2017. A review of organophosphate esters in the environment from biological effects to distribution and fate. *Bulletin of Environmental Contamination and Toxicology*, 98(1), 2–7.

Harwood, J. D., Symondson, W. O. C. and Sunderland, K. D., 2001. Monoclonal antibodies to quantify the effects of alternative prey on aphid predation by spiders. *Antenna-London-Royal Entomological Society*, *25* (4), 257-259.

Haughton, A. J., Bell, J. R., Wilcox, A. and Boatman, N. D., 2001. The effect of the herbicide glyphosate on non-target spiders: Part I. Direct effects on *Lepthyphantes tenuis* under laboratory conditions. *Pest Management Science: formerly Pesticide Science*, 57(11), 1033-1036.

Hukusima, S. and Kondo, K., 1962. Further evaluation in the feeding potential of the predaceous insects and spiders in association with aphids harmful to apple and pear growing and the effect of pesticides on predators. *Jpn. J. Appl. Entomol. Zool*, *6*(4), 274-280.

Ikonov, A. R., Vujić, V. D., Büchs, W., Prescher, S., Sivčev, I. M., Sivčev, L. I., Gotlin-Čuljak, T., Juran, I., Tomić, V. T. and Dudić, B. D., 2019. Does Application of Ppyrethroid Insecticides Induce Morphological Variations of *Oedothorax apicatus* Blackwall, 1850 (Araneae: Linyphiidae)?. *Acta Zoologica Bulgarica*, 71(4): 557-566.

Korenko, S., Niedobová, J., Kolářová, M., Hamouzová, K., Kysilková, K. and Michalko, R., 2016. The effect of eight common herbicides on the predatory activity of the agrobiont spider *Pardosa agrestis*. *BioControl*, 61(5), 507-517.

Kumar, S., Kaushik, G., and Villarreal-Chiu, J. F., 2016. Scenario of organophosphate pollution and toxicity in India: A review. *Environmental Science and Pollution Research*, 23, 9480–9491.

Lacava, M., García, L. F., Viera, C. and Michalko, R., 2021. The pest-specific effects of glyphosate on functional response of a wolf spider. *Chemosphere*, 262, 127785.

Leccia, F., Kysilková, K., Kolářová, M., Hamouzová, K., Líznarová, E. and Korenko, S., 2016. Disruption of the chemical communication of the European agrobiont ground-dwelling spider *Pardosa agrestis* by pesticides. *Journal of Applied Entomology*, 140(8), 609-616.

Lengwiler, U. and Benz, G., 1994. Effects of selected pesticides on web building behaviour of *Larinioides sclopetarius* (Clerck) (Araneae, Araneidae). *Journal of Applied Entomology*, 117 (1-5), 99-108.

Li, H. P., Chang, J., Tao, F. E. N. G. and Gao, X. W., 2015. Differential effects of insecticides on mitochondrial membrane fluidity and ATPase activity between the wolf spider and the rice stem borer. *Journal of Integrative Agriculture*, 14(12), 2574-2580.

Lukaszewicz-Hussain, A., 2010. Role of oxidative stress in organophosphate insecticide toxicity–Short review. *Pesticide biochemistry and physiology*, 98(2), 145-150.

Mansour, F. and Nentwig, W., 1988. Effects of agrochemical residues on four spider taxa: laboratory methods for pesticide tests with web-building spiders. *Phytoparasitica*, 16(4), 317-325.

Mansour, F. and Whitecomb, W. H., 1986. The spiders of a citrus grove in Israel and their role as biocontrol agents of *Ceroplastes floridensis* [Homoptera: Coccidae]. *Entomophaga*, 31(3), 269-276.

Mansour, F., 1984. A malathion-tolerant strain of the spider *Cheiracanthium mildei* and its response to chlorpyrifos. *Phytoparasitica*, 12(3-4), 163-166.

Mansour, F., 1988. Spiders as biocontrol agents in a citrus grove and their susceptibility to different pesticides. In *Citriculture: proceedings of the Sixth International Citrus Congress: Middle-East, Tel Aviv, Israel,* March 6-11, 1988/scientific editors, R. Goren and K. Mendel, editor, N. Goren. Rehovot, Israel: Balaban, 1989.

Mansour, F., Rosen, D., Plaut, H. N. and Shulov, A., 1981. Effect of commonly used pesticides on *Cheiracanthium mildei* and other spiders occurring on apple. *Phytoparasitica*, 9(2), 139-144.

Marc, P., Canard, A. and Ysne, F., 1999. Spiders (Araneae) useful for pest limitation and bioindication. *Invertebrate Biodiversity as Bioindicators of Sustainable Landscapes*, 74: 229-273.

Mattson, M. P., 2008. Hormesis defined. *Ageing research reviews*, 7(1), 1-7.

Meng, X., Li, C., Bao, H., Fang, J., Liu, Z. and Zhang, Y., 2015. Validating the importance of two acetylcholinesterases in insecticide sensitivities by RNAi in *Pardosa pseudoannulata*, an important predatory enemy against several insect pests. *Pesticide biochemistry and hysiology*, 125, 26-30.

Michalková, V. and Pekár, S., 2009. How glyphosate altered the behaviour of agrobiont spiders (Araneae: Lycosidae) and beetles (Coleoptera: Carabidae). *Biological Control*, 51(3), 444-449.

Muhammad, G., 2017. Practical aspects of treatment of organophosphate and carbamate insecticide poisoning in animals. *Matrix Science Pharma (MSP)*, 1(1), 10–11.

Nyffeler, M. and Benz, G., 1978. Zur ökologischen Bedeutung der Spinnen der Vegetationsschicht von Getreide-und Rapsfeldern bei Zürich (Schweiz) 1. *Zeitschrift für angewandte Entomologie*, 87(1-4), 348-376.

Nyffeler, M. and Sunderland, K. D., 2003. Composition, abundance and pest control potential of spider communities in agroecosystems: a comparison of European and US studies. *Agriculture, Ecosystems & Environment*, 95(2-3), .579-612.

O'Brien, R. D., 2016. *Toxic phosphorus esters: Chemistry, metabolism, and biological effects*. New York, NY: Elsevier.

Ossamy, S., Elbanna, S., M., Orabi, G., M. and Semida, F., M., 2016. Assessing the potential role of spiders as bioindicators in Ashtoum el Gamil Natural Protected Area, Port Said, Egypt. *Indian Journal of Arachnology* 5: 101.

Paudyal, B. P., 2008. Organophosphorus poisoning. *Journal of the Nepal Medical Association* 47 (172), 251-258.

Peck, S. L., McQuaid, B. & Campbell, C. L. 1998. Using ant species (Hymenoptera: Formicidae) as a biological indicator of agroecosystems condition. *Environmental Entomology*, 27: 1102–1110.

Pedersen, L. F., Dall, L. ., Sørensen, B. C., Mayntz, D. and Toft, S., 2002. Effects of hunger level and nutrient balance on survival and acetylcholinesterase activity of dimethoate exposed wolf spiders. *Entomologia experimentalis et applicata*, 103(3), 197-204.

Pekár, S. and Beneš, J., 2008. Aged pesticide residues are detrimental to agrobiont spiders (Araneae). *Journal of Applied Entomology*, 132(8), 614-622.

Pekár, S. and Haddad, C. R., 2005. Can agrobiont spiders (Araneae) avoid a surface with pesticide residues? *Pest Management Science (formerly Pesticide Science)*, 61(12), 1179-1185.

Pekár, S., 1999. Foraging mode: a factor affecting the susceptibility of spiders (Araneae) to insecticide applications. *Pesticide Science*, 55(11), 1077-1082.

Pekár, S., 2002. Susceptibility of the spider *Theridion impressum* to 17 pesticides. *Anzeiger für Schädlingskunde= Journal of pest science*, 75(2), pp.51-55.

Pekár, S., 2012. Spiders (Araneae) in the pesticide world: an ecotoxicological review. *Pest management science*, 68(11), 1438-1446.

Pekár, S., 2013. Side effect of synthetic pesticides on spiders. In *Spider Ecophysiology*, 415-427. Springer, Berlin, Heidelberg.

Peng, Y., Shao, X. L., Hose, G. C., Liu, F. X. and Chen, J., 2010. Dimethoate, fenvalerate and their mixture affect *Hylyphantes graminicola* (Araneae: Linyphiidae) adults and their unexposed offspring. *Agricultural and Forest Entomology*, 12(4), 343-351.

Perkins, M., J., Inger, R., Bearhop, S, and Sanders, D., 2018. Multichannel feeding by spider functional groups is driven by feeding strategies and resource availability. *Oikos* 127: 23–33.

Rademacher, W., 2015. Plant Growth Regulators: Backgrounds and Uses in Plant Pproduction. *Journal of plant growth regulation* 34 (4): 845-872.

Riechert, S. E. and Lockley, T., 1984. Spiders as biological control agents. *Annual review of entomology*, 29(1), 299-320.

Samu, F. and Szinetár, C., 2002. On the nature of agrobiont spiders. *Journal of Arachnology*, 30 (2), 389-402.

Samu, F., Szinetár, C., Szita, É., Fetykó, K. and Neidert, D., 2011. Regional variations in agrobiont composition and agrobiont life history of spiders (Araneae) within Hungary. *Arachnologische Mitteilungen*, 40, 105-109.

Serra, A. A., Nuttens, A. Larvor, V. Renault, D., Couée, I., Sulmon, C., and Gouesbet, G., 2013. Low environmentally relevant levels of bioactive xenobiotics and associated degradation products cause cryptic perturbations of metabolism and molecular stress responses in *Arabidopsis thaliana*. *Journal of Experimental Botany*, 64, 2753–2766.

Sidhu, G. K., Singh, S., Kumar, V., Dhanjal, D., S., Datta, S. and Singh, J., 2019. Toxicity, monitoring and biodegradation of organophosphate pesticides: a review. *Critical reviews in environmental science and technology*, (13), 1135-1187.

Singh, B. K., Walker, A., 2006. Microbial degradation of organophosphorus compounds. *FEMS Microbiology Reviews* 30 (3), 428-471.

Singh, P., and Prasad, S. M., 2018. Antioxidant enzyme responses to the oxidative stress due to chlorpyrifos, dimethoate and dieldrin stress in palak (*Spinacia oleracea* L.) and their toxicity alleviation by soil amendments in tropical croplands. *Science of the Total Environment*, 630, 839–848.

Skalský, M., Ouředníčková, J., Michalko, R. and Bartošková, A., 2019. Synergistic effects of glyphosate formulation herbicide and tank-mixing adjuvants on *Pardosa* spiders. *Environmental Pollution*, 249, 338-344.

Stalmach, M., Wilczek, G., Homa, J. and Szulinska, E., 2015. Antioxidative and immunological responses in the haemolymph of wolf spider *Xerolycosa nemoralis*

(Lycosidae) exposed to starvation and dimethoate. *Environmental Pollution*, 206, 551-559.

Sun, L., Xu, W., Peng, T., Chen, H., Ren, L., Tan, H., Fu, Z., 2016. Developmental exposure of zebrafish larvae to organophosphate flame retardants causes neurotoxicity. *Neurotoxicology and Teratology*, 55, 16–22.

Tanaka, K., Endo, S. and Kazano, H., 2000. Toxicity of insecticides to predators of rice planthoppers: spiders, the mirid bug and the dryinid wasp. *Applied Entomology and Zoology*, 35(1), 177-187.

Tietjen, W. J. and Cady, A. B., 2007. Sublethal exposure to a neurotoxic pesticide affects activity rhythms and patterns of four spider species. *The Journal of Arachnology*, 35 (2), 396-406.

Tietjen, W. J., 2006. Pesticides affect the mating behavior of *Rabidosa rabida* (Araneae, Lycosidae). *The Journal of Arachnology,* 34(2), 285-288.

Toft, S. and Jensen, A. P., 1998. No negative sublethal effects of two insecticides on prey capture and development of a spider. *Pesticide Science*, 52(3), 223-228.

Tristram D. W., 2003. *Pheromones and animal behaviour: communication by smell and taste*. Cambridge University press.

Uetz, G. W., Halaj, J. and Cady, A. B., 1999. Guild structure of spiders in major crops. *Journal of Arachnology*, 270-280.

Vagi, M. C., Petsas, A. S., Pavlaki, M. D., Smaragdaki, N. M., and Kostopoulou, M. N., 2018. *Toxic Effects of the Organophosphorus Insecticide Fenthion on Growth and Chlorophyll Production Activity of Unicellular Marine Microalgae Tetraselmis suecica: Comparison between Observed and Predicted Endpoint Toxicity Data.*

Vale, A. and Lotti, M., 2015. Organophosphorus and carbamate insecticide poisoning. *Handbook of clinical neurology*, *131*, 149-168.

Van Erp, S., Booth, L., Gooneratne, R. and O'Halloran, K., 2002. Sublethal responses of wolf spiders (Lycosidae) to organophosphorous insecticides. *Environmental Toxicology: An International Journal*, 17(5), 449-456.

Wang, A., Cockburn, M., Ly, T. T., Bronstein, J. M., and Ritz, B., 2014. The association between ambient exposure to organophosphates and Parkinson's disease risk. *Occupational and Environmental Medicine*, 71, 275–281.

Wang, J., Li, Q., Peng, Y., Yang, H., Wang, Z., Luo, L., Song, Q. and Stanely, D., 2019. Methamidophos Influences Midgut Proteinase Activity and Subcellular Structures in the Wwolf Spider *Pardosa pseudoamulata* (Araneae: Lycosidae). *Journal of economic entomology*, 112(1), 335-340.

Wang, R., Tang, J., Xie, Z., Mi, W., Chen, Y., Wolschke, H., Ebinghaus, R., 2015. Occurrence and spatial distribution of organophosphate ester flame retardants and plasti cizers in 40 rivers draining into the Bohai Sea, North China. *Environmental Pollution*, 198, 172–178.

Wang, Z., Fu, X., Song, D. and Yan, H., 2006. Effect of low-dose chemical pesticides on distribution and activity of carboxylesterase and acetylcholinesterase in the wolf spider, *Pardosa pseudoannulata* (Araneae: Lycosidae). *Acta Entomologica Sinica*, 49(2), 260.

Wang, Z., Song, D. and Zhu, M., 2006. Functional response and searching behavior to the brown planthopper, *Nilaparvata lugens* by the wolf spider, *Pardosa pseudoannulata* under low-dose chemical pesticides. *Acta Entomologica Sinica*, 49(2), 295.

Wilczek, G., 2005. Apoptosis and biochemical biomarkers of stress in spiders from industrially polluted areas exposed to high temperature and dimethoate. *Comparative Biochemistry and Physiology Part C: Toxicology & Pharmacology*, 141(2), 194-206.

Wilczek, G., Babczyńska, A. and Wilczek, P., 2013. Antioxidative responses in females and males of the spider *Xerolycosa nemoralis* (Lycosidae) exposed to natural and anthropogenic stressors. *Comparative Biochemistry and Physiology Part C: Toxicology & Pharmacology*, 157(2), 119-131.

Wilczek, G., Mędrzak, M., Augustyniak, M., Wilczek, P. and Stalmach, M., 2016. Genotoxic effects of starvation and dimethoate in haemocytes and midgut gland cells of wolf spider *Xerolycosa nemoralis* (Lycosidae). *Environmental Pollution*, 213, pp.370-378.

Wilson, C., Tisdell, C., 2001. Why farmers continue to use pesticides despite environmental, health and sustainability costs. *Ecological Economics* 39 (3), 449-462.

Wise, D. H., 1995. *Spiders in ecological webs*. Cambridge University Press.

Wrinn, K. M., Evans, S. C. and Rypstra, A. L., 2012. Predator cues and an herbicide affect activity and emigration in an agrobiont wolf spider. *Chemosphere*, 87(4), 390-396.

Yadav, I. C., Devi, N. L., Zhong, G., Li, J., Zhang, G., and Covaci, A., 2017. Occurrence and fate of organophosphate ester flame retardants and plasticizers in indoor air and dust of Nepal: *Implication for human exposure. Environmental Pollution*, 229, 668–678.

Yang, L., Zhao, Y. H., Zhang, B. X., Yang, C. H. and Zhang, X., 2005. Isolation and characterization of a chlorpyrifos and 3, 5, 6-trichloro-2-pyridinol degrading bacterium. *FEMS Microbiology Letters*, *251* (1), 67-73.

Zhang, Y., Shao, Y., Jiang, F., Li, J. and Liu, Z., 2014. Identification of two acetylcholinesterases in *Pardosa pseudoannulata* and the sensitivity to insecticides. *Insect biochemistry and molecular biology*, 46, 25-30.

Zhi, W., Daxiang, S., Xiuqin, F. and Yunlong, L., 2006. Effect of methamidophos on midgut proteinase activity in the wolf spider, *Pardosa pseudoannulata* (Araneae: Lycosidae) assayed with piezoelectric bulk acoustic wave impedance analysis method. Kun Chong xue bao. *Acta Entomologica Sinica*, 49(4), 700-704.

Zobiole, L., H. S., Kremer, R. J., de Oliveira, R. S. Jr., and Constantin, J., 2012. Glyphosate effects on photosynthesis, nutrient accumulation, and nodulation in glyphosate-resistant soybean. *Journal of Plant Nutrition and Soil Science*, 175, 319–330.

Chapter 12

Organophosphates as Flame Retardants

Nebojša I. Potkonjak*

University of Belgrade, "Vinča" Institute of Nuclear Sciences,
National Institute of the Republic of Serbia, Belgrade, Serbia

Abstract

The application of organophosphorus flame retardants (OPFRs) is continuously increasing over the years. There is a growing concern for OPFRs' safety due to their high production. Unavoidable, large quantities of OPFRs enter the environment. Sewage and solid waste (particularly e-waste) treatment plants are the major sources of OPFRs released into the environment. OPFRs are extensively detected in air, dust, water, soil, sediment, and sludge. This chapter discusses and compares the most advanced analytical procedures of OPFRs, including extraction, purification, gas, and liquid chromatography determination techniques. The environmental occurrence of OPFRs was analyzed. Finally, the accumulation and metabolic pathway of OPFRs on living organisms are overviewed.

Keywords: organophosphates, flame retardants, analytical methods, environmental occurrence, accumulation, metabolism

Introduction

Flame retardants (FRs) can be defined as functional additives possessing specialized functions to reduce flammability and prevent the starting and

* Correspondging Author's Email: npotkonjak@vin.bg.ac.rs.

In: Organophosphates
Editor: Tamara Lazarević-Pašti
ISBN: 978-1-68507-652-8

spreading of fires. It is achieved by their introduction to potentially combustible materials (Pantelaki and Voutsa, 2019). Due to fire safety, flame retardants are used as chemical modifiers of organic materials whose main function is to create a flame retardant barrier and protect against the damage of good by fire (Matsukami et al., 2007). FRs are divided into two main classes depending on usages (Van der Veen and De Boer, 2012):

- additive FRs;
- reactive FRs.

FRs show a flame retardant effect by their presence as flame retardants in combustibles; on the other hand, reactive FRs are introduced to improve the material's flame resistance. Chemical reactions introduce these FRs into the material. Numerous halogenated chemicals, including BFRs and polychlorinated biphenyls (PCBs), have been established to be persistent, bioaccumulative, and have a toxic effect on humans, animals, and the environment. It was the focus of concern for public health, resulting in the banned production of PCBs at the beginning of the seventies (Aresta et al., 2003). It was followed by the slow restriction of BFRs, which resulted in its final phasing out in 2011. Because of the limitation on the use of brominated flame retardants (BFRs) such as PBDEs, organophosphorus flame retardants (OPFRs)compounds have been increasingly used as main flame retardants (Rauert et al., 2018). According to the European Flame Retardants Association (EFRA) (CEFIC, 2007), the total consumption of FRs in Europe in 2006 was 465 000 t (Table 1). The OPFRs, used for almost 150 years (Andrae, 2007), are considered to be appropriate replacements for BFRs. Since the need for vapor-phase activity, a number of volatile OPFRs, including tributyl phosphate (TBP), triphenyl phosphate (TPHP), and triphenylphosphine oxide (TPPO), have been recognized as feasible substitutes for BFRs used in textile back-coatings (Horrocks et al., 2007).

If OPFRs are to be used as a replacement for BFRs, it is significantly important to avoid those compounds proven to be more persistent, bio-accumulative, and toxic to humans and the environment than BFRs. The German Federal Environmental Agency carried out a large research project on the substitution of hazardous BFRs. The examined FRs, the halogenated PFR, tris(1,3-dichloro-2-propyl)phosphate (TDCPP) and tetrakis (2-chloroethyl) dichloro isopentyl diphosphate are targeted to be the right choice for substitutes of BFRs (van der Veen and de Boer 2012).

Table 1. Distribution of certain FR categories in Europe consumption (2006) presented as percentage (%)

FR category	**2006**
OPFRs	20%
BFRs	10%
Metal hydroxides	54%
Chlorinated paraffin	7%
Borates and stannates	3%
Antimony trioxide	3%
Melamine based FRs	3%

Table 2. Share of certain FR categories in worldwide consumption presented as percentage (%)

FR category	**2008**	**2017**
OPFRs	11%	32%
BFRs	23%	19%
Aluminum Trihydrate	40%	25%
Others	25%	24%

In 2017, the worldwide consumption of FRs reached 2,530,000 t, with over 30% being OPFRs and 20% being BFRs (Wang, 2020), presented in Table 2. OPFRs presence in worldwide production increases at a rate of 15% per year. This chapter aims to give recent advances in analytical methods, environmental occurrences, toxicity and risks assessment of OPFRs. In addition, some novel major studies are incorporated and reviewed regarding the occurrence and analysis of OPFRs.

Three different general structures of these OPFRs can be recognized:

- the organophosphate esters (OPEs),
- the phosphonates,
- phosphinates.

The third group is the widely used group of halogenated OPFRs, combining the properties of both the halogen and the phosphorus components. The presence of the halogen raises the lifetime of the FRs in the end-product by decreasing its mobility in the polymer (van der Veen and de Boer 2012)). Examples of halogenated PFRs are TCPP and TCEP. Within these three groups, two basic types of PFRs can be recognized. The first type is reactive

FRs, which are reactive components chemically built into a polymer molecule (WHO, 1997). Due to the chemical binding, losses during the product's lifetime are limited (van der Veen and de Boer 2012).

Organophosphate esters (OPEs) belong to the group of phosphoric acid derivatives (Greaves and Letcher, 2017; Yang et al., 2018; van der Veen and de Boer, 2012; Wei et al., 2015). These compounds can be divided into sub-groups which depend on substituent group:

- chlorinated OPEs;
- alkyl OPEs;
- aryl OPEs.

OPEs sub-groups show similar physicochemical properties:

- water solubility;
- log octanol-water partition coefficient and vapor pressure;
- distribution; and ultimate fate in the environment.

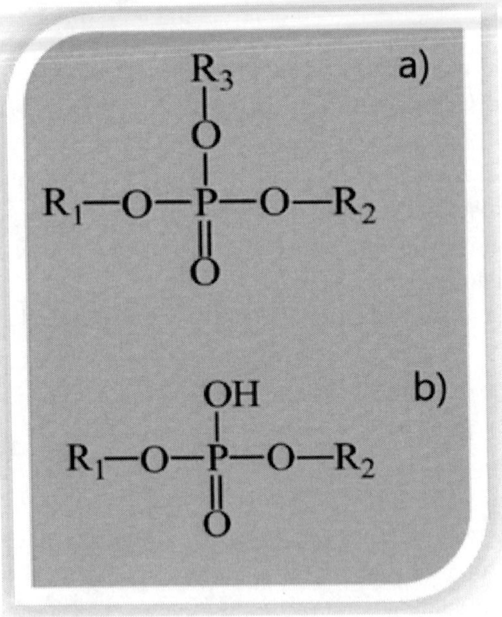

Figure 1. Structures of OPEs: OP triesters (upper) and OP diesters (lower).

All OPEs have identical phosphate base units, a phosphate molecular placed in a center and various heterogeneous substituents (Greaves and Letcher, 2017; Yang et al., 2018), Figure 1. They occur as:

- organophosphate (OP) triesters,
- organophosphate OP diesters (the decomposition products of OP triesters),
- polyphosphates.

Organophosphate trimesters can be classified as halogenated/non-halogenated alkyl or aryl phosphates based on the difference in substituents (Greaves and Letcher, 2017). Because of their high resistance fire potential and flexibility, OPEs are extensively used as FRs and plasticizers in furniture, textiles, electrical/electronic equipment, and polyurethane foam.

The chlorinated and brominated OPEs are most frequently used as flame retardants, organophosphorus flame retardants (OPFRs) (Yin et al., 2019). Among chlorinated OPEs used as FRs tris(2-chloroethyl) phosphate (TCEP), tris(2-chloro iso-propyl) phosphate (TCPP), and tris(1, 3-dichloro-2-propyl) phosphate (TDCPP) are mostly used. On the other hand, non-chlorinated OPEs such as tri-n-butyl phosphate (TBP) and triphenyl phosphate (TPHP) are mostly used as plasticizers, lubricants, extreme pressure additives, antiwear agents in hydraulic fluids. Triphenyl phosphate is used as FRs and plasticizers in PVC, cables and water pipes. Tris(2-butoxy ethyl) phosphate (TBEP) is usually added to rubber, and tritolyl phosphate (TTP) is used in hydraulic fluids, PVC, cellulose, plastic, and transmission fluid.

The non-halogenated OPEs are mostly used as plasticizers (Andresen et al., 2004). Considering their side-chain substituents, OPFRs can be divided as:

- chlorinated-OPFRs (Cl-OPFRs);
- alkyl-OPFRs;
- brominated-OPFRs (Br- OPFRs);
- aryl-OPFRs.

Rough projection implies that more than forty different OPEs have been used as flame retardants or plasticizers additives in a crowd of products, such as electronics, building material, textile, plastics, furniture, lubricant, varnish etc. (Miyake et al., 2018; García-López et al., 2008; 7. Möller et al., 2012).

Analytical Procedures

Aqueous samples can be collected either in a stainless steel container or in polypropylene or polyethylene bottles (Gustavsson et al., 2018; Lee et al., 2018; Wolschke et al., 2015; Hu et al., 2014). Sampling equipment should be rinsed with ethanol, Milli-Q water, and acetone, respectively and before outfield use and rinsed three times with a sample before collection (Gustavsson et al., 2018; Wang et al., 2015). Collected samples should be transported at 4°C until analysis. In order to remove suspended particles, filtration through glass fiber filters should be employed upon arrival at the laboratory(Yang et al., 2017). Regarding the presence of OPFRs in solid samples such as sediments and sludge, several studies were published. In these studies, sediment samples were collected using a box-core sampler or stainless steel shovels. Thereupon, samples were stored in amber glass bottles. Plastic containers should be avoided due to possible interference of OPFRs. Samples were freeze-dried during transportation to the laboratory. Before sample treatment in the laboratory, samples should be sieved by stainless-steel sieves (<400 μm) and stored at −20 °C (Wang et al., 2018; Hu et al., 2017; Gao et al., 2016; Pang et al., 2016).

The most commonly used techniques for the OPFRs recovery from water/wastewater samples are liquid-liquid extraction (LLE) and solid-phase extraction (SPE). When LLE is used, it implies the extraction with dichloromethane (Martínez-Carballo et al., 2007; Wang et al., 2015; Wolschke et al., 2015). The advantage of LLE is the low-cost equipment. Demand for a large amount of solvents, possible contamination (from glassware), and analyte losses are the main disadvantages of LLE (Pantelaki and Voutsa, 2019). If compared with LLE, SPE is a common extraction technique. It requires a small amount of organic solvents, having a short time for analysis and automated procedure (Ren et al., 2019). Most frequently, techniques of SPE used for the extraction of OPFRs are Oasis HLB and ENVI-18. Recommended extractive solvents are ethyl acetate, acetonitrile, methanol or dichloromethane/hexane. Both extraction techniques, SPE and LLE, work with high efficient recoveries for utmost OPFRs. However, one must remember that natural organic matter could be present in water samples, which could undesirably compromise the efficiency of the extraction (Gustavsson et al., 2017). In order to avoid side matrix effects and improve the efficiency of the extraction, the microwave-assisted head-space SPME (MA-HS-SPME) technique gives significant results (Tsao et al., 2011). These authors have successfully managed to conduct the rapid and solvent-free analysis of OPEs in aqueous samples via one-step microwave-assisted head-space solid-phase micro-extraction (MA-HS-SPME) followed by gas chromatography-mass spectrometry (GC–MS).

The recovery of OPFRs from sediments can be performed by soxhlet extraction and ultrasonic extraction by using dichloromethane/hexane, acetone/hexane or ethyl acetate/hexane as solvents (Lee et al., 2018; Hu et al., 2017; Cristale and Lacorte, 2013). The main disadvantage of soxhlet extraction is a time-consuming process of extraction; the duration of the extraction typically lasts between 16–24 h with a high amount of solvents used (Hu et al., 2017; Lee et al., 2018; Woudneh et al., 2015). Application of ultrasonic extraction needs minor volumes of solvents and shorter times, typically from 10 to 30 min with 2–3 repeated cycles (Cristale and Lacorte, 2013; Wang et al., 2018). Moreover, microwave-assisted extraction (MAE) and pressurized liquid extraction (PLE) have also been used (Giulivo et al., 2017; García-López et al., 2009). Better infiltration of solvent in the sample due to higher temperatures and pressures in these techniques results in efficient extraction of OPFRs in shorter times with fewer solvents volume. The OPFRs extraction from sediment with MEA at 150°C with two consecutive steps of 15 min duration showed good results (García-López et al., 2009). The volume of solvent (acetone and acetonitrile) was5 mL, and the sample weight was 0.5 g. The recovery of OPFRs from sediments can be successfully made by using pressurized liquid extraction (PLE) by using hexane/acetone (1:1) solvent mixture at 1500 psi and 373 K (Giulivo et al.) (2017). In order to reduce matrix effects, a clean-up step usually follows the extraction process of environmental samples. The particular role of the clean-up step is to remove matrix components that might disturb the analytical method performances but not the targeted material. It could be seen that Gustavsson et al. (2017) have tested the efficiency of three sorbents: 10% florisil (deactivation), 6% alumina (deactivation) and acidified silica (40% H_2SO_4 w/w) in a water sample (river). In this study, alumina displayed improved efficiency for most OPFRs.

There is a growing need for advanced and reliable analytical methods that enable rapid, sensitive, and selective determination of emerging pollutants in environmental samples at trace levels. Numerous analytical methodologies for the OPFRs determination in surface and wastewaters are accessible in the literature summarized in an article by Rodil et al., 2009. Since OPFRs have an extensive range of physical and chemical properties, the analytical method development should focus on manipulating all features to improve sensitivity and accurate quantification (Rodil et al., 2009). These studies have shown that triesters of OPFRs are mostly determined (aryl phosphates, alkyl phosphates and chlorinated alkyl phosphates) by using gas or liquid chromatography (GC or LC) combined by mass spectrometry or electrochemical detectors (Rodriguez et al., 2006; Wolschke et al., 2015; Pang et al., 2016; Hao et al., 2018).

The gas chromatography instrumental technique has been applied in the determination of non-polar compounds. Non-polar stationary phases used in most cases are 5% phenyl methyl-polysiloxane (HP5, DB-5MS). In GC measurements, the column lengths diverge from 15 m to 30 m. Typically, GC columns having dimensions 30 m × 0.25 mm of internal diameter and 0.25 μm of film width are used (Jin et al., 2016). Separation and detection of certain OPFRs components can be demanding, depending on the stationary phase material. Some results suggest that when the efficiency of two capillary columns with the same dimensions (30 m × 0.25 mm i.d., 0.25 μm film thickness) but with changed coated phases: DB-5, (5%-phenyl)-methylpolysiloxane) and SPB-1701, 14%-cyanopropyl-phenyl)-methyl-polysiloxane) are compared, good results for most OPFRs were obtained (Rodriguez et al., 2006). However, results have suggested poor separation of TCEP and an isomer of TClPP (PB-1701 column). Furthermore, temperature-dependent selectivity was observed for TBOEP and TPHP in the DB-5 column. Determination of OPFRs in various water samples has been done using a nitrogen phosphorous detector (NPD) (Jin et al., 2016). The flame photometric detector (FPD) instrumentation methods can be used as an alternative to NPD. FDP offers satisfying sensitivity and selectivity, comparable to NPD (Gao et al., 2013). Nevertheless, one must admit that the use of mass spectrometry offers firm advantages in identifying OPFRs. The GC–MS and the GC–MS/MS are the techniques that are frequently used under electron bombardment ionization. If compared, some investigations report suggests that GC–MS and GC–MS/MS have similar detectability for most OPFRs (Gustavsson et al., 2017). Some other investigations reported improved detection limits in GC–MS/MS compared with GC–MS due to better selectivity (Cristale et al., 2012). GC connected with inductively coupled plasma mass spectrometry (GC-ICP-MS) can be observed as an interesting analytical technique for the determination of OPFRs. This method was found to provide very good sensitivity and selectivity for P-containing molecules determination in environmental matrices. It can be explained by the effect of the combination of low radio frequency power, pulsed split-less injection and difficult extraction conditions (García-López et al., 2009).

Liquid chromatography coupled with mass spectrometry (LC-MS)can also be used to determine OPFRs that are not volatilized enough for GC analysis, for example, TEHP, EHDPP TPHP, TMPP, and TNBP (Lorenzo et al., 2016). It can be noticed that when the reversed-phase separation of OPFRs is used, mobile phases are made from mixtures of acidified water (formic acid) with acetonitrile or methanol in gradient elution. Triple quadrupole (QqQ) mass analyzers in LC-MS/MS or UHPLC-MS/MS methods were most commonly used to OPFRs determine in environmental samples. It is due to their high sensitivity and selectivity. OPFRs analysis was undertaken *via* electrospray ionization (ESI) in both ionization modes (positive or negative), depending on the compound and acquisition mode. The coupling of chromatographic methods and mass spectrometry provided high accuracy in identifying and quantifying OPFRs in environmental samples (Gustavsson et al., 2017). Yet, enhancement or suppression of the signal by matrix effects is often observed. It can be suppressed by internal standard (isotopically labeled) and isotopic dilution quantification methods. Matrix-matched calibration and standard addition methods can also be used to avoid limited availability and relatively high cost of commercial standards. All chromatographic methods have good recoveries and low quantification limits (0.02–30 ng l^{-1}) for most OPFRs (van der Veen and de Boer, 2012). Though, possible contamination due to the omnipresence of OPFRs in laboratory air and laboratory equipment must be avoided. Evaluating the quality of the results of OPFRs in numerous matrices, blank problems for different OPFRs (TNBP, TIBP, TBOEP and TClPP) suggest that these OPFRs were present (Brandsma et al., 2013). The main sources of interferences of OPFRs are plastic containers. To avoid contamination of OPFRs samples by those present in plastic containers, the use of glassware was proposed (Pang et al., 2016). Generally, in order to eliminate contamination of OPFRs samples, all laboratory equipment must be carefully checked for trace amounts of these pollutants. It means that laboratory equipment must be OPFRs-free. It can be done by careful cleaning, heated at a temperature from 100 to 400°C for 4–6 h and rinsed with polar and non-polar organic solvents (Regnery and Püttmann, 2010; Giulivo et al., 2016; Gustavsson et al., 2018).

Environmental Occurrence

OPFRs belong to the group of semi-volatile organic compounds which are not chemically bound to the polymer materials (van der Veen and de Boer, 2012;

Addington et al., 2020, Aničijević et al. 2021.). Therefore, OPPFs can easily escape into the environment. The particle-bound concentrations of OPEs in the total concentrations in bulk air have been reported in a vast number of research articles (Abdollahi et al., 2017; Möller et al., 2012; Sühring et al., 2016). All these studies have shown that the distribution of OPEs in the air can be observed as the particle matter (PM) phase distribution rather than gas phase distribution (Möller et al., 2011; Salamova et al. 2014; Shoeib et al., 2014; Liu et al., 2016; Ren et al., 2016; Li et al., 2019), Table 3. It can be seen that the percentage of OPEs present as PM in the air is in a pretty high amount if compared with the gas-phase distribution.

Nevertheless, it should be noticed that difference in the detection limits of the methods used in these studies, especially because these differences in methodology can be of essential importance on results obtained regarding the distribution of OPEs concentrations in air probes. Specifically, results obtained for OPEs concentrations in the air may be misjudged because the detection limits can range from 1 to 96 pg m^{-3}, presented in Table 3 (Li et al., 2019; Möller et al., 2011; Shoeib et al., 2014). OPEs concentrations in the air are under a strong anthropogenic influence. The total amount of OPEs ($OPEs_{tot}$) in the air of industrial and traffic centers can reach 16.6 ng m^{-3} (Ren et al., 2016). On the other hand, OPEstotal in rural areas was considerably lower: 0.148 ng m^{-3} arctic air in Canada (Sühring et al., 2016) or 0.37 ng $m^{-3}$the the North Sea in Germany (Möller et al., 2011).

Table 3. Percentage of organophosphate esters (OPEs) present as the particle matter (PM) in the air

OPEs as PM in the air / %	Sources
86%	Möller et al., 2011
90%	Li et al., 2019b
91%	Liu et al., (2016), Ren et al., (2016)
95%	Salamova et al., (2014)
100%	Shoeib et al., (2014)

Various studies have shown that chlorinated OPEs (Cl-OPEs) are predominately present in the total amount of OPEs in the air of urban areas: 50% to 80% in the Great Lakes region, USA (Salamova et al. 2014), >78% in Beijing, China (Liu et al. 2016) and ~43% New York, USA (Li et al., 2019). Contrary, non-chlorinated OPEs are present in major 75% of the air of rural areas (Salamova et al., 2014).

Regarding the different OPEs composition profiles, tris(isobutyl) phosphate (TIBP) and tri(n-butyl)phosphate (TNBP) were found to be mostly present in the air of Corunna, Spain (Quintana et al., 2007). Tri(n-butyl)phosphate has a high ability for skin penetration and irritation effects on skin and eyes for animals and humans, and an insignificant inhibitory effect on cholinesterase of human plasma (Quintana et al., 2007). Sun et al. (2016) have reported neurodevelopmental toxicity in zebrafish statically exposed to TNBP at 3125 $\mu g\ l^{-1}$ or 11.7 μM, but no effect on the AChE level. When exposed to TNBP, phaeodactylumtricornutum showed growing inhibition and cell apoptosis in a range from 0.2 to 1.6 $mg\ l^{-1}$, or from 0.75 to 6.0 μM, which indicates its evolving toxicity (Liu et al., 2019). *In vitro* experiments demonstrated the disruption effects of TNBP on the endocrine system (Kojima et al., 2013). Triphenyl phosphate (TPHP) was the be main OPEs, above 40% in suburban air in Canada (Abdollahi et al., 2017; Shoeib et al., 2014) and 37% in Shanghai (Ren et al., 2016). Tri(n-butyl)phosphate is extensively used in electronic, plastic and foam furniture industries (He et al., 2017; Tollbäck et al., 2006). Numerous investigations described the neurobehavioral effects of TPHP at exposure doses < 0.64 (Noyes et al., 2015), < 0.4 (Jarema et al., 2015) and <0.3 μM (Shi et al., 2018). Shi et al., 2018 have shown that TPHP inhibited AChE activity. These OPEs were found to disrupt the reproductive performance in zebrafish (Liu et al., 2013). Disruption effects of TPHP to metabolic (Du et al., 2016) and endocrine (Kojima et al., 2013; Zhang et al., 2014) systems have also been established equally *in-vitro* and *in-vivo*. Studies of OPEs that cover atmospheric air above oceans (counting Pacific, the Arctic, Indian, and the Southern Ocean), the median contribution of three important Cl-OPEs (i.e., TCEP, TCPP and TDCP) was found to be 86% (Möller et al., 2012). Significantly higher concentrations of Cl-OPEs were also detected in arctic air in Canada (Sühring et al., 2016) and the North Sea in Germany (Möller et al., 2011). These reports have shown the persistent characteristics of the Cl-OPEs. Tris(2-chloroethyl)phosphate (TCEP) is extensively used in polyurethane foams (flexible and rigid), furniture, rubber and the textile industry (Cui et al., 2017). The presence of TCEP in the Antarctica ice sheet suggests long-range transport possibilities in the environment on its persistent characteristics (Cheng et al., 2013). However, TCEP is not considered a bio-accumulative pollutant (Arukwe et al., 2018). Yet, TCEP is transported faster across the skin than other OPEs (Frederiksen et al., 2018). The neurobehavioral toxicity of TCEP on zebrafish development under static exposure was analyzed in several research papers (Sun et al., 2016). Decreased swimming activity of zebrafish larvae exposed to the TCEP at 31.4 μM and

100 μM in the dark (Dishaw et al., 2014), and 6250 μg l^{-1} or 21.9 μM in the light (Sun et al., 2016). It was found that TCEP triggered a hyperactive photo-motor response in zebrafish embryos (24 hpf, hpf: hours post fertilization) at 0.0064 and 0.064 μM (Noyes et al., 2015), and reduced response in zebrafish larvae (120 hpf) at 64 μM. Acetylcholinesterase (AChE) activity in the earthworm, exposed to dry soil containing 0.1–10 mg kg^{-1} TCEP, was significantly increased, with related genes expression reduced in mRNA levels (Yang et al., 2018). *In vitro* experiments imply that exposure to TCEP at 50 μM promoted differentiation of cholinergic phenotype (Dishaw et al., 2011).

In indoor air, OPFRs were found at dozens to hundreds of ng m^{-3}. $OPFRs_{tot}$ in indoor air (40.2 ng m^{-3}) was found to be eight times higher than that in outdoor air (5.13 ng m^{-3}) Zhou et al. (2017). More than 100 times higher of $OPFRs_{tot}$ in indoor air (340 ng m^{-3}) verse outdoor air (3.1 ng m^{-3}) was observed in Stockholm, Sweden (Wong et al., 2018). These investigations suggest that OPFRss are mainly released from indoor to the outdoor environment. These compounds are relatively easy to release into the surrounding environment from household products by volatilization, leaching, and abrasion. They are mixed with the host material instead of chemically bonding to the other constituents (García-López et al., 2008). Studies on occurrences of OPEs in indoor microenvironments showed that $OPEs_{tot}$ in the air from private houses was significantly lower than those found for public buildings (offices and shops) and vehicles (Zhou et al., 2017). The compositional profiles of OPEs in air vary across different indoor microenvironments due to their diverse applications in consumer products and finishing materials in buildings (Wong et al., 2018). Overall, TCPP, TIBP, TNBP and TBEP are the most abundant OPE analogs in indoor air.

Studies on the occurrence of OPEs in dust mainly concentrated on the indoor environment because it is a source of pollution to the outdoor environment as well as human exposure (Wang et al., 2019).

Accumulation and Metabolism

OPFRs accumulation and its concentration fluctuate for different species. The investigation for the eight female herring gulls in multiple tissues showed that the highest concentrations of OPFR striesters are in fat samples. It was followed by muscle and red blood cells, while the levels of OPFRs were below the detection limit in the liver, blood plasma and brain samples (Greaves and Letcher, 2017). It indicates that OPFRs are most likely to accumulate in tissues having high lipid storage. OPFRs were also detected in aquatic organisms. Reports suggest that the tissue-specific concentration distribution forms followed the order (Bekele et al., 2018; Greaves and Letcher, 2017):

The analytical methods for OPFRs determination in biological tissues are still to be developed, and a unified and complete methodology needs to be established. Generally, OPFRs appear to be dispersed in metabolically active parts of tissues if we talk about fish. In the case of birds, it is more likely that OPFRs accumulate in inactive tissues (Strobel et al., 2018), indicating that OPFRs distribution in the tissue may differ among different species. Some results suggested that the chemical structure of OPFRs might be the principal factor defining the tissue-specific accumulation of OPFRs in fish (Bekele et al., 2018, Wang et al., 2017).

Concerning the accumulation of OPFRs in plants, as a representative plant, fast wheat uptake of roots and accumulates less *in vivo* (Liu et al., 2019). It was found from the kinetic analysis that the concentrations of OPFRs in roots increase fast and reach equilibrium and then decrease subsequently (Liu et al., 2019). The three main pathways related to root uptake are (Liu et al., 2019):

- transmembrane;
- symplastic;
- apoplastic.

It was revealed that the mean root concentration factor values for aryl-OPFRs were higher than those for halogenated OPFRs and alkyl-OPFRs in roots (Wang et al., 2019). Results suggest that the hydrophilicity of a OPFRs is the main factor determining the capacity of accumulation and distribution in plants. It appears that roots absorb OPFRs through two mechanisms (Liu et al., 2019):

1. Absorption by the transpiration stream for OPFRs with low hydrophobicity.
2. Combining with proteins for OPFRs with higher hydrophobicity. Subsequently, the OPFRs are carried by proteins to root organic constituents (membranes and lipids) and accumulate.

The accumulated pollutants are removed in two ways from roots: biodegradation and mass transfer from roots to leaves stem and surrounding solution (Liu et al., 2019). The uptake of OPFRs by shoots occurs via two major pathways (Liu et al., 2019):

- the absorption of volatile contaminants by leaves from the air;
- translocation from roots.

Utmost OPFRs having low vapor pressures have effort volatilizing to air at room temperature so that translocation might be the chief source of OPFRs in the shoot (Liu et al., 2019). These substances are transferred from the roots to the shoots by a transpiration stream. Furthermore, it appears that hydrophilic OPFRs tend to be translocated to the shoots by the transpiration stream. Although hydrophobic pollutants cannot be effortlessly translocated within the plant tissue, they are transported by binding with the proteins (Liu et al., 2019; Gong et al., 2020). The accumulation of OPFRs increased with increasing lipid content in shoots. It suggested that OPFRs could be affected by extra factors in addition to lipid content. OPFRs are eliminated rapidly in shoots; 50% of the OPFRs can be removed in 72 h. The elimination of OPFRs in shoots is done throughout three major pathways (Liu et al., 2019; Gong et al., 2020):

- evaporation through pores,
- biodegradation,
- translocation.

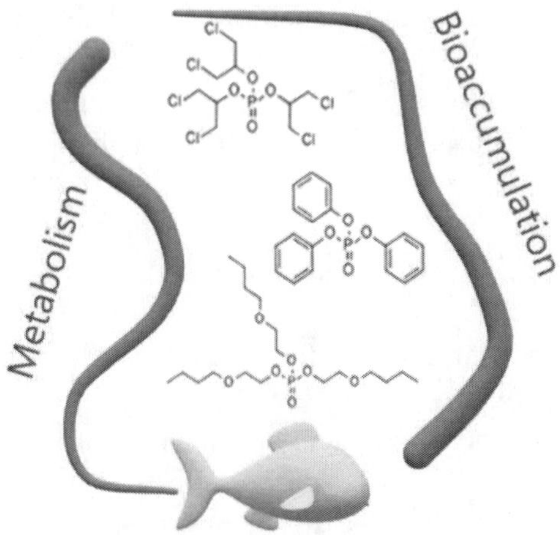

Figure 2. OPFRs cycling between bioaccumulation and metabolism.

The volatilization process is unlikely to go through the pores since most of the OPRFs have low volatility. Consequently, translocation and biodegradation are the key pathways by which OPFRs are eliminated (Liu et al. 2019). This research study implies that TCIPP and TBOEP can be transferred by phloem (from shoots to roots). This result showed that OPFRs are transferred upward or downward through phloem, xylem and protein.

The metabolic machinery for OPFRs in fish is very complex and still is the subject of research. Improved metabolic effects for TNBP and TBOEP are found in the liver microsomes than in the intestinal of fish, Figure 2.

The liver microsome has higher enzyme activity than the intestinal microsome (Hou et al., 2018). Results have indicated that enzyme activity plays an imperative role in the elimination of OPFRs. Wang et al. (2017a) also showed that metabolites are produced mainly in the liver and then released to the intestine via bile. It further proves that the metabolism of OPFRs primarily occurs in the liver. BBOEP, one isomer of bis(2-butoxy ethyl) hydroxyethyl phosphate (BBOEHEP), and three isomers of bis(2-butoxy ethyl) hydroxyl-2-butoxy ethyl phosphate (1-OH-, 2-OH-, and 3-OH-TBOEP) were detected in liver microsomes incubated with TBOEP (Hou et al., 2018; Van den Eede et al., 2015). In the TNBP metabolism assay, only DNBP and 3-OH-TNBP were well separated. These results showed that CYP enzymes, related to the production of hydroxylated and dealkylated metabolites of alkyl-OPFRs, have different affinities for substrates. The metabolic pathways for alkyl- and aryl-OPFRs showed that CYP-mediated hydroxylation is more important than hydrolysis to dialkyl phosphates (DAPs) in fish liver microsomes (Su et al., 2015; Hou et al., 2018). Hydrolysis to DAPs is important in the metabolism of chlorinated OPFRs (Greaves et al., 2016b; Van den Eede et al., 2016). In general, the pathways for OPFRs are related to the structure and physicochemical properties of these substances in the *in vitro* metabolism of fish. Further studies are needed to explore the metabolic pathways and metabolites in liver microsomes *in vitro*.

Diet ingestion is the most significant pathway of human exposure to OPFRs (Ding et al., 2018; Poma et al., 2017). Contamination in food results from the transfer of OPFRs through the food chain and industrial processing and packaging of food products (Wang and Kannan, 2018). Numerous food products were investigated: meat, fish/seafood, dairy product, cereal, vegetable, fruit, beverages, etc. (Poma et al., 2017; Wang and Kannan, 2018). The intermediate levels of $OPEs_{tot}$ in most food groups were typically at a few or less than 1 ng g^{-1}(w:w) (about 1 $ngmL^{-1}$ in beverages as foodstuff), with projected contact doses being between 25 and 85 ng kg^{-1} per day for adults and much higher for children (32–135 ng kg^{-1} per day) (Poma et al., 2017; Wang and Kannan, 2018). High concentrations of OPEs were found in rice from China, and the estimated exposure doses through rice ingestion were 539 and 601 ng kg^{-1} per day for males and females, respectively (Zhang et al., 2016). Ingestion of meat (47.3%) and dairy product (51.5%) mostly contributed to dietary consumption of OPFRs for adults and toddlers, respectively (Wang and Kannan, 2018). While OPFRs are reported to occur in indoor dust at a high level (usually $\mu g g^{-1}$), diet consumption is a more significant exposure lane owing to its considerably larger consumption rate (1000 to 100,000-fold higher) than from dust (Poma et al., 2017).

Developmental neurotoxicity is one of the greatest projecting toxic effects of OPFRs on experimental animals and the human population. The molecular toxicity mechanism is still unknown. Application of an electrochemical biosensor-based assay can be used for investigation of the inhibitory effects of OPFRs on O-linked N-acetyl glucosamine transferase (OGT) activity (Gu et al., 2018). Generally, the electrochemical biosensor-based assay involves various voltammetric methods with the use of the three-electrode electrolytic cell (Gorjanovic et al., 2010; Milic et al., 2011; Potkonjak et al., 2014; Potkonjak 2018; Pagnacco et al., 2018).

Conclusion

Flame retardants are functional additives capable of reducing flammability and spreading fires. Organophosphorus flame retardants compounds have been progressively used as foremost flame retardants. In 2017, the global consumption of FRs reached 2,530,000 t, with over 30% of OPFRs. A careful approach to the analytical procedure for determination of OPFRs includes thorough clean-up of sampling equipment (stainless steel container or in a polypropylene or polyethylene bottles), adequate selection of OPFRs recovery from samples throughout liquid-liquid extraction and solid-phase extraction and the right choice of instrumental technique (gas or liquid chromatography). OPFRs pollute the environment, accumulate and enter metabolic pathways in plants, animals, and humans.

Acknowledgment

This work was supported by the Ministry of Education, Science and Technological Development of the Republic of Serbia (Grant No. 451-03-68/2020-14/200017).

References

Abdollahi, A., Eng, A., Jantunen, L. M., Ahrens, L., Shoeib, M., Parnis, J. M. andHarner T. 2017. Characterization of polyurethane foam (PUF) and sorbent impregnated PUF (SIP) disk passive air samplers for measuring organophosphate flame retardants. *Chemosphere*, 167:212-219.

Addington, C. K., Phillips, K. A. and Isaacs, K. K. 2020. Estimation of the Emission Characteristics of SVOCs from Household Articles Using Group Contribution Methods. *Environmental Science & Technology* 2020: 110-119.

Adler, L., Greulich, K., Kampa, G. and Vieth B. 2006. Residue Analysis of 500 High Priority Pesticides – better by GC-MS or LC-MS/MS? *Mass Spectrometry Reviews* 25: 838-865.

Aresta, M., Caramuscio, P., De Stefano, L. and Pastore T. 2003. Solid state dehalogenation of PCBs in contaminated soil using $NaBH_4$. *Waste Management* 23: 315-319.

Andresen, J. and Kai B. 2006. Elimination of organophosphate ester flame retardants and plasticizers in drinking water purification. *Water Research* 40: 621-629.

Aničijević, V., Jelić, M., Jovanović, A., Potkonjak, N., Pašti, I. and Lazarević-Pašti, T. 2021. Organophosphorous pesticide removal from water by graphene-based materials–Only adsorption or something else as well? *Journal Serbian Chemical Society* 86: 699-710.

Arukwe, A., Carteny, C. C., Eggen, T. and Moeder, M. 2018. Novel aspects of uptake patterns, metabolite formation and toxicological responses in Salmon exposed to the organophosphate esters-Tris(2-butoxy ethyl)- and tris(2-chloroethyl) phosphate. *Aquatic Toxicology* 196: 146-153.

Bekele, T. G., Zhao, H. X., Wang, Y., Jiang, J. Q. and Tan. F. 2018. Measurement and prediction of bioconcentration factors of organophosphate flame retardants in common carp (Cyprinus Carpio). *Ecotoxicology and Environmental Safety* 166: 270-276.

Brandsma, S. H., de Boer, J., Cofino, W. P., Covaci, A. and Leonards,P. E. G. 2013. Organophosphorus flame-retardant and plasticizer analysis, including recommendations from the first worldwide interlaboratory study. *Trends in Analytical Chemistry* 43: 217-228.

Cheng, W., Sun, L., Huang, W., Ruan, T., Xie, Z., Zhang, P., Ding, R. and Li, M. 2013. Detection and distribution of Tris(2-chloroethyl) phosphate on the East Antarctic ice sheet. *Chemosphere* 92: 1017-1021.

Cristale, J., Quintana, J., Chaler, R., Ventura, F. and Lacorte S. 2012. Gas chromatography/ mass spectrometry comprehensive analysis of organophosphorus, brominated flame retardants, by-products and formulation intermediates in water. *Journal of Chromatography A* 124: 1-12

Cristale, J. and Lacorte S. 2013. Development and validation of a multiresidue method for the analysis of polybrominateddiphenyl ethers, new brominated and organophosphorus flame retardants in sediment, sludge and dust. *Journal of Chromatography A* 1305: 267-275.

Cui, K., Wen, J., Zeng, F., Li, S., Zhou, X. and Zeng, Z. 2017. Occurrence and distribution of organophosphate esters in urban soils of the subtropical city, Guangzhou, China. *Chemosphere* 175: 514-520.

Deng, M., Kuo, D. T. F., Wu, Q., Zhang, Y., Liu, X. Liu, S. Hu, X. Mai, B. Liu, Z. and Zhang, H. 2018. Organophosphorus flame retardants and heavy metals in municipal landfill leachate treatment system in Guangzhou, China. *Environmental Pollution* 236: 137-145.

Dishaw, L. V., Hunter, D. L., Padnos, B., Padilla, S. and Stapleton, H. M. 2014. Developmental exposure to organophosphate flame retardants elicits overt toxicity and alters behavior in early life stage zebrafish (Daniorerio). *Toxicological Sciences* 142: 445-454.

Du, Z., Zhang, Y., Wang, G., Peng, J., Wang, Z. and Gao S. 2016. TPhP exposure disturbs carbohydrate metabolism, lipid metabolism, and the DNA damage repair system in zebrafish liver. *Scientific Reports* 6: 21827.

Frederiksen, M., Stapleton, H. M., Vorkamp, K., Webster, T. F., Jensen, N. M., Sorensen, J. A., Nielsen, F., Knudsen, L. E., Sørensen, L. S., Clausen, P. A. and Nielsen, J. B. 2018. Dermal uptake and percutaneous penetration of organophosphate esters in a human skin ex vivo model. *Chemosphere* 197: 185-192.

Gao, Z., Deng, Y., Hu, X., Yang, S., Sun, C. and He H. 2013. Determination of organophosphate esters in water samples using anionic liquid-based sol–gel fiber for head-space solid-phasemicroextraction coupled to gas chromatography-flamephotometric detector. *Journal of Chromatography A* 1300: 141-150.

Gao, L., Shi, Y., Li, W., Liu, J. and Cai Y. 2016. Occurrence and distribution of organophosphate triesters and diesters in sludge from sewage treatment plants of Beijing, China. *Science of the Total Environment* 544:143-149.

García, M., Rodríguez, I. and Cela, R. 2007. Microwave-assisted extraction of organophosphate flame retardants and plasticizers from indoor dust samples. *Journal of Chromatography A* 1152: 280-286.

García-López, M., Rodríguez, I. and Cela, C. 2009. Pressurized liquid extraction of organophosphate triesters from sediment samples using aqueous solutions. *Journal of Chromatography A* 1152, 1216: 6986-6993 .

Giulivo, M., Capri, E., Kalogianni, E., Milacic, R., Majone, B., Ferrari, F., Eljarrat, E. and Barceló, B., 2017. Occurrence of halogenated and organophosphate flame retardants in sediment and fish samples from three European river basins. *Science of the Total Environment* 586: 782-791.

Gong, X. Y., Wang, Y., Pu, J., Zhang, J. J., Sun, H. W. and Wang, L. 2020. The environment behavior of organophosphate esters (OPEs) and di-esters in wheat

(Triticumaestivum L.): uptake mechanism, in vivo hydrolysis and subcellular distribution. *Environment International* 135: 105405.

Gorjanovic S. Z., Novaković, M. M., Potkonjak N. I. and Sužnjević, D. Z. 2010. Antioxidant Activity of Wines Determined by a Polarographic Assay Based on Hydrogen Peroxide Scavenge. *Journal of Agricultural and Food Chemistry* 58: 4626-4631.

Greaves A. K. and R. J. Letcher R. J. 2017. A review of organophosphate esters in the environment from biological effects to distribution and fate. *Bulletin of Environmental Contamination and Toxicology* 98: 2-7.

Gu, Y., Yang, Y., Wan, B., Li, M. and Guo, L. H. 2018. Inhibition of O-linked N-acetyl glucosamine transferase activity in PC12 cells - a molecular mechanism of organophosphate flame retardants developmental neurotoxicity. *Biochemical Pharmacology* 152:21-33.

Gustavsson, J., Ahrens, L., Nguyen, M. A., Josefsson, S., Berggren Klejac, D. and Wiberg, K. 2017. Influence of natural organic matter on the extraction efficiency of flame retardants from surface waters. *Journal of Chromatography A* 1524: 74-86.

Gustavsson, J., Ahrens, L., Nguyen, M. A., Josefsson, S. and Wiberg, K. 2017. Development and comparison of gas chromatography–mass spectrometry techniques for analysis of flame retardants. *Journal of Chromatography A* 1481: 116-126.

Gustavsson, J., Wiberg, K., Ribeli, E., Nguyen, M. N., Josefsson, S. and Ahrens, L. 2018. Screening of organic flame retardants in Swedish river water. *Science of the Total Environment* 625: 1046-1055.

Hao, C., Helm, P. A., Morse, D. and Reiner, E. J. 2018. Liquid chromatography-tandem mass spectrometry direct injection analysis of organophosphorus flame retardants in Ontario surface water and wastewater effluent. *Chemosphere* 191 (2018) 288-295.

He, M. J., Yang, T., Yang, Z. H., Li, Q. and Wei S. Q. 2017. Occurrence and distribution of organophosphate esters in surface soil and street dust from Chongqing, China: implications for human exposure. *Archives of Environmental Contamination and Toxicology* 7: 1-13.

Horrocks, A. R., Davies, P. J., Kandola, B. K. and Alderson A. 2007. The potential for volatile phosphorus-containing flame retardants in textile back-coatings. *Journal of Fire Sciences* 25: 523-540.

Hou, R., Huang, C., Rao, K. F., Xu, Y. P. and Wang, Z. J. 2018. Characterized *in vitro* metabolism kinetics of alkyl organophosphate esters in fish liver and intestinal microsomes. *Environmental Science & Technology* 52 3202-3210.

Hu, M., Li, J. Zhang, B., Cui, Q., Wei, S. and H. Yu, H. 2014. Regional distribution of halogenated organophosphate flame retardants in seawater samples from three coastal cities in China. *Marine Pollution Bulletin* 86: 569-574.

Hu, Y. X., Sun, Y. X., Li, X., Xu, W. H., Zhang, Y., Luo, X. J., Dai, S. H., Xu, X. R. and Mai B. X. 2017. Organophosphorus flame retardants in mangrove sediments from the Pearl River Estuary, South China. *Chemosphere* 181: 433-439.

Jarema, K. A., Hunter, D. L., Shaffer, R. M., Behl, M. and Padilla, S. 2015. Acute and developmental behavioral effects of flame retardants and related chemicals in zebrafish. *Neurotoxicology and Teratology* 52: 194-209.

Jin, T., Chenga, J., Cai, C., Cheng, M., Wu S. and Zhou, H. 2016. Graphene oxide based sol-gel stainless steel fiber for the headspacesolid-phase microextraction of organophosphate ester flameretardants in water samples. *Journal of Chromatography A* 1457: 1-6.

Kojima, H., Takeuchi, S., Itoh, T., Iida, M., Kobayashi, S. and Yoshida, T. 2013. *In vitro* endocrine disruption potential of organophosphate flame retardants via human nuclear receptors. *Toxicology* 314: 76-83,

Lee, S., Cho, H. J., Choi, W. and Moon, H. B. 2018. Organophosphate flame retardants (OPFRs) in water and sediment: occurrence, distribution, and hotspots of contamination of Lake Shihwa, Korea. *Marine Pollution Bulletin* 130: 105-112.

Li, W., Wang, Y. andKannan K. 2019. Occurrence, distribution and human exposure to 20 organophosphate esters in air, soil, pine needles, river water, and dust samples collected around an airport in New York state, United States. *Environment International* 131: 105054.

Liu, R., Lin, Y., Liu, R., Hu, F.,Ruan, T. and Jiang G. 2016. Evaluation of two passive samplers for the analysis of organophosphate esters in the ambient air. *Talanta* 147: 69-75.

Liu, Q., Wang, X. L., Yang, R. Y., Yang, L. P., Sun, B. B. and Zhu, L. Y. 2019. Uptake kinetics, accumulation, and long-distance transport of organophosphate esters in plants: impacts of chemical and plant properties. *Environmental Science & Technology* 53: 4940-4947.

Lorenzo, M., Campo, J. and Picó Y. 2016. Ultra-high-pressure liquid chromatography tandem mass spectrometry method for the determination of 9 organophosphate flame retardants in water samples. *Methods X* 3: 343-349.

Martínez-Carballo, E., González-Barreiro, C., Sitka, A., Scharf, S. andGans, O. 2007. Determination of selected organophosphate esters in the aquatic environment of Austria. *Science of the Total Environment* 388: 290-299.

Matsukami, H., Tue, N. M., Suzuki, G., Someya, M., Tuyen, L. H., Viet, P. H. Takahashi, S., Tanabe, S. and Takigami H. 2015. Flame retardant emission from e-waste recycling operation in northern Vietnam: environmental occurrence of emerging organophosphorus esters used as alternatives for PBDEs. *Science of the Total Environment* 514: 492-499.

Milić, S. Z., Potkonjak, N. I., Gorjanović, S. Z., Veljović-Jovanović, S. D., Pastor F. T. and Sužnjević, D. Z. 2011. A Polarographic Study of Chlorogenic Acid and Its Interaction with Some Heavy Metal Ions. *Electroanaysis* 23: 2935-2940.

Miyake, Y., Tokumura,M.,Wang, Q., Amagai, T., Takegawa, Y.,Yamagishi, Y., Ogo, S., Kume, K., Kobayashi, T., Takasu, S., Ogawa, K. andKannan K. 2018. Identification of novel phosphorus-based flame retardants in curtains purchased in Japan using Orbitrap mass spectrometry. *Environmental Science & Technology Letters*, 5: 448-455.

Möller, A., Sturm, R., Xie, Y., Cai, M., He, J. and Ebinghaus R. 2012. Organophosphorus flame retardants and plasticizers in airborne particles over the northern Pacific and Indian Ocean toward the polar regions: evidence for global occurrence. *Environmental Science & Technology* 46: 3127-3134.

Möller, A., Xie, Z., Caba, A., Sturm, R. and Ebinghaus, R. 2011. Organophosphorus flame retardants and plasticizers in the atmosphere of the North Sea. *Environmental Pollution* 159: 3660-3665.

Noyes, P. D., Haggard, D. E., Gonnerman, G. D. and Tanguay, R. L. 2015. Advanced morphological-behavioral test platform reveals neurodevelopmental defects in embryonic zebrafish exposed to comprehensive suite of halogenated and organophosphate flame retardants. *Toxicological Sciences* 145: 177-195.

Pagnacco, M. C., Maksimović, J. P., Potkonjak, N. I., Božić, B. Ð. and Horváth, A. K. 2018. Transition from Low to High Iodide and Iodine Concentration States in the Briggs-Rauscher Reaction: Evidence on Crazy Clock Behavior. *Journal of Physical Chemistry A* 122: 482– 491.

Pang, L., Yuan, Y., He, H., Liang, K., Zhang, H. and Zhao J. 2016. Occurrence, distribution, and potential affecting factors of organophosphate flame retardants in sewage sludge of wastewater treatment plants in Henan Province, Central China. *Chemosphere* 152: 245-251.

Pantelaki, I. and Voutsa, D. 2019. Organophosphate flame retardants (OPFRs): a review on analytical methods and occurrence in wastewater and aquatic environment. *Science of the Total Environment* 649: 247-263.

Poma, G., Glynn, A., Malarvannan, G., Covaci, A. andDarnerud, P. O. 2017. Dietary intake of phosphorus flame retardants (PFRs) using Swedish food market basket estimations. *Food and Chemical Toxicology* 10: 1-7.

Potkonjak, N. I. 2018. Consideration about a voltammogram as the bifurcation diagram of oscillating electrochemical systems: a case study of the copper|1 M trifluoroacetic acid oscillator. *Reaction Kinetics, Mechanisms and Catalysis* 123: 155-163.

Potkonjak N. I., Nikolić, Z. M., AnićS. R. and Minić, D. M. 2014. Electrochemical oscillations during copper electrodissolution/passivation in trifluoroacetic acid induced by current interrupt method. *Corrosion Science* 83: 355-358.

Quintana, J. B., Rodil, R., López-Mahía, P.,Muniategui-Lorenzo, S. and Prada-Rodríguez D. 2007. Optimisation of a selective method for the determination of organophosphorous triesters in outdoor particulate samples by pressurised liquid extraction and large-volume injection gas chromatography–positive chemical ionisation–tandem mass spectrometry. *Analytical and Bioanalytical Chemistry* 388: 1283-1293.

Rauert, C., Schuster, J. K., Eng, A. and T. Harner T. 2018. Global atmospheric concentrations of brominated and chlorinated flame retardants and organophosphate esters. *Environmental Science & Technology* 52: 2777-2789.

Regnery, J. and Püttmann, W. 2010. Occurrence and fate of organophosphorus flame retardants and plasticizers in urban and remote surface waters in Germany. *Water Research* 44: 4097-4104.

Ren, G., Chen, Z., Feng, J., Ji, W., Zhang, J., Zheng, K., Yu, Z. and Zeng X. 2016. Organophosphate esters in total suspended particulates of an urban city in East China. *Chemosphere* 164: 75-83.

Ren, J., Wang, X.. Gong, P. and Wang C. 2019. Characterization of tibetan soil as a source or sink of atmospheric persistent organic pollutants: seasonal shift and impact of global warming. *Environmental Science & Technology* 53: 3589-3598.

Rodil, R., Benito Quintana, J., Lopez-Mahia, P., Muniategui-Lorenzo, S. and Prada-Rodriguez, D. 2009. Multi-residue analytical method for the determination of emerging pollutants in water by solid-phase extraction and liquid chromatography–tandem mass spectrometry. *Journal of Chromatography A* 1216: 2958-2969.

Rodriguez, I., Calvo, F., Quintana, J. B., Rubi, E., Rodil, R. and Cela, R. 2006. Suitability of solid-phase microextraction for the determination of organophosphate flame retardants and plasticizers in water samples. *Journal of Chromatography A* 1108: 158-165.

Salamova, A., Ma, Y., Venier, M. and Hites, R. A. 2014. High levels of organophosphate flame retardants in the Great Lakes Atmosphere. *Environmental Science & Technology Letters* 1 8-14.

Shi, Q., Wang, M., Shi, F., Yang, L., Guo, Y., Feng, C., Liu, J. and Zhou, B. 2018. Developmental neurotoxicity of triphenyl phosphate in zebrafish larvae. *Aquatic Toxicology* 203: 80-87.

Shoeib, M., Ahrens, L., Jantunen, L. and Harner, T. 2014. Concentrations in air of organobromine, organochlorine and organophosphate flame retardants in Toronto, Canada. *Atmospheric Environment* 99: 140-147.

Strobel, A., Willmore, W. G., Sonne, C., Dietz, R. and Letcher, R. J. 2018. Organophosphate esters in East Greenland polar bears and ringed seals: adipose tissue concentrations and *in vitro* depletion and metabolite formation. *Chemosphere* 196: 240-250.

Sun, L., Xu, W., Peng, T., Chen, H., Ren, L., Tan, H., Xiao, D., Qian, H. and Fu, Z. 2016. Developmental exposure of zebrafish larvae to organophosphate flame retardants causes neurotoxicity. *Neurotoxicology and Teratology* 55: 16-22.

Sühring, R., Diamond, M. L., Scheringer, M., Wong, F., Pućko, M., Stern, G., Burt, A., Hung, H., Fellin, P., Li, H. and Jantunen L. M. 2016. Organophosphate esters in Canadian Arctic air: occurrence, levels and trends. *Environmental Science & Technology* 50: 7409-7415.

Tollbäck, J., Tamburro, D., Crescenzi, C. and Carlsson H. 2006. Air sampling with Empore solid phase extraction membranes and online single-channel desorption/liquid chromatography/mass spectrometry analysis: determination of volatile and semi-volatile organophosphate esters. *Journal of Chromatography A* 1129: 1-8.

Tsao, Y. C., Wang, Y. C., Wu, S. F. and Ding, W. H. 2011. Microwave-assisted head-space solid-phase microextraction for the rapid determination of organophosphate esters in aqueous samples by gas chromatography-mass spectrometry, *Talanta* 84:406-410.

Van der Veen I. and de Boer J. 2012. Phosphorus flame retardants: Properties, production, environmental occurrence, toxicity and analysis. *Chemosphere* 88: 1119– 1153.

Wang, R., Tang, J., Xie, Z., Mi, W., Chen, Y., Wolschke, H., Tian, C., Pan, X., Y. Luo, Y. and Ebinghaus R. 2015. Occurrence and spatial distribution of organophosphate ester flame retardants and plasticizers in 40 rivers draining into the Bohai Sea, north China. *Environmental Pollution* 198: 172-178.

Wang, G. W., Shi, H. H., Du, Z. K., Chen, H. Y., Peng J. B. andGao, S. X. 2017. Bioaccumulation mechanism of organophosphate esters in adult zebrafish (Daniorerio). *Environmental Pollution* 229: 177-187.

Wang, Y. Sun, H., Zhu, H., Yao, Y., Chen, H., Ren, C., Wu, F. and Kannan, K. 2018. Occurrence and distribution of organophosphate flame retardants (OPFRs) in soil and outdoor settled dust from a multi-waste recycling area in China. *Science of the Total Environment* 625: 1056-1064.

Wang, Q. Z., Zhao, H. X., Xu, L. and Wang, Y. 2019. Uptake and translocation of organophosphate flame retardants (OPFRs) by hydroponically grown wheat (*Triticumaestivum* L.). *Ecotoxicology and Environmental Safety* 174: 683-689.

Wang, X., Zhu, Q., Yan, X., Wang, Y., Liao C. and Jiang G.2020. A review of organophosphate flame retardants and plasticizers in the environment: Analysis, occurrence and risk assessment. *Science of the Total Environment* 731: 139071.

Wei, G. L., Li, D. Q., Zhuo, M. N., Liao, Y. S., Xie, Z. Y., Guo, T. L., Li, J. J., Zhang, S. Y. and Liang, Z. Q. 2015. Organophosphorus flame retardants and plasticizers: Sources, occurrence, toxicity and human exposure. *Environment Pollution* 196: 29–46.

Wolschke, H., Sühring, R., Xie, Z. and Ebinghaus R. 2015. Organophosphorus flame retardants and plasticizers in the aquatic environment: a case study of the Elbe River, Germany. *Environmental Pollution* 206: 488-493.

Wang, Y. and Kannan, K. 2018. Concentrations and dietary exposure to organophosphate esters in foodstuffs from Albany, New York, United States. *Journal of Agricultural and Food Chemistry* 66: 13525-13532.

Wong, F., de Wit, C. A. and Newton, S. R. 2018. Concentrations and variability of organophosphate esters, halogenated flame retardants, and polybrominateddiphenyl ethers in indoor and outdoor air in Stockholm, Sweden. *Environmental Pollution* 240: 514-522.

Woudneh, M. B., Benskin, J. P., Wang, G., Grace, R., Hamilton, M. C. and Cosgrove, J. R. 2015. Quantitative determination of 13 organophosphorous flame retardants and plasticizers in a wastewater treatment system by high performance liquid chromatography tandem mass spectrometry. *Journal of Chromatography A* 1400: 149-155.

Yang, C., Li, Y., Zha, D., Lu, G. Q., Sun, Q. and Wu D. 2017. A passive sampling method for assessing the occurrence and risk of organophosphate flame retardants in aquatic environments. *Chemosphere* 167: 1-9.

Yang, Y., Xiao, Y., Chang, Y., Cui Y., Klobučar, G. and Li, M. 2018. Intestinal damage, neurotoxicity and biochemical responses caused by tris (2-chloroethyl) phosphate and tricresyl phosphate on earthworm. *Ecotoxicology and Environmental Safety* 158: 78-86.

Yin, H., Wu, D., You, J., Li, S., Deng, X., Luo, Y. and Zheng, W. 2019. Occurrence, distribution, and exposure risk of organophosphate esters in street dust from Chengdu, China. *Archives of Environmental Contamination and Toxicology* 76: 617-629.

Zhang, X., Suehring, R., Serodio, D., Bonnell, M., Sundin, N. andDiamond, M. L. 2016. Novel flame retardants: estimating the physical-chemical properties and environmental fate of 94 halogenated and organophosphate PBDE replacements. *Chemosphere* 144: 2401-2407.

Zeng, X., He, L., Cao, S., Ma, S., Yu, Z., Gui, H., Sheng, G. and Fu. J. 2014. Occurrence and distribution of organophosphate flame retardants/plasticizers in wastewater treatment plant sludges from the Pearl River Delta, China. *Environmental Toxicology and Chemistry* 33: 1720-1725.

Zhou, L., Hiltscher, M., Gruber, D. and Puettmann, W. 2017. Organophosphate flame retardants (OPFRs) in indoor and outdoor air in the Rhine/Main area, Germany: comparison of concentrations and distribution profiles in different micro-environments. *Environmental Science and Pollution Research* 24: 10992-11005.

Editor's Contact Information

Tamara Lazarević-Pašti, PhD

"Vinča" Institute of Nuclear Sciences

National Institute of the Republic of Serbia

University of Serbia

Belgrade, Serbia

Email: tamara@vinca.rs

Index

D

E

F

R

S

T

U

V

W

X

Z